Editors
Mahendra Sethi
National Institute of Urban Affairs
New Delhi, Delhi
India

Jose A. Puppim de Oliveira
Getulio Vargas Foundation (FGV/EAESP and FGV/EBAPE)
Sao Paulo and Rio de Janeiro
Brazil

ISSN 2367-0045 ISSN 2367-0053 (electronic)
Exploring Urban Change in South Asia
ISBN 978-981-10-5815-8 ISBN 978-981-10-5816-5 (eBook)
https://doi.org/10.1007/978-981-10-5816-5

Library of Congress Control Number: 2017948639

Printed on acid-free paper

This Springer imprint is published by Springer Nature
The registered company is Springer Nature Singapore Pte Ltd.
The registered company address is: 152 Beach Road, #21-01/04 Gateway East, Singapore 189721, Singapore

Foreword

Climate Change is a global challenge and a local one too, which can only be successfully overcome through a global collaborative and cooperative effort. India is prepared to play its role as a responsible member of the international community and make its own contribution, and is taking part in multilateral negotiations under the UN Framework on Climate Change. The outcomes must be effective, fair and equitable. Every citizen of this planet must have an equal share of the planetary atmospheric space. It also has to be ensured that despite India's developmental imperatives, per capita greenhouse gas emissions should not exceed the expected levels and also in no case should they exceed the present levels, at the very least. Climate change is one of the most important global environmental challenges with implications for food production, water supply, health, energy, and so on, which otherwise may have an adverse impact on the day-to-day life of urban dwellers. Addressing climate change requires a good scientific understanding as well as coordinated actions at the settlement level.

India is still a developing country. Per capita energy consumption is about 500 kJ against the world average of 1800 kJ, and there are estimated 400 million Indians who do not have access to commercial energy in significant measure. Development imperatives are huge and we are determined to meet them with a sense of environmental responsibility. In the overall global scenario, unfortunately many poor cities in the country are proceeding with a growth structure that leads to unsustainable practices with regard to water and waste management, and biodiversity, including air quality. The rigid separation of housing, employment, commercial and recreational activities creates a dependency on mobility which if not planned well, in turn contributes to increasing environmental pollution. The phenomenon of the urban heat island that exacerbates climate change is by now well known. Pavements, buildings and other structures typical of urban area replace natural vegetation and eliminates the cooling provided by vegetation through both shade and evapo-transpiration. In urban areas, temperatures may be raised from this process by between 2 and 10 °C; this contributes to the formation of ground level ozone which is detrimental to human health. Furthermore, the compounding effects of urban heat islands and increased temperatures from climate change result in increased demand for air-conditioning which in turn means increased generation and consumption of electricity, which if allowed to be not used well as it is now, adversely affects the climate. On the other hand, there are new technologies that convert waste into economic returns, with zero-landfill and redeploying the waste heat productivity.

Climate change risk in India's urban centres can be seen in the perspective of the expected transition in city growth. In addition to 500 million people, an estimated 10,000 plus urban settlements are expected by 2050. This will involve environmental transition related to water, sanitation, environmental, health, air and water pollution, and climate change. Urban challenges for India for the twenty first century are dual. On the one side there are calls for priority interventions such as providing housing and basic services to all, infrastructure to support an ever-growing population and addressing poverty. On the other side there is the challenge of coping with the various kinds of pollution, whether air, noise or water pollution or environmental degradation, and climate change-induced extreme events and disasters.

India has embarked on a comprehensive national action plan for meeting the challenges of climate change. The approach of the action plan is to promote developmental objectives through pathways that also yield co-benefits for addressing climate change. In principle, it includes inclusive and sustainable development, cost-effective strategies for induced demand-side management, combating growth with ecological sustainability, adoption of appropriate technology for adaptation and mitigation of greenhouse gas emissions extensively, engineering new and innovative market's regulatory and voluntary mechanism, effective programme implementation through linkages with the local bodies, and the people and public-private partnerships, and welcoming international cooperation or R&D and transfer of technologies. The National Action Plan for Climate Change, which is being closely monitored by the Prime Minister himself, has eight missions, namely: the National Solar Mission, National Mission for Enhanced Energy Efficiency, National Mission on Sustainable Habitat, National Water Mission, National Mission for Sustaining the Himalayan Eco System, National Mission for a Green India, National Mission for Sustainable Agriculture, and National Mission on Strategic Knowledge for Climate Change.

The responsibility for spearheading the National Mission for Sustainable Habitat has been entrusted to the Ministry of Urban Development (MoUD). The mission covers a broad spectrum of issues related to both mitigation and adaptation. As far as mitigation is concerned, the focus would be on energy efficiency, in residential and commercial sectors, urban water management, urban planning, bringing about a modal shift in the area of urban transportation, and so on. As far as adaptation is concerned, the focus would be on ensuring universal access to water, protection against floods and cyclones, establishment of disaster warning systems, and measures taken for pre- and post-disaster preparations. India has already initiated action on many of these issues. Its cities require and in fact the *participants* can give feedback as to how things are being done and how things could be done even better.

India's cities require integrated interventions that involve urban development and growth coupled with environmental safety and sustainability. Any framework needs to first identify the city-level vulnerability and risk and then redirect investments and programmes in the city-planning process. This calls for reducing city social structural vulnerability. Service level benchmarks have also been developed that will help to design the specific projects and programmes. It involves redirecting investment towards climate sensitive planning and linking the different levels of institutional interventions. This means equipping cities in two ways: building resilience coping mechanisms against disaster, extreme events and planning city development in an environmentally friendly manner that is sustainable in the long run.

The strategies may include promoting low-rise high density mix land use development to reduce demand for travel and land integrating sustainability norms in master plans or development plans of cities and towns. A lot of debate is going on, whether there should be high-rise buildings or not, and perhaps the *workshop* could deliberate on this issue also, in the context of sustainability. Besides increasing and restoring the green cover in and around cities, use of building materials and surfaces that reduce urban heat islands and improve the storm water drainage system will also help. Strategies could also involve interventions at the regional level like proper resource management, improving weather forecasting to prepare for extreme events, having a disaster management plan in place for environmentally sensitive drought and flood prone areas, and preservation of the threatened areas.

It is noticeable that India often gets the sudden cloudburst kind of rainfall now and this is creating another kind of challenge. When planners and city authorities are questioned, they go back to their database and they say that this never happened before, therefore they never planned for it. They say that in the last 30 years it never rained like this in a short spell. However, what we are viewing as a nuisance, as a challenge, is actually a resource. If you plan for this sort of rainfall, the water can be stored well, and it can change the face of the cities. This is a challenge because not all climate change issues need to be reactive; we can proactively prepare ourselves too. Make good use of whatever is appearing to be a challenge or a nuisance. This could be achieved through specific mitigation and adaptation plans that are built into city planning, ascertaining institutional backup to enable operationalization of these plans.

But a huge knowledge gap exists in terms of possible climate change impacts at regional, local and city levels. The nature of responses should be designed to take care of location-specific impacts. Hence, there is a need to look at the R&D needs, specific to design of urban responses against climate change of the cities. The role of town planners assumes significance as they have to sensitize the numerous agencies and the citizen towards the adverse impacts of climate change and as pointed out above, not all climate change issues need to be a threat; they are opportunities as well, provided we appreciate them seriously, carefully. While undertaking preparation of master plans, zonal developmental plans and layout plans, it has to be ensured that the adverse impacts on quality of air and water are minimized. I hope that the state town and country planning departments, urban development and planning authorities, municipal bodies, schools of planning and architecture and related professional and statutory bodies will play a proactive role in taking up the issue of climate change and accordingly take the necessary steps and studies for incorporating the environmental concerns of cities to reduce the adverse impact of these threats.

I also hope that the strategies adopted for the purpose focus on sustainability and an inclusive approach. The MoUD have revised the Urban Development Plan Formulation and Implementation Guidelines. The discourse on climate issues could suitably influence these guidelines. Since United Nations University (UNU) has got a global presence, it can bring the global knowledge and awareness into India successfully. India is a very large country and the issues (that are found in different parts of the world are probably found in this country) are such that it deserves to have a chapter of UNU in India. The book deliberates on all the relevant issues, themes and comes up with excellent suggestions that would be useful for one and all who are concerned. I wish the book every success!

Dr. Sudhir Krishna
Former Secretary, Ministry of Urban Development,
Government of India

Preface

The twenty-first century is considered to be the greatest opportunity to attain sustainable development. However, a failure to transform our patterns of development can lead to irreversible impacts with consequences for the people and planet. In this context, the future of sustainable development objectives is intrinsically linked to the development processes in our cities, as evident with the inclusion of a Cities Goal in the Sustainable Development Goals (SDGs), though urban development affects most of the SDGs. Rapid urbanization puts increasing pressure on the global environment and exacerbates inequalities as cities both produce and consume material and energy resources. The idea of generating partnerships for co-benefits is embedded in how we can integrate the solutions to global challenges and the pressure for local development objectives in cities in the short and long term. Cities in the developing world face multiple challenges to achieve sustainability goals. These challenges include jobless labour, slums and health risks; lagging access to basic levels of infrastructure, such as adequate housing, transport, electricity; losing prime green and blue land uses to industries, buildings and roads; facing the onslaught of local pollution and global climate change impacts. Nevertheless, cities have historically been and will continue to be centres of hope, knowledge, innovation, social and political reforms that catalyse transformations.

In this sense, collaborative approaches between various functional sectors and levels of governance are key to create the local and global innovations, institutions and mechanisms to generate co-benefits as, for example, reducing the causes and impacts of climate change and air pollution and improving human health. However, in view of adhering to international commitments like the Paris Agreement, SDGs and the recently agreed New Urban Agenda in Post-Habitat III and, for the case of India, attaining ambitious national goals for sustainable urbanization under the Smart Cities Mission, AMRUT (Atal Mission for Rejuvenation and Urban Transformation), HRIDAY (National Heritage City Development and Augmentation Yojana) and the Swachh Bharat Mission, we need a comprehensive discourse to transform broad concepts into practical results for the implementation of such commitments. The objective of this book is to discuss how we can develop policies and instruments to boost the capacity of societies to generate climate, environmental and development-oriented co-benefits in cities to synergistically achieve the local and global goals of sustainable development for India and beyond.

This book is based on the research efforts propounded by United Nations University—Institute for the Advanced Study of Sustainability (UNU-IAS), Japan under its Sustainable Urban Futures Programme between 2010 and 2015. Under its urban development with co-benefits project supported by the Ministry of the Environment, Japan (MOEJ), UNU-IAS instigated a debate on this theme in India by holding with TERI (The Energy Research Institute) an International Conference on Urban Development with Climate Co-Benefits: Aligning Climate, Environmental and Other Development Goals in Cities on 9 December 2013 in New Delhi. This research took concerted efforts to build and expand on its deliberations, invite new ideas from different domain experts, and discuss pathways to identify and generate large urban co-benefits; that is, positive impacts for integrating climate change with local sustainable development goals in cities. The co-benefits could be vital to articulate policy, make social and technological transitions and can promote more sustainable urban initiatives that cut across sectors and regions. This book presents outcomes of empirical research carried out at UNU-IAS and similar efforts in India that interprets and tests prevailing policies and further evaluates the potential urban co-benefits, particularly needful to converge global goals in the post-Paris, SDG and Habitat III era with India's national urban missions.

India, like many other rapidly developing countries, has the unprecedented chance of following a different path of urban development. This book will look at the different aspects of climate co-benefits relevant for Indian cities. There are tremendous opportunities to harness massive urban climate co-benefits, which will be the only way to improve the well-being of the existing and growing number of urban dwellers without threatening the health of the planet and its inhabitants. We bring specialists from various sectors to discuss the possibilities of changing the urbanization path in India based on previous domestic and international experiences, as well as assessing new tools and frameworks.

For simplicity in comprehension, the book is presented in four main parts or themes. Part 1, 'Introduction to the Concept and Theory of Co-benefits' introduces the reader to the concept and theories of co-benefits. It also reports a plethora of tools that facilitate in evaluating co-benefits in cross-cutting sectors. Part 2, 'Contextualizing Co-benefit Issues: Across Spatial Scales and Sectors' demonstrates how the theory can be applied to diverse operating levels, regional to local. It presents discussions on the South Asian context and the challenges to align global environmental issues with urban development policies. Part 3, 'Co-Benefits in Energy, Transport, Buildings, Waste and Biodiversity' forms the main body of the book. With empirical investigation and analysis, it demonstrates how potential co-benefits can be accrued from different development sectors. While some may argue the use of a sector-driven approach, it serves multiple research and practical purposes. The chapters relate well with established domains and their experts and identified co-benefits (and gaps thereof) could be directly and indirectly integrated with a particular sector. The evaluation or quantification of co-benefits around the sectors provides irrefutable evidence to strategize needful innovations, policy shifts and governance reforms. This brings us to Part 4, 'Promoting Co-benefits in the Urban Context: Innovations and Reforms', the last section, which explores how co-benefits could be introduced into the prevailing milieu through urban innovations and institutional reforms and how co-benefits can effectively be mainstreamed.

The book is perhaps the first of its kind providing comprehensive research on cities, climate change and co-benefits research and policy in India. As evident, it mainstreams the Indian state of affairs with the international discourse on the subject. While some outcomes could be unique to India, the reader would certainly find universal lessons for problem statement, theory and analytical techniques, and their application to policy and governance. The editors are hopeful that this 'easy to read and refer' book would generate an equal interest and liking among policy-makers, the private sector, environmental or non-government organizations, academicians, researchers and students.

New Delhi, India — Dr. Mahendra Sethi
Sao Paulo, Brazil — Dr. Jose A. Puppim de Oliveira

Acknowledgements

Conducting this exhaustive research and disseminating the learning in the form of a book has been a strenuous and time consuming task for the entire research team. It is significant to acknowledge all the individuals and institutions that have been instrumental in shaping this endeavour. We hold immense appreciation for United Nations University—Institute for the Advanced Study of Sustainability (UNU-IAS), Japan, that provided the necessary ground for this research to germinate, and the Ministry of the Environment Japan (MOEJ) for the support to carry out the initial research on urban co-benefits that resulted in the book. Special thanks are due to Professor Govindan Parayil, Dean of Patel College of Global Sustainability (PCGS), the University of South Florida, USA, former Vice-Rector of the United Nations University (UNU) and former Director, United Nations University Institute for the Advanced Study of Sustainability, Japan in ensuring that necessary dialogue is initiated with Indian researchers and communities. The support of the Energy and Resources Institute (TERI), New Delhi, as a local partner was noteworthy in facilitating this exchange, and the School of Planning and Architecture Bhopal for the partnership in the research project. Last but not the least, we would like to acknowledge the hard work and patience of all the authors, without their able contribution this body of work would not have been possible. The first editor (Mahendra Sethi) owes special gratitude for the UNU-IAS PhD Fellowship award/grant bestowed upon him that inspired to take up this initiative.

Contents

Contributors

Shomit Bade Department of Planning, School of Planning and Architecture (SPA), Bhopal, India

Amit Chatterjee Department of Planning, School of Planning and Architecture (SPA), Bhopal, India

Binayak Choudhury Department of Planning, School of Planning and Architecture (SPA), Bhopal, India

Meenakshi Dhote Department of Environmental Planning, School of Planning and Architecture, New Delhi, India

Magali Dreyfus CNRS, CERAPS, Lille Cedex, France

Sudhir Gota Independent Consultants, Bangalore, India

Bhola Ram Gurjar Department of Civil Engineering and Centre for Excellence in Transportation Systems, Indian Institute of Technology Roorkee (IIT-R), Roorkee, India

Aastha Joshi National Institute of Urban Affairs, New Delhi, India

Manmohan Kapshe Department of Architecture and Planning, Maulana Azad National Institute of Technology (MANIT), Bhopal, India

Mukesh Khare Department of Civil Engineering, Indian Institute of Technology Delhi, New Delhi, India

Priyanka Kulshrestha Lady Irwin College, University of Delhi, New Delhi, India

Rama Krishna Reddy Kummitha School of Management, Politecnico di Milano, Milano, Italy

Alvin Mejia Metro Manila, Philippines

Shilpi Mittal Indian Society for Applied Research & Development, New Delhi, India; GD Goenka University, Gurgaon, India

Debojyoti Mukherjee Consultant, IPE Global Ltd., New Delhi, India

Ajay Singh Nagpure Center for Science, Technology, and Environmental Policy, Hubert H. Humphrey School of Public Affairs, University of Minnesota, Twin Cities, USA

A. Narender Consultant to Various Organizations, Hyderabad, India

Toshimasa Ohara Regional Atmospheric Modeling Section, Asian Environment Research Group, National Institute for Environmental Studies, Tsukuba, Japan

Jose A. Puppim de Oliveira Getulio Vargas Foundation (FGV/EAESP and FGV/EBAPE), Sao Paulo and Rio de Janeiro, Brazil

Usha P. Raghupathi National Institute of Urban Affairs, New Delhi, India

Mahendra Sethi National Institute of Urban Affairs, New Delhi, India

Divya Sharma Urban Policy and Planning, Oxford Policy Management Limited, Oxford, UK

Richa Sharma National Institute of Urban Affairs, New Delhi, India

Raina Singh Center for Urban Planning and Governance, The Energy and Resources Institute, New Delhi, India

Vandana Tyagi Department of Civil Engineering and Centre for Excellence in Transportation Systems, Indian Institute of Technology Roorkee (IIT-R), Roorkee, India

List of Figures

Chapter 1

Chapter 2

Chapter 3

Chapter 4

Chapter 5

Chapter 6

Chapter 7

Chapter 8

Chapter 9

Chapter 11

Chapter 12

Chapter 13

List of Tables

Chapter 6

Chapter 7

Chapter 8

Chapter 9

Chapter 12

Chapter 13

Chapter 14

Chapter 15

Part I
Introduction to the Concept and Theory of Co-benefits

Chapter 1
Cities and Climate Co-benefits

Mahendra Sethi and Jose A. Puppim de Oliveira

Abstract The New Urban Agenda (NUA) was launched during Habitat III, the United Nations Conference on Housing and Sustainable Urban Development, which took place in Quito, Ecuador, from 17 to 20 October 2016. This was one of the major global policy deliberations after the negotiation of the sustainable development goals (SDGs) and the Paris Agreement. A key challenge remains how to actualize development goals without exacerbating the causes and consequences of climate change. Cities in developing countries particularly face an uphill task to become sustainable. Most of the climate research and action in the developed world like the US, Europe and Japan focused on mitigation strategies. Cities in developing countries until recently have had a limited research on climate response, that too pre-occupied with adaptation agenda. The co-benefit approach tries to bridge this gap, with integration of mitigation and adaptation agendas, vertical and horizontal coordination between sectors and scales of intervention. In this regard, India and other rapidly developing countries have the historical chance of following a different path of urban development. There are tremendous opportunities to harness massive urban climate co-benefits, which will be the only way to improve the well-being of the existing and growing number of urban dwellers without threatening the health of the planet and its inhabitants. This research will present a conceptual framework to investigate the different aspects of climate co-benefits relevant for Indian cities.

Keywords New urban agenda · SDG · Paris agreement · IPCC UNFCCC · Co-benefits · Urbanization · India · Climate policy · ULBs

---

M. Sethi (✉)
National Institute of Urban Affairs, New Delhi, India
e-mail: mahendrasethi@hotmail.com

J.A. Puppim de Oliveira
Getulio Vargas Foundation (FGV/EAESP and FGV/EBAPE),
Sao Paulo and Rio de Janeiro, Brazil
e-mail: japo3@yahoo.com

M. Sethi and J.A. Puppim de Oliveira (eds.), *Mainstreaming Climate Co-Benefits in Indian Cities*, Exploring Urban Change in South Asia,
https://doi.org/10.1007/978-981-10-5816-5_1

1 Introduction

The 'New Urban Agenda' (NUA) was launched during Habitat III, the United Nations Conference on Housing and Sustainable Urban Development, which took place in Quito, Ecuador, from 17 to 20 October 2016[1] (UN 2016). The Agenda includes three transformative commitments: (i) Sustainable Urban Development for Social Inclusion and Ending Poverty; (ii) Sustainable and Inclusive Urban Prosperity and Opportunities for All; and (iii) Environmentally Sustainable and Resilient Urban Development. It tries to change the urbanization patterns to enhance its opportunities for human development and reduce its negative impacts. There is a subtle message, which is to avoid the many past mistakes of urbanization based on high levels of material consumption and concrete building to focus on quality of life, resource conservation and sufficiency.

Climate change is an important component of the NUA. Although it is not amongst the explicit commitments, as the UN has a specific convention on climate change for this purpose, yet it is mentioned in several parts of the text. The main goals of the Agenda include many of the pressing needs of Indian cities, which are the provision of jobs and economic opportunities, energy, housing, transportation and other urban services. However, a key challenge remains on how to achieve those goals without exacerbating the causes and consequences of climate change. Urban development that generates climate co-benefits can also tackle inequality, another key point of the New Urban Agenda. Current urbanization trends are core drivers of inequities and inequalities, and climate change exacerbates them. Climate co-benefits can reduce inequality, while addressing the core of the problem, that is, patterns of unsustainable urbanization in the North, and more recently in the South. One of the major problems is the capacity of cities to concentrate and consume extraordinarily, and expel their unwanted residues to other places, causing several consequences for part of the urban population elsewhere, particularly in rural areas. Achieving the multiple goals of the Agenda will require us to take a different view of the role of cities and their main long-term development objectives, moving urban development from being based on buildings to improving human well-being. Resources and services will have to be better shared, generating climate co-benefits and reducing inequalities in access to them and giving more opportunities to all.

In this regard, India and other rapidly developing countries have the historical chance of following a different path of urban development. There are tremendous opportunities to harness massive urban climate co-benefits, which will be the only way to improve the well-being of the existing and growing number of urban dwellers without threatening the health of the planet and its inhabitants. This book will look at the different aspects of climate co-benefits relevant for Indian cities. Specialists from various sectors discuss the possibilities of changing the urbanization path in India based on previous domestic and international experiences, as well as assessing new tools and frameworks.

---

[1]Habitat III is the third bi-decennial UN summits on urbanization. The first conference took place in Vancouver, Canada, in 1976 and the second in Istanbul, Turkey, in 1996.

2 Climate Change and the Differentiated Contributions from Sectors and Regions

Climate change is characterized by increased average temperature in the atmosphere of Earth due to accumulation of greenhouse gases (GHGs), including those produced by anthropogenic activities. Anthropogenic activities primarily relating to production, distribution and consumption of various goods and services across the world have resulted in emission of GHGs,[2] of which the carbon dioxide emissions constitute a major share. The need to compile and assess scientific knowledge in this area led to the formulation of the Intergovernmental Panel on Climate Change (IPCC) in 1988. The initial task for the IPCC, as outlined in UN General Assembly Resolution 43/53, was to prepare a comprehensive review and recommendations with respect to the state of knowledge of the science of climate change; the social and economic impact and possible response strategies and elements for inclusion in a possible future international convention on climate. The United Nations Framework Convention on Climate Change (UNFCCC), adopted in 1992, acknowledged that change in the Earth's climate and its adverse effects are a common concern of humankind and expressed concern that human activities have been substantially increasing the atmospheric concentrations of GHGs, that these increases enhance the natural greenhouse effect, and that this will result on average in an additional warming of the Earth's surface and atmosphere and may adversely affect natural systems and humankind (UN 1992).

The IPCC in its fourth assessment report (AR4) in 2007 had already predicted that the global GHG emissions would continue to grow over the next few decades and continued GHG emissions at or above current rates would cause further warming and induce many changes in the global climate system during the twenty first century that would very likely to be larger than those observed during the twentieth century (IPCC 2007a). The fifth assessment report (AR5), the latest and by far the most comprehensive analysis by the IPCC, has confirmed the previous predictions and made the following key observations (IPCC 2014):

- Anthropogenic greenhouse gas emissions have increased since the pre-industrial era, driven largely by economy and population growth, and are now higher than ever. This has led to atmospheric concentrations of carbon dioxide, methane and nitrous oxide that are unprecedented in at least the last 800,000 years.
- Recent climate changes have had widespread impacts on human and natural systems. The atmosphere and ocean have warmed, the amounts of snow and ice have diminished, and sea level has risen.

---

[2]Carbon dioxide (CO_2), methane (CH_4), nitrous oxide (N_2O), hydrofluorocarbons (HFCs), perfluorocarbons (PFCs) and sulphur hexafluoride (SF_6) are the six greenhouse gases covered by the first commitment period of the Kyoto Protocol (UNFCCC 2015, http://unfccc.int/kyoto_protocol/items/3145.php). There are other gases, such as water vapour, which are not included in the Protocol.

- Changes in many extreme weather and climate events have been observed since about 1950. Some of these changes have been linked to human influences, including a decrease in cold temperature extremes, an increase in warm temperature extremes, an increase in extreme high sea levels and an increase in the number of heavy precipitation events in a number of regions.
- Scenarios leading to CO_2-equivalent concentrations in 2100 of about 450 ppm or lower are *likely* to maintain warming below 2 °C over the twenty first century relative to pre-industrial levels (*high confidence*). In these scenarios, temperature peaks during the century and subsequently declines.
- Continued emission of greenhouse gases will cause further warming and long-lasting changes in all components of the climate system, increasing the likelihood of severe, pervasive and irreversible impacts for people and ecosystems. Limiting climate change would require substantial and sustained reductions in greenhouse gas emissions which, together with adaptation, can limit climate change risks.
- Surface temperature is projected to rise over the twenty first century under all assessed emission scenarios. It is *very likely* that heat waves will occur more often and last longer, and that extreme precipitation events will become more intense and frequent in many regions. The ocean will continue to warm and acidify, and global mean sea level will continue to rise.
- Climate change will amplify existing risks and create new risks for natural and human systems. Risks are unevenly distributed and are generally greater for disadvantaged people and communities in countries at all levels of development.
- Increasing magnitudes of warming increase the likelihood of severe, pervasive and irreversible impacts for people, species and ecosystems. Continued high emissions would lead to mostly negative impacts for biodiversity, ecosystem services and economic development and amplify risks for livelihoods and for food and human security.

The contribution to global GHG emissions from different economic sectors and world regions varies considerably. AR5 has calculated the global GHG emissions by sector and found that the energy sector is the largest emitter, with electricity and heat production comprising a quarter of contributions (IPCC 2014a). This is primarily from coal, diesel and gas based thermal power plants that spew, besides carbon dioxide, polluting gases such as sulphur dioxide and oxides of nitrogen. Energy emissions are followed by industry (21%) and transport (14%) that akin to energy sector burn fossil-fuels. Refer to Fig. 1 for the respective contributions from each sector.

Scientific and policy literature has also brought out the regional variations in greenhouse gas emissions across world regions and countries. According to UN Habitat (2011), there are striking differences in GHG emissions between regions and countries. While 18% of the world's population lives in developed countries, they account for 47% of global CO_2 emissions. On the other hand, 82% of the world's population living in developing countries account for the remaining 53%. The US and Canada alone account for 19.4% of global GHG emissions, while

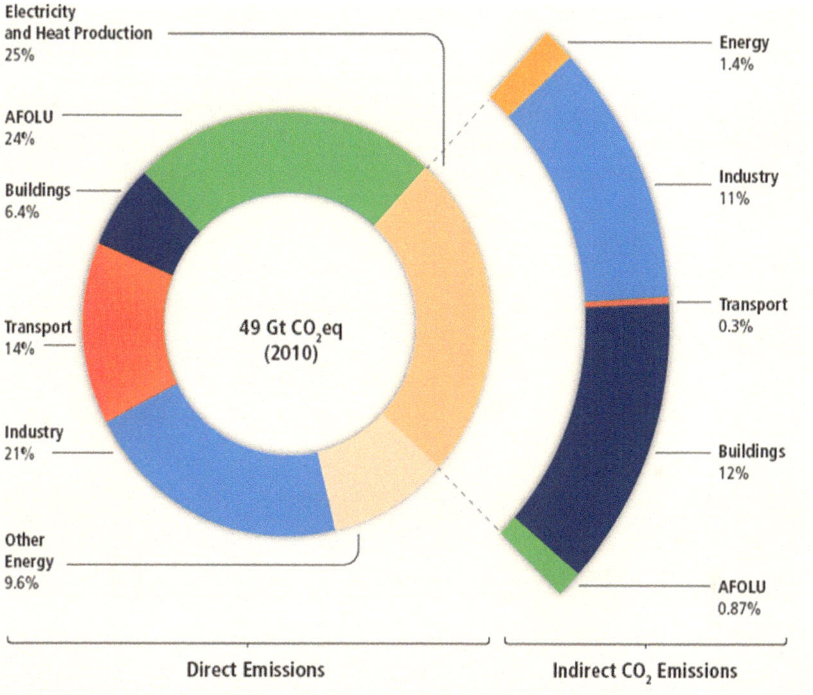

Fig. 1 Total anthropogenic greenhouse gas (GHG) emissions (gigatonne of CO_2-equivalent per year, $GtCO_2$-eq/yr) from economic sectors in 2010. The circle shows the shares of direct GHG emissions (in % of total anthropogenic GHG emissions) from five economic sectors in 2010. The pull-out shows how shares of indirect CO_2 emissions (in % of total anthropogenic GHG emissions) from electricity and heat production are attributed to sectors of final energy use. *Source* IPCC 2014a

South Asia accounts for 13.1% and Africa just 7.8%. China and the US contribute a significant portion of the total carbon emissions in the world, followed by India, the Russian Federation and Japan in the range of 4–5% each (see Table 1 for a list of the largest CO_2 emitters, based on data feeds from Oliver et al. 2016). The per capita carbon emission (another significant indicator used for international comparisons) from different world regions and countries are equally diverse in magnitude. For some countries like the US, China, Canada, Russian Federation, South Korea, Saudi Arabia, Spain, Ukraine and other mainly developed countries, it is above the global average. But for most of the world's population residing in developing and least developed countries, particularly in Asia, Africa and Latin America it is in the range of 2.37–2.77 t/capita which is half of the global norm. In the case of Africa, the per capita emission is still less at 0.77 t/capita which is about one-sixth of the world average.

It has been argued that by enhancing their efforts in keeping with historical responsibility, the developed and resource rich countries could reduce the burden of their action from being borne by developing countries that carry the additional

Table 1 CO_2 emissions from major world regions and countries in 2015

Rank	Country	Carbon emissions (in million tonnes)	Change in carbon emissions, 1990–2015 (%)	Per capita carbon emissions (tonnes) in 2015
1	China	10,720	355	7.7
2	United States	5,180	3	16.1
3	India	2,470	272	1.9
4	Russian Federation	1,760	−26	12.3
5	Japan	1,260	8	9.9
6	Germany	780	–24	9.6
7	Canada	680	23	19.0
8	Iran	630	214	8.0
9	South Korea	620	129	12.1
10	Saudi Arabia	510	201	16.0
11	Indonesia	500	214	2.0
12	Brazil	490	120	2.3
13	Mexico	470	63	3.7
14	Australia	450	60	18.6
15	South Africa	420	47	7.7
16	UK	400	−31	6.2
17	Turkey	360	132	4.5
18	Italy	350	−17	5.9
19	France	330	−14	5.1
20	Poland	300	−19	7.6

Source Data feeds from Oliver et al. (2016)

responsibility of finding resources to meet their development needs and strive to improve their Human Development Indexes (MoEF 2015). The variability in total emissions, per capita emissions and circumstances of different countries in terms of their economy, sector contributions, level of technology, vulnerability to climate change, and so on, could affect their choice for a more preventive, mitigation-oriented approach towards climate change or rather an accommodating and adaptive one.

Traditionally, disparities in carbon footprints of different countries were evaluated and negotiated from a purely economic or 'state of development' perspective. While global climate governance has always been in policy paralysis over differences in access and allocation of the carbon space, research shows how (a) carbon inequities thrive within the prevailing climate regime; (b) the role of ethics, fairness and justice in climate governance is growing; and (c) methods vary in evaluating empirical access and allocation of global carbon space (Sethi 2015b). Moreover, new research analysing data from more than 200 countries over five decades shows that rates of urbanization correlate more with carbon emissions than wealth (gross domestic product per capita). As countries urbanize, their cities' contributions of carbon emissions and greenhouse gases start to become disproportionately high in comparison to their population and wealth (Sethi and Puppim de Oliveira 2015).

Box 1.1: COP21 and the Paris Agreement

Adopted on 12 December 2015 by 195 nations attending the United Nations' twenty first Conference of the Parties (COP21), the Paris climate agreement has been termed a monumental triumph, the turning point for mankind and the greatest opportunity to save our future Earth by some great world leaders like US President Obama. The agreement commits for the signatory countries to limit the rise of global temperatures to well below 2.0 °C above pre-industrial levels while pursuing efforts to keep the temperature increase to just 1.5 °C above pre-industrial levels. This new climate deal will come into force in 2020. The 184 signatories also agreed to reach peaking of GHGs as soon as possible and to undertake rapid reductions thereafter, by means of submitting their Intended Nationally Determined Contribution (INDC) documents every five years to progressively reduce their carbon dioxide emissions using the highest possible ambition. Just prior to the talks, 154 countries responsible for over 85% of global annual emissions of greenhouse gases, and representing over 90% of global gross domestic product (GDP), in 2012 had already submitted INDCs, including pledges to reduce annual national emissions after 2020.

The Paris agreement also creates a $100 billion annual Green Climate Fund under which developed countries shall provide financial resources to assist developing country parties with respect to both mitigation and adaption. Also for the first time at a COP meeting, education was brought to the table. The text acknowledges the role of education, training, public awareness and access to information in the transition to a climate responsible world. The Paris Agreement saw the finalization of not only a fortnight of talks, but of more than 23 years of international diplomacy under the UN to forge collective action on the global problem. Since 1992, climate negotiations were marked by discord and failure, the refusal of every party to be on board and abide to commitments. The successful agreement at Paris hence restores faith in the collective UN process. Along with the usual euphoria and hope associated with global environmental negotiations, there are also corresponding reservations and criticisms on the agreement. Moreover, the new American president, Donald Trump, has declared that America should withdraw from the Paris Agreement, creating concerns about its future.

Observers skeptical of catastrophic anthropogenic climate change have condemned the agreement calling it 'a fraud, with no action, but just an aspirational document. By proponents' own admission, the first emissions-reduction plans don't even come close to reducing the volume of greenhouse gas emissions necessary to limit warming to 2.0 °C. Secondly it could be seen that the agreement's language was crafted to devoid it of essential legal teeth. There are no sanctions for creating sub-standard plans or penalties for non-compliance. Similarly, there is also no penalty for countries that miss making GCF payments. As then-Secretary of State John Kerry admitted 'it doesn't have mandatory targets for [temperature] reduction and it doesn't have an enforcement, compliance mechanism.'

It is also essential to point out the fact that, 'Unless and until the individual countries actually enact the [emissions] targets through domestic law, they aren't even binding within the legal system of any individual country.' This was evident with countries not ratifying the previous agreement, the Kyoto Protocol in their national legislatures. Though there has been progress compared with hypothetical 'business as usual' global emissions pathways, there still remains a gap between the submitted INDCs, and a pathway that is consistent with a reasonable chance of limiting the rise in global average temperature to no more than 2 °C above pre-industrial levels. Some of the emissions reduction pledges put forward in countries' INDCs ahead of COP21 in Paris are not sufficiently ambitious; in addition they say little about countries' ability for credible policy implementation.

The Paris Agreement lays emphasis on enhancing linkages and creating synergy between, inter alia, mitigation, adaptation, finance, technology transfer and capacity building, thus affirming belief in the co-benefits approach. The agreement also calls to make INDCs with quantifiable information and targets that are measurable, transparent, verifiable in reporting. This underlines the use of quantitative methods tools in assessment of emissions. In addition, the agreement acknowledges the role of non-party stakeholders like civil society, the private sector, cities and other sub-national authorities, local communities, etc. to mobilize stronger and more ambitious climate action. As such, a pro-active role of cities in generating co-benefits through measurable and verifiable reporting tools in limiting emissions cannot be undermined.

In this regard, there was an historic moment for mayors at the Paris COP with 'resilience pledges' to dedicate 10% of city budgets to climate resilience and innovation signalling a new level of recognition of the tremendous role to be played by cities and regional authorities. The direct responsibilities of sub-national and territorial authorities over the management of land, water, air, waste and natural resources, and their roles in adaptation brings to mind the significance of local governance to climate futures.

Other major outcomes during the agreement were that India and France led 120 countries in announcing an International Solar Alliance supporting solar energy deployment in developing countries. More than 20 developed and developing countries launched Mission Innovation, pledging to double public investment in clean energy research and development over five years. All through the year, France encouraged non-state actors to demonstrate their action and support by entering pledges into the NAZCA Portal set up under the Lima-Paris Action Agenda. By the time of Paris, the portal listed nearly 11,000 commitments from 2250 cities, 150 regions, 2025 companies, 424 investors, and 235 civil society organizations.

Source Excerpt from UNFCCC 2015

Box 1.2: What Are 'Co-benefits'?

As the study of co-benefits is still progressing, there is no standard or globally accepted definition. But normative meanings are evolving in respective sectors or targeted areas of policy/study. An overview of some important definitions:-

The United States Environmental Protection Agency (USEPA) provides a basic definition of co-benefits as 'all of those positive outcomes associated with multiple, simultaneous emissions reductions' (USEPA 2005).

The IPCC's 4th Assessment Report quotes co-benefits as the non-climate benefits of GHG mitigation policies that are explicitly incorporated into the initial creation of mitigation policies. Thus, the term co-benefits reflects that most policies designed to address GHG mitigation also have other, often at least equally important, rationales involved at the inception of these policies (e.g., related to objectives of development, sustainability, and equity). This definition captures the intended positive benefits of a policy and distinguishes it from unintended positive side effects or ancillary benefits (IPCC 2007b).

The Institute for Global Environmental Strategies (IGES) defines a co-benefits approach as a 'win-win strategy aimed at capturing both development and climate benefits in a policy or measure' (IGES 2010).

IPCC's 5th Assessment Report refers to co-benefits as a government policy or a measure intended to achieve one objective that often affects other objectives, either positively or negatively. For example, mitigation policies can influence local air quality. When the effects are positive they are called 'co-benefits', or 'ancillary benefits'. Negative effects are referred to as 'adverse side effects'. Some measures are labeled 'no or low regret' when their co-benefits are sufficient to justify their implementation, even in the absence of immediate direct benefits. Co-benefits and adverse side effects can be measured in monetary or non-monetary units (IPCC 2014b).

Finally, we have to define climate co-benefits. Doll and Puppim de Oliveira (2017) define as following: "Climate co-benefits can be generated when one outcome of an intervention is a benefit in terms of climate change mitigation or adaptation. On the one hand, climate co-benefits could occur as the benefits of certain development interventions in tacking climate change as compared to other viable development alternatives (including doing nothing or business as usual). On the other hand, the co-benefits can arise when purposely mitigating climate change leads to development benefits at the local or regional level, such as reduction in air pollution, creation of jobs or energy cost savings. The latter may be considered to be developmental co-benefits generated by climate action, and the former the climate co-benefits generated by a climate friendly development option." (Doll & Puppim de Oliveira, 2017, p.2)

3 Climate Mitigation, Adaptation or Co-benefits?

Deciding upon mitigation or adaptation alternatives to tackle climate change impacts has always been a tricky policy decision for societies and nations. It is known that anthropogenic GHG emissions are mainly driven by population growth, economic activities, energy use, change in land use patterns and technology.

Through technology and policy measures, mitigation options are known to be available in every sector. These could be made economical by following a planned and integrated approach that decarbonizes energy supply, reduces energy use, net emissions, greenhouse gas intensity of end-users and increases sequestration in carbon sinks. It has been argued with *high confidence* that without additional mitigation efforts beyond those in place today, and even with adaptation, warming by the end of the twenty first century will lead to a high to very high risk of severe, widespread and irreversible climatic impacts globally with huge social and economic negative consequences (IPCC 2014b).

While mitigation can lead to co-benefits and risks due to adverse side effects, these risks do not involve the same possibility of severe, widespread and irreversible impacts as risks from climate change, increasing the benefits from near-term mitigation efforts. Furthermore, adaptation and mitigation are complementary strategies for reducing and managing the risks of climate change. Substantial emission reductions over the next few decades can reduce climate risks in the twenty first century and beyond, increase prospects for effective adaptation, reduce the costs and challenges of mitigation in the longer term and contribute to climate-resilient pathways for sustainable development (IPCC 2014a). But, effective adaptation and mitigation response is contingent upon policies and actions across multiple scales of intervention, that is, international, regional, national and sub-national. After several scientific and policy-based discussions on the various adaptation and mitigation options available in addressing climate change, it is now widely being regarded that no single option is complete by itself. Effective response depends on a range of policies and cooperation in all activity sectors and scales which could be enhanced through integrated approach like climate co-benefits.

The IPCC's recent assessment underscores that since AR4, it is being reported with high confidence that there has been an increased focus on policies designed to integrate multiple objectives, increase co-benefits and reduce adverse side effects (IPCC 2014a). In addition, national governments often refer to co-benefits in climate and sector-specific plans and strategies. It has been reported that sector-based policies have been more widely used than economy-wide policies. Although most economic theory suggests that economy-wide policies for mitigation would be more cost-effective than sector-specific policies, as administrative and political barriers may make economy-wide policies harder to design and implement. The latter may be better suited to address barriers or market failures specific to certain sectors and may be bundled in packages of complementary policies. Nonetheless, specific studies are significant to evaluate gaps in research, policy and its application. Indeed there is a perceptible and growing interest in climate co-benefits.

Co-benefits of mitigation could affect other objectives, such as those related to energy and food security, air quality, human health, biodiversity, livelihoods, income distribution, labour supply and employment and urban sprawl (Puppim de Oliveira 2013). In the absence of complementary policies, however, some mitigation measures may have adverse side effects (at least in the short term), for example on biodiversity, food security, energy access, economic growth and income distribution. Similarly, co-benefits of adaptation policies may include improved access to infrastructure and services, extended education and health systems, reduced disaster losses, and better governance. Whether or not and to what extent side effects materialize is case- and site-specific, and depend on local circumstances and the scale, scope and pace of implementation. Many co-benefits and adverse side effects have not been well quantified.

Significant co-benefits, synergies and trade-offs exist between different mitigation and adaptation actions spanning over diverse sectors and operational scales. Increasing efforts to deal with climate impacts implies an increasing complexity of interactions among climate, water, energy, land use and biodiversity, but tools to understand and manage these interactions remain limited. Examples of actions with co-benefits include: (i) improved energy efficiency and cleaner energy sources, leading to reduced emissions of health-damaging, climate-altering air pollutants; (ii) reduced energy and water consumption in urban areas through greening cities and recycling water; (iii) sustainable agriculture and forestry; and (iv) protection of ecosystems for carbon storage and other ecosystem services (IPCC 2014a). The effect of co-benefits and adverse side effects from climate policies on overall social welfare has not yet been quantitatively examined, with the exception of a few recent multi-objective studies.

It has been argued that co-benefits are contingent upon local circumstances (Puppim de Oliveira 2013). Comprehensive strategies in response to climate change that are consistent with sustainable development take into account the co-benefits, adverse side effects and risks that may arise from both adaptation and mitigation options. The assessment of overall impacts is complicated by variables like climate change itself, actions undertaken, climate policy alternatives, along with the pre-existing development-oriented policies. For example, in terms of air quality and human health, the value of the extra tonne of carbon dioxide (CO_2) reduction that occurs with climate change mitigation through reduced fossil fuel combustion depends greatly on the stringency of CO_2 control policies. If prevailing instruments are weak, the CO_2 reductions would be substantial, but if they are already strict, in spite of pitching tremendous funds and efforts, the response is bound to be negligible.

The just concluded Paris Agreement recognizes the social, economic and environmental value of voluntary mitigation actions and their co-benefits for adaptation, health and sustainable development (UNFCCC 2015). It has entrusted a Technology Executive Committee to prepare a summary for policy-makers, with information on specific policies, practices and actions representing best practices and with the potential to be scalable and replicable, and on options to support their implementation, as well as on relevant collaborative initiatives, and publishing the summary at least two months in advance of each session of the Conference of the Parties. This would include a technical paper on the mitigation benefits and co-benefits of policies, practices and actions for enhancing mitigation ambition, as well as on options for supporting their implementation, information on which should be made available in a user-friendly online format. The agreement underscores the relevance of the co-benefits approach particularly in bridging with actions needful in adaptation arena. It emphasizes the need to identify actions that could significantly enhance the implementation of adaptation actions, including actions that could enhance economic diversification and have mitigation co-benefits and promote cooperative action on adaptation. This would be fundamental to the implementation of actions to achieve the commitments of the New Urban Agenda.

4 Role of Cities in Contributing to Climate Change

As growing evidence from scientific literature points out, cities contribute significantly to climate change and they are also affected by it (IPCC 2007a; UN-Habitat 2011; IPCC 2014b). Figure 2 represents the agents that contribute to climate change in the atmosphere (including ozone depletion) and the ones that in turn impact the settlements. While the discourse on the former focuses on the study of GHG emissions and is integral to define mitigation strategies, the latter part of the inter-relation attains greater understanding in the realm of climate change adaptation.

A noteworthy research on the subject (World Bank 2010), estimates the magnitude of this relation by underscoring the urban stakes of economic production, consumption and emissions. It states that the 50 largest world cities combined, if they were one country, would rank third in both population and GHG emissions,

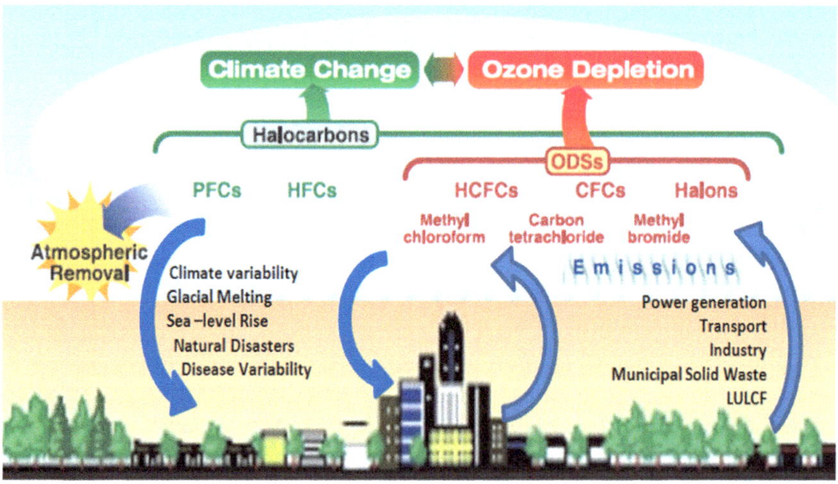

Fig. 2 Cities contribute to climate change and in turn get affected by it

and second in GDP. Cities meet approximately 72% of their total energy demand from coal, oil and natural gas—the main contributors to GHGs. Meanwhile, World Energy Outlook 2008 and studies from the UN claim that cities are responsible for 75% of global energy consumption and 80% of GHGs (UN 2007; IEA 2008). On a temporal scale, GHG emissions show a strong association with the process of industrialization and urbanization. Just 200 years ago a mere 3% of the Earth's inhabitants lived in cities, but in 2009, for the first time in history, city dwellers outnumbered rural inhabitants (UN DESA 2011). As per UN estimates, people living in cities across the globe are expected to double by 2050 (see Fig. 3), while the built-up area is expected to triple during the same period (Angel et al. 2005).

The building up of GHG emissions in the atmosphere has been escalating steadily over the last century. Emissions from fossil fuel burning, which form the biggest bunch of GHG basket, have mounted seven-fold from about 1000 mt CO_2e in the 1920s to over 6500 mt CO_2e in 2000 (see Fig. 4). As the United Nations, in their review of linkages between climate change and urbanization points out, the two pose to be the greatest challenges currently facing humanity in the twenty first century, whose effects are converging in dangerous ways (UN-Habitat 2011). The future scenarios in GHG emissions discussed earlier, if seen against the trends in global urban expansion (refer to Box 1.3 for a detailed perspective), prompt towards a comprehensive and methodical research enquiry to systematically and strategically guide national and sub-national urbanization, urban policy and governance, with India as the case study in this book.

Box 1.3: The Global Urban Expansion

Urbanization is a global phenomenon that is transforming human settlements. Today, 52% of the global population, or 3.6 billion persons, is urban, compared to only 3% in 1800 and 13% in 1900. By 2050, the global urban population is expected to increase between 2.5 and 3 billion, corresponding to 64–69% of the world population. Each week the urban population is increasing by approximately 1.3 million. Future trends in the levels, patterns, and regional variation of urbanization will be significantly different from those of the past. Most of the urban population growth will take place in small to medium-sized towns in developing countries. Another key dimension of urbanization is the increase in built-up area and urban land cover. Worldwide, urban land cover occupies a small fraction of global land surface, with estimates ranging between 0.28 and 3.5 million km^2, or between 0.2 and 2.7% of ice-free terrestrial land. A meta-analysis of 326 studies using satellite data shows a minimum global increase in urban land area of 58,000 km^2 between 1970 and 2000, or roughly 9% of the 2000 urban extent. At current rates of declining densities among developing country cities, a doubling of the urban population over the next 30 years will require a tripling of built-up areas. The kinds of towns, cities, and urban agglomerations that ultimately emerge over the coming decades will have a critical impact on energy use and carbon emissions (Excerpts from UN DESA 2012; IPCC 2014b).

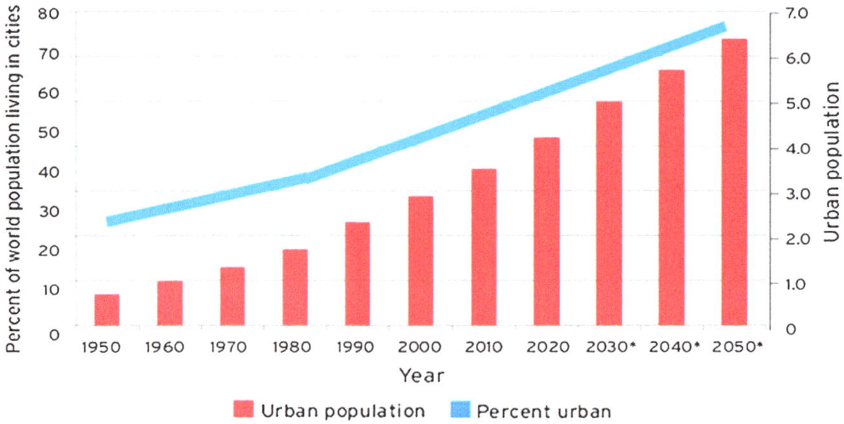

Fig. 3 People living in cities (percentage of world population and total). *Source* UN DESA 2011

Box 1.4: What Is Urban?

There is no universally accepted definition of *urban*. Hence, in its assessments, the United Nations follows respective national definitions. Interestingly, a global analysis of countries shows different criterion or a mix of methods being used by governments to define urban. For instance, 105 countries base it on *administrative jurisdiction*, while for 83 it is the sole criterion; 100 countries define cities on *density or population size* ranging it from 200 to 50,000 inhabitants, in which 57 use this as the sole criterion; 25 countries use *economic characteristics* like population engaged in non-agrarian activities; 18 countries count availability of *urban infrastructure* like paved streets, water supply and sewerage to classify cities (UN Habitat 2006, 2007). The Census of India recognizes as urban all those settlements which either have a *statutory* status like municipal committee/corporation/notified area committee/cantonment board, estate office and so on, or that are *census towns* that fulfil all of the following three conditions simultaneously: (i) a population of more than 5000; (ii) more than 75% of the male working population engaged in non-agricultural activities; and (iii) a density of population of more than 400 people per square kilometre. Accordingly, India had an urban population of 377 million in 2011, which is 31.16% of the total population, distributed across 7935 towns; that is, 4031 statutory towns and 3894 census towns (Census of India 2011).

Cities are a complex lot in terms of size, economic activities, infrastructure and facilities. A review by the United Nations into their definition reveals their global diversity (see Box 1.4). A critical aspect of cities is that they are meant to absorb additional, often unskilled labour that is expected to migrate from rural areas. The irony is that cities are expected to become engines of growth, while at the same time

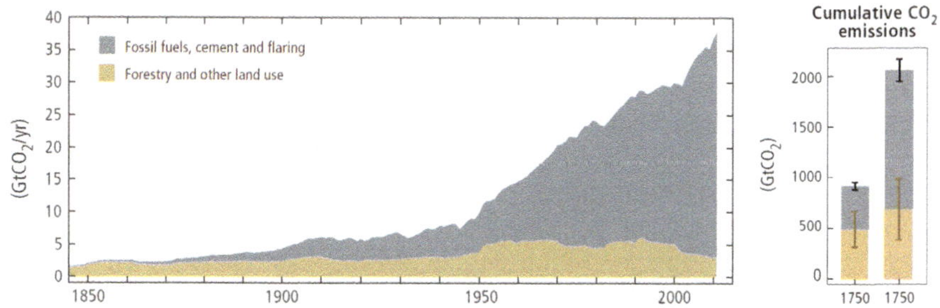

Fig. 4 Global carbon emissions from fossil fuel and cement, 1750–2003. *Source* IPCC 2014a

reduce emissions by pursuing sustainable strategies. Thus, it becomes important to study cities and urbanizing areas for their contribution to GHG emissions for not only the causative part of the problem, but also the solution. The international community unequivocally concurs that though the cause and implication of climate change are global, local action for mitigation and adaptation is paramount (OECD 2009; UN Habitat 2011). Urban authorities and local governments have the potential to implement mitigation programmes effectively, because of the type of responsibilities they hold in relation to land use planning, local public transportation and the enforcement of industrial regulations. Second, the concentration of people and industries in large cities provides the opportunity for technological innovations, such as combined heat and power plants, waste-to-energy generation plants, and mass transit systems for transportation. Third, this concentration also provides the opportunity for the rapid spread and adoption of new ideas and innovations, both as business and behavioural solutions. Finally, reducing greenhouse gas emissions can also lead to various other urban benefits (Dodman 2009; Bulkeley and Bestill 2003; Puppim de Oliveira et al. 2013); elaborated later in this chapter. There is an additional potential for developing countries as they have the 'late development advantage'. This ensures an opportunity to control their urbanization rates and make more sustainable use of space at lesser human and financial cost (UN Habitat 2011), logically determine locations of economic investments like special economic zones, industrial corridors and attempt to regulate high 'locked-in emissions' for the new infrastructure and buildings being planned.

There is limited empirical knowledge on cities' actual contributions to GHG emissions, and appropriate means to mitigate them (Schroeder and Bulkeley 2009; World Bank 2010; UN Habitat 2011), particularly how cities in the 'global south' are responding to climate change and whether and how their local authorities are planning for the impacts of climate change (Betsill and Bulkeley 2007). Taking advantage of this opportunity, however, will require a radical change in the anti-urbanization stance taken by many developing country policy-makers who still try to impede or slow urban growth rather than prepare ahead for it (Martine 2009). In this perspective, Indian development policy and governance generally follows the Gandhian worldview that India lives in its villages, and its climate policy in particular faces a similar dilemma in domestic and international situations. India occupies an intriguing dual position in global climate politics—a poor and developing economy with low levels of historical and per capita emissions, and a large and rapidly growing economy with rising emissions and high technological capabilities in some areas. Indian climate politics has substantially been shaped around the first perspective and increasingly, under international pressure, is being forced to grapple with the second. It has been observed that translating low carbon development into action in India has a high potential as both the knowledge and resources are available, but it poses a considerable governance challenge (Dubash 2012). In terms of urbanity, India being a developing country with a large rural population base (urbanization level of 31.6% as recently reported in Census of India 2011), the general perception is that cities hardly have any carbon footprint. On the other hand, unreeling economic growth and rapid urbanization during the last two decades, coupled with limited data on the subject pressingly seeks a systematic determination of emission from urban areas and its implications in policy and local governance to achieve a better urban environment. Recent research demonstrates that in spite of relatively low levels of national urbanization, Indian cities can be associated with 85% of energy generation from thermal plants (Sethi 2015a) and are responsible for 66.5–70.3% of the national GHG emissions (Sethi and Mohapatra 2013), giving irrefutable evidence about urban causations of global warming.

5 Generating Co-benefits in Urban Areas

Climate change is caused by complex anthropogenic activities and requires multiple strategies for mitigation and adaptation, as discussed earlier in Sect. 3. It is generally accepted now that by linking policies in climate change within various development sectors like energy, buildings, transportation, land use and waste, coordinated efforts at the local level can be fostered as an effective intervention for pursuing co-benefits (Doll and Puppim de Oliveira 2017) . At the same time, these actions are crucial to attain national and international goals of sustainable development. The urban climate co-benefits approach (UNU-IAS 2013) refers to *the implementation of initiatives (policies, projects,* etc.*) that simultaneously contribute to reducing the contribution to man-made global climate change while solving local environmental problems in cities, and in turn potentially having other positive developmental impacts, such as improvements in citizen health, energy security, income generation, etc. Though many actions meant to combat climate change, inadvertently have other local benefits, the co-benefits approach seeks to purposefully multiply and mainstream climate co-benefits into the development process.*

In order to vividly understand the potential co-benefits at the intersection of multiple development and environmental objectives across different spatial scales. A conceptual framework is presented in Fig. 5.

Cities have always been at the forefront in dealing with environmental issues like water pollution, air pollution, noise and urban greenery and have also led the charge in matters of climate change. Being an intensive agglomeration of people, technology, business and progressive ideas, they naturally become laboratories of

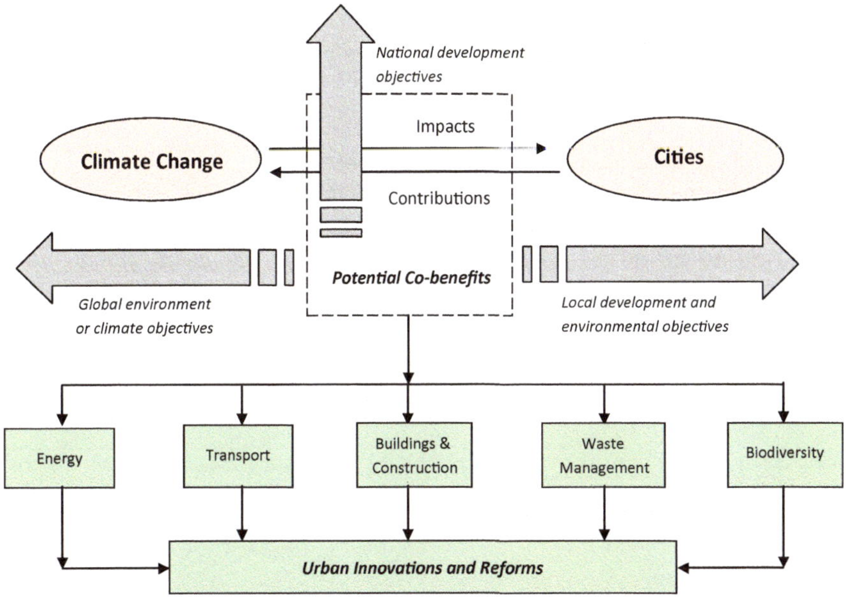

Fig. 5 Conceptual framework to understand potential co-benefits at the intersection of multiple development and environmental objectives across different spatial scales

research and experiment. The co-benefits concept is constantly being tried and tested in urban areas across the world. It has been argued that if climate change policies are to be implemented, they must match the local priorities of countries and cities (Puppim de Oliveira 2009). With strategies of setting up emission targets, relocating industries, promoting public transportation, setting up cap and trade systems, regulating urban sprawl, initiating tree-planting activities, recycling wastewater, investing in green energy, and so on, cities like Curitiba, London, Los Angeles, Toronto, Tokyo, Singapore have been pioneers in bringing a favourable change.

Most of the research and intervention in urban co-benefits pertain to the developed world; for example, the US, Europe and Japan have focused on mitigation strategies. Cities located in developing countries until recently have shown limited empirical research on climate response; that too is pre-occupied with adaptation agenda. Is this probably because the issues in developing cities are devolved from their national narratives on climate change? It is actually an intellectual and policy vacuum that developing countries identify themselves as hardly contributing to the global GHG burden, but rather facing higher climate risks and vulnerabilities. However, the local situation in these countries is gradually but surely changing. In the process, they have to simultaneously deal with local developmental and environmental challenges of migration, slums, sanitation, transport, environmental pollution, waste, parks and biodiversity. In this regard, it would be mindful to observe how other communities and nations have fared up, who have tread this path before. In recent years, cities in developing countries have increased their emissions significantly, through growth in both economic production and materials consumption. This is simultaneous with their rapidly growing urban levels (see Box 1.5 for urbanization trends in developing regions).

Box 1.5: Urbanization in Developing Countries

Urbanization rates in developed regions are high, between 73% in Europe and 89% in North America, compared to 45% in Asia and 40% in Africa. The majority of urbanization in the future is expected to take place primarily in Africa and Asia, and will occur at lower levels of economic development than the urban transitions that occurred in Europe and North America. While its urbanization rate is still lower than that of Europe and the Americas, the urban population in Asia increased by 2.3 billion between 1950 and 2010.

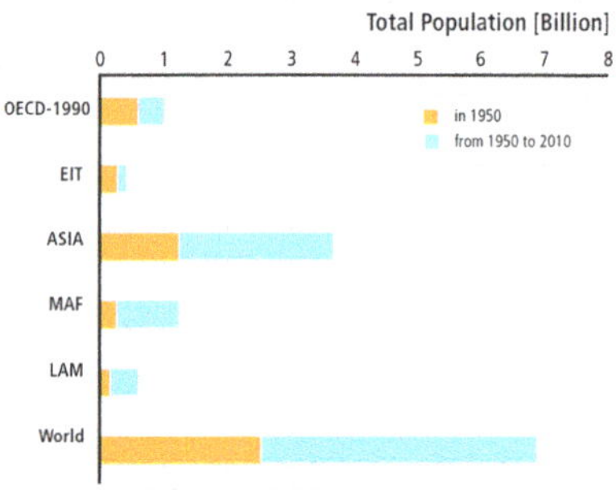

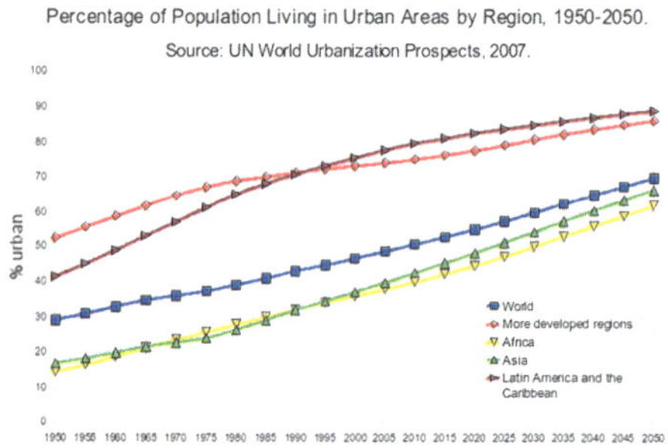

Overall, urbanization has led to the growth of cities of all sizes. Although mega-cities (those with populations of 10 million or greater) receive a lot of attention in the literature, urban population growth has been dominated by cities of smaller sizes. About one-third of the growth in urban population between 1950 and 2010 (1.16 billion) occurred in settlements with populations fewer than 100,000. Currently, approximately 10% of the 3.6 billion urban dwellers live in mega-cities of 10 million or greater. Within regions and countries, there are large variations in development levels, urbanization processes and urban transitions.

Worldwide, populations will increasingly live in urban settlements, particularly in developing countries. By the middle of the century, the global urban population is expected to reach between 5.6 and 7.1 billion, with trends growth varying substantially across regions. The proportion of rural population in the developed regions has declined from about 60% in 1950 to less than 30% in 2010, and will continue to decline to less than 20% by 2050. While highly urbanized North America, Europe, Oceania, and Latin America will continue to urbanize, the increase in urbanization levels in these regions is relatively small. Urbanization will be much more significant in Asia and Africa where the majority of the population is still rural.

Source Figures and excerpts from IPCC 2014b and UNDESA 2007

It has been reported that 89% of the CO_2 emissions growth in cities between 2006 and 2030 is expected to occur in non-OECD countries (IEA 2008). For example, consider the world's nations with the biggest urban populations—China and India. In China, the largest 35 cities contribute 40% of the country's CO_2 emissions, though they have only 18% of the population (Dhakal 2009). Similarly, urban India which is home to about one-third of the national population contributes to about two-thirds of its GHG emissions, if accounted from location or production perspective (Sethi and Mohapatra 2013). Hence it could be argued that the co-benefits approach is highly relevant to transforming societies burgeoning in the developing countries, which have to address several challenges simultaneously with limited technical know-how, financial resources and managerial capacities.

Cities in developing countries particularly face an uphill task to become sustainable. They are stressed with a mounting population and the growing need for goods and services. This natural increase in demand of resources is simultaneous with degradation of the local environment, insanitary conditions, air, water and noise pollution and overlooking of best practices. Environmental issues are least prioritized in cities of developing countries, where provision of basic services, roads, shelter and jobs for the urban population takes centre stage. These cities are engines of growth, propelled by a carbon-spewing energy and manufacturing sector or highly consumptive and inefficient services sector. In both these cases, GHG footprints of urban settlements extend well beyond their administrative boundaries. On top of this, these cities have little resilience to the impact of global warming, heat waves, floods, storm surges, landslides, heat islands, and so on. They also host a large population of poor people living in the most environmentally vulnerable locations (Satterthwaite 2009; IPCC 2014b). Responding to local environmental problems and the test of global change simultaneously, developing country cities are in the right position to effectively assess multiple challenges and their trade-offs associated with rapid growth, sustainable development, mitigating excessive emissions and adapting to climate change.

Thus, the urban challenges for developing countries are significantly different from those of developed countries. While the cities in the latter have reached high standards of living for their inhabitants with increasing levels of GHG emissions and now focus on decarbonizing their activities, the cities in the former still have a large number of development challenges with a growing population. Nevertheless, they cannot follow the same development path of the now rich countries without putting the world, and themselves, in climatic risks and need to find a different way to develop in order to avoid an unsustainable global rise in GHG emissions. Thus, while the question for cities in rich countries is 'how can we reduce our GHG emissions without reducing, or keep increasing, our standards of living?' the question in developing countries is twisted to 'how can we improve the standards of living of our population without putting too much GHG in the atmosphere and adapt to the changing climate?'

The co-benefit approach tries to bridge this gap, with integration of mitigation and adaptation agendas, vertical and horizontal coordination between sectors and scales of intervention. It attempts to internalize the global change policy within prevailing development, environmental portfolios, policy instruments, research discourse and governance mechanisms, through innovative strategies and actions. It also helps accrue benefits from possible locked-in emissions. A well-determined application of co-benefit approach is still at an incipient stage in India. Nonetheless, India has explicitly expressed its ambitious plans to erect new and revive old cities, making them economically competitive and environmentally sustainable through the Smart Cities Mission. This makes it an ideal case to explore possible avenues for urban climate co-benefits in India.

6 International Experiences of Co-benefits in the Developing Context

Cities in several countries have experienced initiatives that generated climate co-benefits as an approach for mainstreaming global environmental issues at the local level. It is essential to encourage the implementation of climate-friendly measures in developing cities which are not obliged to reduce their carbon emissions but where much progress can be effectively made. In the context of urban sustainability, there have been several studies documenting such innovative practices across the world. Most of those examples are of an incremental nature, challenging to assess the contribution they have made in tackling larger environmental problems or whether they have the potential to scale up and stimulate systemic change in urban development processes.

Brazil has experienced one of the world's most drastic processes of socioeconomic and territorial reorganization as a result of rapid urbanization since the 1930s, now being one of the most urbanized countries in the world ($\sim$85% of urban population). There is huge potential for co-benefits as the lack of good public transportation and waste management and disposal are a growing concern among policy-makers and increasing pressure from organized civil society groups and population in general. The City of São Paulo started with a series of actions to address climate change in the energy sector, as well as waste management, whose land-fill gas capturing initiative reduced the GHG emissions by more than 10% (Puppim de Oliveira 2009). In Rio de Janeiro there is also a set of initiatives to reduce urban pollution and tackle climate change at the same time, through a series of transport initiatives such as expansion of the metro lines and Bus Rapid Transit (BRT) system.

Urbanization in China has significantly enhanced economic development and social changes. However, some negative influences have emerged following the rapid urban development. There are unprecedented resources and energy needed to support this kind of urban growth rate and scale, which has put stress on the local environment (like air and water quality) and generated large amounts of GHG emissions because of the dependence on coal for electricity generation. Lately,

China has started to take numerous initiatives to tackle both local environmental pollution and, more recently, GHG emissions, even making international voluntary commitments to improve carbon efficiency. The country has recently become the largest producer and consumer of solar and wind energy. Transformations are also taking place in its cities. Shenyang is one of biggest industrial cities in China and it has planned to lower its carbon emissions and pollution. Besides the transportation sector (Geng et al. 2013), the city has especially addressed the pollution in the industrial district of Teixi in its urban plan by shifting some industries to a different location. The energy efficiency in the building sector in Shanghai was another case addressed in the research (Jiang et al. 2013a). Research has also noted necessary policy measures introduced in Tiexi District of Shenyang and Baoshan industrial district in Shanghai, as they are also going through a series of initiatives that can lead to co-benefits (Jiang et al. 2013b).

Indonesia has witnessed rapid economic growth recently along with an increasing urbanization. Environmental problems have aggravated in several cities owing to the lack of infrastructure or urban services. The national government has shown some interest in reducing GHG emissions by both reducing deforestation and taking initiatives in the urban area. The city of Surabaya implemented waste composting schemes that had a significant impact in reducing organic waste by around 25%. This reduced the load being sent to the final disposal site and GHG emissions (Kurniawan et al. 2013). Yogyakarta also reported some interesting initiatives in community-based solid waste management (CBSWM) with some positive results. For example, the number of CBSWM groups in the city increased from 27 to 93 between 2008 and 2010 and the volume of solid waste taken to the landfill reduced from 91,000 to 65,000 metric tons per year. The city also started a BRT initiative that led to many other initiatives in the transportation sector (Dirgahayani 2013).

7 Urbanization, Climate Change and GHG Emissions in India

Like many developing nations, India too is urbanizing rapidly. For the first time since its independence in 1947, the absolute increase in population in urban areas is more than in the rural areas, on account of net rural urban classification and migration (56%), against natural increase (44%) (Census of India 2011). With a total urban population of 377 million in India, urbanization increased from 27.81% in 2001 to 31.16% in 2011 (MoUD 2011). The variation may seem marginal in percentage points, but urban India is actually adding 92 million every decade, the population equivalent of four Australias put together. It took nearly 40 years (1971–2011) for India's urban population to rise 270 million, but in future it may take half the time to add the same number. As per various estimates, by 2030, India's urban population will be approximately 600 million (Mckinsey 2010; MoUD 2011), that is about 40% of the total and it would break even with the rural population by 2039.

Beyond this phenomenal increase of urbanites, the Indian population stands against a huge risk of climate variability. India has time and again stated that there is no country more vulnerable to climate change than India on so many fronts (MoEF 2015, Foreword in Dubash 2012) due to the following:

- Heavy dependence of the Indian economy and agriculture on monsoons where half of the fluctuations in GDP are related to monsoons
- One of the longest coast lines in the world where 250–300 million people live along the coast and the majority of them depend on climate sensitive livelihoods such as agriculture and fishing
- Threat of melting of Himalayan glaciers which will have a direct impact on water availability across the Gangetic belt
- Dependence on natural resource extraction and mining which could lead to deforestation and loss of biodiversity.

The Government of India brought out a climate change impact assessment report on India titled 'Climate Change Assessment and India: A 4 × 4 Assessment' focusing on a sectoral and regional analysis for 2030. The key findings of the Report (MoEF 2010a) with regard to climate change impact, among others were:

- Rise in annual mean surface air temperature from 1.7 to 2.0 degree centigrade
- Intensifying of extreme daily maximum and minimum temperatures
- Small increase in annual precipitation
- Increase in diseases due to rise in temperature, extreme rainfall and flooding
- Rise in sea level along the Indian coast at the rate of about 1.3 mm per year on an average
- Reduction in forest vegetation
- Decrease in water availability.

An Asian Development Bank Study on assessing the costs of climate change adaptation in South Asia projects the economic damage and losses in India from climate change to be around 1.8% of its GDP annually by 2050 (Ahmed and Suphachalasai 2014). Meanwhile the most recent emission assessment of the Ministry of Environment and Forests (MoEF), Government of India, through its Indian Network of Climate Change Assessment (INCCA) accounts for India's GHG emissions at 1.72 bt CO_2e in 2007 (MoEF 2010b). Recent assessments of over 2 bt CO_2e (including land use change and forestry), position India as the third largest emitter after China and the USA. Interestingly, the volume of emissions from China and the USA are 5–2.5 times respectively that of India. However, it is important to note that India is making significant efforts to reduce carbon emissions and this has resulted in the decline in carbon emissions intensity of India's GDP by more than 30% during the period 1994–2007. India has announced its commitment to further reduce the emissions intensity of GDP by 20–25% between 2005 and 2020 (Foreword to the INCCA-MoEF Report, May 2010 by Jairam Ramesh).

A study by MoEF titled 'India's GHG Emissions Profile' predicts that the emissions over the next two decades, based on an average of 5 climate models (NCAER-CGE, TERI-MoEF, IRADe-AA, TERI-Poznan, and McKinsey), (MoEF 2009), is bound to become three folds, from 1.72 bt CO_2e (2007) to 5.22 bt CO_2e (2030–31) while per capita emissions will also become two and half times from 1.5 t CO_2e (2007) to 5.6 t CO_2e (2030–2031). According to this study, the per capita GHG emissions of India will remain modest, well below those of some developed countries, even if they were to take ambitious emission reduction targets (25–40%) as recommended by the IPCC for the mid-term. In fact, it is argued that India's per capita GHG emissions would be well below the global average 25 years earlier. The comparisons of aggregate emissions with China, with whom India is usually clubbed, are also revealing, in the sense that they are miniscule (for details refer to Box 1.6).

Box 1.6: The Case of China and India

Depending on the scenario and forecast, 55% of the total urban land in 2030 is expected to be built in the first three decades of the twenty first century. Nearly half of the global growth in urban land cover is forecasted to occur in Asia, and 55% of the regional growth will take place in China and India. These forecasts provide first-order estimates of the likelihood that expansion of urban areas will occur in areas of increasing vulnerability to extreme climate events including floods, storm surges, sea level rise, droughts, and heat waves. Urban expansion and associated land clearing and loss of above-ground biomass carbon in the pan-tropics is expected to be 1.38 PgC between 2000 and 2030, or 0.05 PgC/yr.

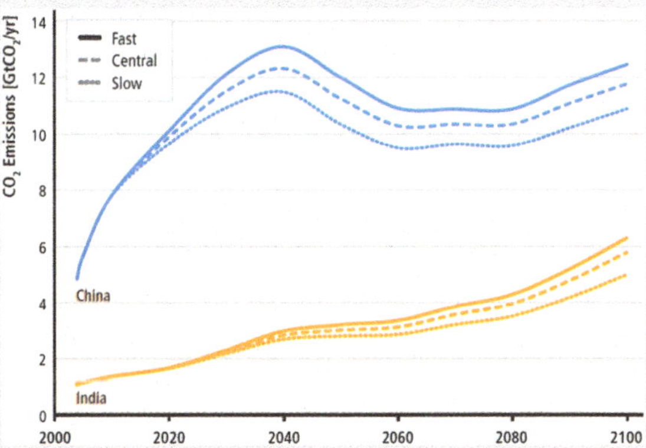

Source IPCC 2014b

According to the INCCA report, GHG emissions from energy, industry, agriculture, and waste sectors constituted 38, 22, 17 and 3% of the net carbon emissions respectively. Data on sub-sectoral contributions to GHG emissions within the energy sector shows that around 65% of GHG comes from electricity generation followed by the transport sector and residential/commercial sectors, which contribute around 13% each. Within the transport sector, the predominant contribution to GHG emissions comes from road sector. The four wheelers and two wheelers are the biggest contributors to GHG emissions. The aviation sector contributes around 10%, followed by railways at 7%. Urban areas have a strong linkage with the road sector and its resultant GHGs. Emissions from waste, though, contribute around 3% of total emissions, but most of this originates from the urban areas (MoEF 2010b).

Taking into account national statistics, a superimposed time-series of urban population and GHG emissions, a scenario spanning over 80 years is plotted in Fig. 6. It clearly indicates that akin to the globally evident phenomenon, rapid economic growth and urbanization point to an immediacy to limit global warming, a situation where cities play a pivotal role in mitigation of climate change. The above narrative and scientific evidence provides an essential base—it is how cities choose to plan and manage their local development and environment that will determine the pace of global warming. In the Indian context, the approach of generating and estimating climate co-benefits in urban areas could be highly potential sum measure for their sustainable and low-carbon future. A detailed study of emissions from India's urban sphere reports huge contributions from different economic sectors, namely the energy (59.26%), iron and steel (9.66%), cement (10.70%) and transport (5.85%) sectors (Sethi and Mohapatra 2013). It further emphasizes that there is an urgent need to internalize climate strategies within the prevailing urban governance, which is only viable through rational and systematic estimation of climate co-benefits in cities.

Almost all the macroeconomic models predict that anticipated needs in the future will be large. Rapid urbanization in the country will be one of the most dominant trends in the coming years. It is expected that over 40% of the population in 2030 would be urban as against 32% currently. As population expands and incomes grow, this shift will likely be realized alongside demographic changes that will exponentially increase the demand for urban amenities like housing, energy, transport, water, waste disposal. It is estimated that more than half of India of 2030 is yet to be built (MoEF 2015). India's official submission to UNFCCC at Conference of Parties (COP) 21 in Paris notes this in clear terms. It states that given the development agenda in a democratic polity, the infrastructure deficit represented by different indicators, the pressures of urbanization and industrialization and the imperative of sustainable growth, India faces a formidable and complex challenge in working for economic progress towards a secure future for its citizens. There are some early signs of how urban India is fostering new climate initiatives, through select policies, programmes and projects (see Box 1.7).

Box 1.7: Major Climate Initiatives in Urban India

Policies

- Intended Nationally Determined Contribution 2015
- National Environment Policy 2006
- Integrated Energy Policy 2006
- National Action Plan for Climate Change 2009
- National Urban Transport Policy 2006

Programmes

- National Mission for Sustainable Habitat
- National Mission on Energy Efficiency
- National Mission for Enhanced Energy Efficiency
- National Mission for a Green India
- National Solar Mission
- Bachat Lamp Yojana
- National Energy Labeling Programme
- Implementation of Energy Conservation Building Code
- Standards & Labeling Programme
- Perform Achieve and Trade scheme

Projects

- UN-Habitat in partnership with ICLEI South Asia's Urban-Low Emission Development Strategies (LEDS) project wherein a total of fifteen cities in India participated. Two Model Cities—Thane and Rajkot—and six Satellite Cities were selected.
- GIZ and ICLEI South Asia's Demonstrating the Urban NEXUS approach to optimize water, energy and land in Nashik, Maharashtra.
- Solar Housing Complex, Kolkata and Magarpatta Housing, Pune showcase local best practices in energy efficiency.
- The Rockefeller Foundation supported Asian Cities Climate Change Resilience Project initiated work in Surat, Indore and Gorakhpur, later expanded to several others.
- Gujrat Urban Development Company's energy efficiency project in street lighting and water pumping across 150 municipalities.
- CDM Project for street lighting in Madhya Pradesh and Chattisgarh.
- SUNYA-Towards Zero Waste in South Asia Project, an EU-funded project being implemented in Shimla and Coimbatore.
- The Climate Project India (TCPI) is to inspire individuals on the impacts and solutions to climate change by presenting relevant information and knowledge in an interactive manner.
- GIZ supported (i) Development and management of nationally appropriate mitigation actions in India and (ii) Climate Change Adaptation in planning and design of Industrial Areas in Andhra Pradesh.

- Resource efficiency and secondary raw materials management as a contribution to climate change mitigation commissioned by the German Federal Ministry for the Environment.
- WWF-India is engaged with Kanpur and Meerut municipalities to develop local action plan on climate change.

8 Climate Policy in India

It has been universally accepted that 'the global nature of climate change calls for the widest possible cooperation by all countries and their participation is an effective and appropriate international response, in accordance with their common but differentiated responsibilities and respective capabilities and their social and economic conditions' (UN 1992 p1). The same has been reinstated by India in its Intended Nationally Determined Contribution (INDC) submitted towards the COP Paris 2015 agreement.

India's environment policy is anchored in the Constitution of India. Article 48-A of the Constitution states that 'The State shall endeavour to protect and improve the environment and to safeguard the forests and wildlife of the country'. In 2006, the

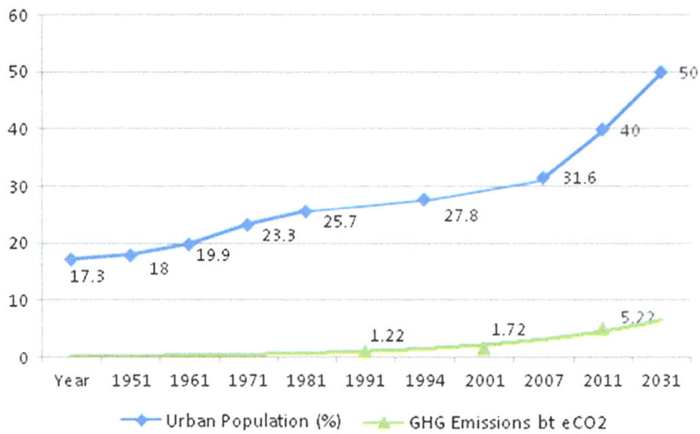

Fig. 6 Scenarios of India's urbanization and GHG emissions. *Source* The time series analysis is generated by the authors based on the following data sources: *1* Urban Population data from TCPO (2012), 'Data Highlights (Urban) based on Census of India', Town & Country Planning Organisation, Government of India, New Delhi. Projections for 2039 from MoUD 2011, 'India's Urban Demographic Transition'. *2* GHG emissions data (1994 & 2007) from MoEF (2010b): 'India: Greenhouse Gas Emissions 2007', Indian Network for Climate Change Assessment, MoEF, India and forecasts for 2030/2031 averaged from five models from MoEF (2009): 'India's GHG Emissions Profile', Climate Modelling Forum, MoEF, India)

Government of India's policy framework, called the National Environment Policy (NEP), was released. Climate Change finds a narrow reference in the document, which basically clarifies India's position in the international climate change debate rather than to pave a nationwide integrated approach on the subject (MoEF 2006). It upholds the principle of common but differentiated responsibilities and respective capabilities of different countries. Pertaining to mitigation, the policy emphasizes multilateral approaches, rights to equal per capita entitlements of global environmental resources, priority to rights to development and encouraging Indian industry to participate in clean development mechanisms through capacity building. With regards to energy, NEP recommends an integrated approach to energy conservation and adoption of renewable energy technologies, including hydropower, by appropriately linking efforts to improve conversion, transmission, distribution, end-use efficiency and, also, research, development and dissemination. It further suggests removing policy, legal and regulatory barriers to setting up decentralized generation and distribution systems for power and other secondary energy forms, based on local primary energy resources.

Large developing countries such as India were not in favour of any legal binding on carbon emission targets and a stipulated time frame for peak carbon emissions as they felt it would jeopardize their poverty reduction and economic development efforts. Despite these differences, most of the countries have voluntarily committed targets for reduction in carbon emissions and have achieved significant success. There are several policy debates on the stand that India could pursue globally and domestically with regard to reduction of carbon emissions. Dubash (2012) has classified these policy debates into three groups: (a) the growth-first realist; (b) the sustainable development realist; and (c) the sustainable development internationalist.

The 'growth-first realist' approach supports a high and unconstrained growth strategy for India on the ground that the developed countries were historically responsible for much of the carbon emissions in their pursuit for high economic growth and now it is India's turn. The national ambition to provide *Electricity for All* could realistically remain unaccomplished without India's reliance on its thermal power plants. This strategy implies no or minimum commitments by India for reduction in carbon emissions. The 'sustainable development realist' approach considers that climate change and its likely impacts pose a serious threat to India and advocates a shift in India's growth strategy in favour of more environmental sustainable and equitable development by pursuing 'co-benefits' at home. However, this approach argues for per capita based burden sharing and for not linking legally domestic efforts in the reduction of carbon emissions to the international process. This appears to be the current official stand of the Government of India and also several NGOs and academics. Meanwhile, the 'sustainable development internationalist' approach shares the thinking of 'sustainable development realist' on domestic strategy but calls for a more aggressive and proactive role for India in international processes aligning India interests with a strong global climate regime to bring development and climate co-benefits.

The Government of India through the Prime Minister's Council on Climate Change formulated the 'National Action Plan on Climate Change' (NAPCC) in (2008) which appears to be based on the 'sustainable development realist' approach. The key objectives of the NAPCC (2008) are:

- Protecting the poor and vulnerable sections of society through an inclusive and sustainable development strategy sensitive to climate change
- Achieving national growth objectives through a qualitative change in direction that enhances ecological sustainability leading to further mitigation of greenhouse gas emissions
- Devising efficient and cost-effective strategies for end-use demand side management
- Deploying appropriate technologies for both adaptation and mitigation of greenhouse gas emissions extensively as well as at an accelerated pace
- Engineering new and innovative forms of market, regulatory and voluntary mechanisms to promote sustainable development
- Effecting implementation programs through unique linkages including with civil society and local government institutions and through public private partnership
- Welcoming international cooperation for research, development, sharing and transfer of technologies enabled by additional funding and a global intellectual property rights regime that facilitates technology transfer to developing countries under the UNFCCC.

It has been observed that the NAPCC has drawn largely from the prevailing international discourse, from ideas of co-benefits propagated by IPCC AR4, IEA, OECD, UNFCC and the World Bank. In doing so, the NAPCC articulated for a directional shift in the development pathway and outlined a number of steps to simultaneously advance India's development and climate change related objectives of adaptation and mitigation. As a way forward the National Action Plan suggested for preparation of the eight national missions of which the National Mission on Sustainable Habitat (NMSH) directly pertains to urban settlements, while the National Solar Mission (NSM) and the National Mission for Enhanced Energy Efficiency (NMEEE) are also two other relevant sub-missions. The National Mission for Sustainable Habitat broadly covers the following aspects: extension of the energy conservation building code—which addresses the design of new large commercial buildings to optimize their energy demand; better urban planning and a modal shift to public transport—which makes long-term transport plans to facilitate the growth of medium and small cities in such a way that ensures efficient and convenient public transport; recycling of material and urban waste management—a special area of focus will be the development of technology for producing power from waste. The National Mission will include a major R&D programme, focusing on biochemical conversion, waste water use, sewage utilization and recycling options wherever possible (CSE 2014). Accordingly, States and Union Territories have put in place the State Action Plan on Climate Change (SAPCC), attempting to mainstream climate change concerns in their planning process. But actualization of

NAPCC and SAPCC poses a big challenge considering the fact that the urban sector and cities in India actually exhibit a complex architecture of several agencies and actors. In its present form, it needs to be cautioned that the co-benefits approach does not become a tool for the governments to hyperbole existing policies with emerging global concerns. A review of some SAPCC shows that most of them fell considerably short in embracing the idea of co-benefits due to lack of visionary leadership and enterprise in states, coupled with little guidance and handholding by the Centre. Any initiative meant to internalize climate smart development and co-benefits within cities has to undeniably acknowledge the current state of India's urban governance (see Box 1.8).

The Planning Commission set up an Expert Group on 'Low Carbon Strategies for Inclusive Growth' in 2011 to suggest strategies to reduce carbon emissions and mitigate and adapt to climate change. The Expert Group submitted a Final Report highlighting various strategies for climate mitigation and co-benefits. The low carbon strategies in the power sector include reducing electricity demand by use of more efficient appliances, introduction of more fuel efficient power plants and changes in the mix of power plants. The strategies for the transport sector include promoting goods transport by railways, mass transport for passenger movement, facilitating non-motorized transport and increasing fuel efficiency of vehicles. Among industries, the strategies focus on the possibilities of reducing emissions through change in technology in the steel, cement, oil and gas sectors. The scope for reducing energy needs of commercial buildings is also examined. The report observed that the emission intensity of India's GDP can be brought down between 23–25 and 33–35% over the 2005 levels under different scenarios. The report has articulated sectoral strategies for five sectors, namely power, transport, industry, buildings and forestry. In order to generate co-benefits in urban areas, inroads will have to be made into several national policies promulgated by different government ministries/departments in various sectors. This internalization has become even more important after doing away with the erstwhile Planning Commission and its five year plan regime.

Accordingly, in its submission to UNFCCC at COP 21 Paris, India has now declared that despite having no binding mitigation obligations as per the Convention, it aspires to achieve the ambitious target of reducing the emissions intensity of its GDP by 33–35% by 2030 from the 2005 level. It has identified a series of mitigation measures like clean and efficient energy systems, promoting waste to wealth conversion, safe smart and sustainable green transportation network, planned afforestation, abatement of air pollution, citizen and private sector contribution in combating climate change and adaptation strategies in agriculture, water, health, coastal regions and islands, disaster management, and so on. The launch of a number of urban sector schemes like the Smart Cities Mission, Atal Mission for Rejuvenation and Urban Transformation (AMRUT) and National Heritage City Development and Augmentation Yojana (HRIDAY) would act as the terra-firma for these mitigation and adaptation strategies to germinate. It makes an even more pressing case to explore the innovation and reforms based means and methods of attaining climate co-benefits in urban areas.

Box 1.8: A Review of India's Urban Governance

Genesis: India is a democratic, sovereign republic with a federal governance structure bearing a centre-state system. The division of powers is defined in the Seventh Schedule, Article 246 of the Indian Constitution along with its subsequent amendments. Accordingly, 97 subjects of national importance (such as defence, foreign affairs, ports, airports, rail and national highways, atomic energy, interstate disputes, banking, income tax) are under the Union List, 66 subjects (such as public order, public health, police, local government, land, agriculture, trade and commerce, taxes on land, buildings, goods, agriculture, minerals, professions, vehicles, tolls) are in the State List, while 47 subjects (such as criminal law, transfer of properties, economic and social planning, social security, labour welfare, electricity) are in the Concurrent List. Subjects relevant to cities have been interpreted from the State List, namely public health and sanitation, hospital and dispensaries, land, works, taxes on land and buildings and from the Concurrent List, namely economic and social planning, electricity, and so on. There is no separate constitutional provision to make rules and govern for the bottom tier of urban local bodies (ULBs).

Legislative Powers: Out of 7935 census towns (numerically defined by Census of India on the basis of a settlement's population size, density and proportion of male workers in non-agricultural activities, see Box 1.4 for details) in India, there are 3842 statutory towns. These include larger urban areas called *nagar nigam* or municipal corporations (population of 1 million +) smaller urban areas called *nagar palika* or municipality/municipal council/ municipal board or transitional areas like *nagar panchayats* or city/town/ notified area councils or the erstwhile town/notified area committees) all exercise different authority, powers and civic functions delegated by the states. While municipal corporations and municipalities are fully representative bodies, the notified or town area committees are either fully or partially nominated bodies. In order to decentralize power, the Government of India took a historic decision as it legislated to confer constitutional authority to the ULBs through the 74th Constitutional Amendment Act (CAA) in 1992, making them pillars for tier-III governance, below the tier of centre and the states. It added Article 243 W (also known as the Twelfth Schedule) defining 18 functions that include urban planning, regulation of land use and construction of buildings, planning for economic and social development, water supply for domestic, industrial and commercial purposes, public health, sanitation conservancy and solid waste management, urban forestry; protection of the environment and promotion of ecological aspects, slum improvement and upgradation, urban poverty alleviation; parks, gardens, playgrounds; promotion of cultural, educational and aesthetic aspects; public amenities including street lighting, parking lots, bus stops and public conveniences, and so on. In addition to the above, depending on certain specific needs of defence, industry, infrastructure, socio-economic activities and shelter, centre and state governments have constituted (by statutes) development authorities, cantonment boards, special economic zones, industrial estates, export regions or particular urban functions like city improvement trusts, housing/shelter boards, heritage trusts, power corporations, water and sewerage boards.

Functional Mandate: As evident, most of the above provisions endow reasonable legal authority on the mitigating capacities of a city. While provision of energy or generation of power is not a direct mandate of ULBs, they have extensive command to protect the environment and ecology in day-to-day affairs and land-use planning, water supply in prospective scenarios that can considerably influence mitigation of emissions and local pollution abatement from thermal power plants. But without functions, jurisdictions, and so on actually devolved to ULBs by their state governments, from line departments, para-statal authorities, boards, or trusts already functioning in urban areas, the CAA remains a law on paper. Only 10 states out of 29 have transferred functions mandated under the 74th CAA to the ULBs until 2015.

Financial Autonomy: In spite of the clear directions of the 74th CAA and recommendations made by various commissions, the states have not provided much local autonomy to the ULBs to deal with sources of funds or activities that contribute to GHG emissions and hence influence climate change. The Act also provides for constitution of the State Finance Commissions (SFCs), every five years, to review financial administration by local governments and suggest further transfers. However, as observed by the 13th Central Finance Commission, there are long delays in constitution and delivering of final reports of the respective SFCs.

Policy and Plan Preparation: While controlling industrial emissions are within the competence of the municipal bodies and the metropolitan authorities, the increase in air pollution is related significantly to vehicular growth. It could be argued that ULBs may substantially generate climate co-benefits from the demand-side by influencing usage of resource and consumption behaviour, but regulating emissions from the supply-side poses a major governance challenge. In addition, there are external challenges like growing population, horizontal coordination, political allegiance and transparency, hence putting mitigation of climate change on the backburner. Jawaharlal Nehru Urban Renewal Mission (JNNURM) was a major initiative taken by the national government in 2005 to invest more than Rs.1,00,000 crore in 65 cities for upgrading physical infrastructure, urban transport, housing for urban poor and good governance. The funding was on the basis of a strategic planning document called the City Development Plan. This was in addition to the development planning process in cities that produces a Master Plan. It is striking that any aspect pertaining to climate change mitigation or vulnerability was not a prerequisite for this strategic vision.

Local Governance: There is growing talk about changes in the traditional role of ULBs, from provider to facilitator. It is gradually being realized that policy and regulatory functions can be with the government, and the delivery functions can be assigned to the private sector. But, the crux of the matter is that without giving formidable powers to the ULBs to collect tax and user charges for financial sustenance, the current situation of urban governance in India is bound to remain unchanged.

Source Excerpt from Sethi and Mohapatra 2013

9 Research Design and Reporting

As in most developing countries, co-benefits have to play a significant role in urban India in order to allow the country to achieve human development goals and constrain the GHG emissions and pollution in its cities. Collaborative approaches between various functional domains or sectors and across working scales can facilitate the creation of local and global innovations, business opportunities, institutions and mechanisms to generate co-benefits as, for example, climate mitigation, air pollution and improvement of human health. However, we need to transform broad concepts into practical results for their implementation. The objective of this book is to discuss how we can develop policies and instruments to boost the capacity of societies, particularly in India, to generate climate, environmental and development oriented co-benefits in cities to synergistically achieve the local and global goals of sustainable development.

This edited book aims to build on the deliberations in this multidisciplinary area, and discuss pathways to identify and generate large urban co-benefits. These co-benefits can help us to understand how we make economic, policy and technological transitions that can promote more sustainable urban initiatives that cut across sectors and regions and lead to radical impacts on the way our societies use environmental resources and distribute its benefits. Most of the chapters in the book are based on empirical research conducted to understand the policies behind co-benefits in urban India and also align it with the current international discourse.

Accordingly, our analysis was structured into four main thematic areas: (1) concepts and theories behind cities and climate co-benefits; (2) contextualizing co-benefit issues across spatial scales and sectors; (3) sectoral analyses of co-benefits in energy, transport, buildings, waste and biodiversity; and (4) innovations and reforms needed to promote co-benefits in the urban context. The same is represented in Fig. 7. The contents within these thematic areas are briefly explained below.

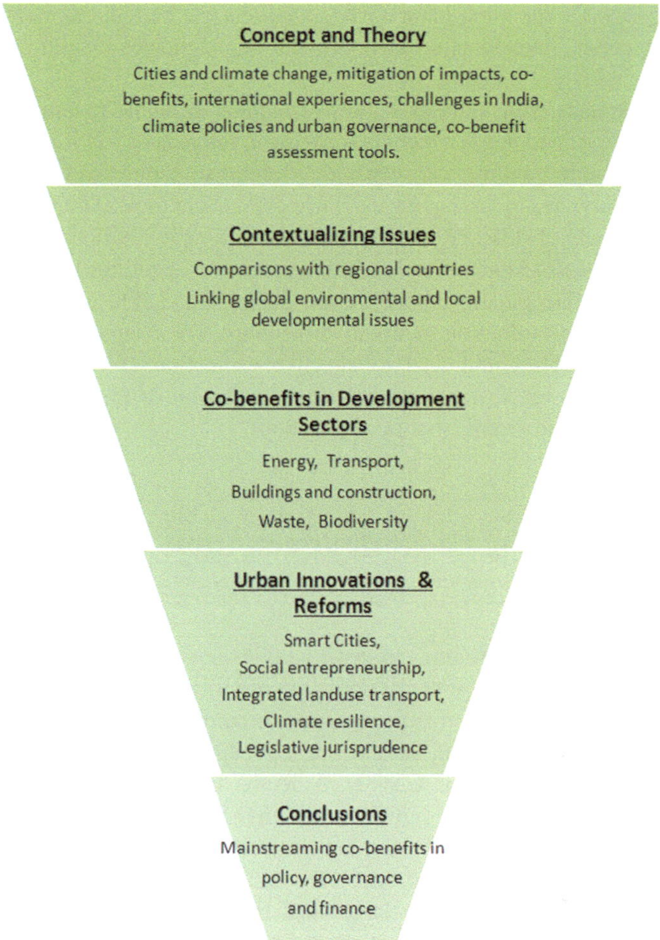

Fig. 7 Reporting in this book is based on four major research themes

9.1 Concept and Theory

Research in the cause and impact of climate change has witnessed widespread growth since the 1980s and 90s, while cities have recently started taking an interest in this inter-disciplinary arena. A co-benefits approach helps to guide interventions that are at the crossroads of climate action, development needs and improvement of local environment. Different international initiatives inform us that there are multiple opportunities for co-benefits in the urban context in developing countries. Experiences around the world show that development interventions for improving the quality of life of cities' inhabitants can also lead to climate change mitigation in several sectors. On the other hand, GHG emissions can be reduced with initiatives that also generate development outcomes. But this mandates measurable and verifiable data based evaluation.

Chapter 2 deals with numerous tools or models available to estimate climate, local environment and development co-benefits in urban areas. This chapter aims to have a fair understanding on generating and estimating co-benefits in this inter-disciplinary area by reviewing available assessment tools. The study reports in detail on 44 tools of different kinds—database, evaluation and simulation, across both mitigation and adaptation areas. In the process, it identifies the grey zones or research gaps—conceptual, methodological, empirical and of the policy-governance kind in systematic evaluation of urban co-benefits. The chapter investigates how future research tools could lead to effective estimation of co-benefits through assimilation of concepts, methodical analysis, inclusion of available, reliable and comparable data, and science-policy applications.

9.2 Contextualizing Co-benefit Issues Across Spatial Scales and Sectors

There are immense opportunities to generate climate co-benefits in urban areas in India, maybe more than most of the Asian countries, as cities in India lack several services and an adequate infrastructure and are burdened with air and water pollution and lack of employment, but they have not sprawled or 'overdeveloped' like cities in the West or China. Chapter 3 demonstrates urban air pollution and co-benefits, appreciating the case of India within the context of South Asia. Most of the South Asian countries are developing or emerging economies, and India is the fastest growing economy among them. The Indian subcontinent is comprised of hundreds of densely populated large and medium-size cities including five megacities (each with a population of more than 10 million). Constantly increasing energy-intensive activities in burgeoning cities and biomass burning in their hinterland are responsible for a large share in the unacceptably high emissions of health endangering and environment polluting gaseous and particulate pollutants. The air pollution problem in India and the rest of South Asia came into the limelight by the Atmospheric Brown Cloud (ABC), which is said to be responsible for 100,000 premature deaths annually in the South Asian region. The analysis presented in this chapter is expected to help initiate appropriate policy measures and suitable action plans to limit emissions and adopt ways based on a co-benefits approach that promotes sustainable development.

Aligning global and local environmental concerns with the immediate development goals of urban areas in developing countries is a daunting task. Chapter 4 attempts to examine how environmental issues in Indian cities align with local development issues. It reviews the status of a climate co-benefits approach in Indian cities, illustrated through various *sectoral case studies*. It discusses the context of climate and environmental co-benefits in urban India and relates it to urban policies. Implementation of the Asian Cities Climate Change Resilience Network (ACCCRN) programme in the three cities—Indore, Surat and Gorakhpur—has been discussed to highlight climate co-benefits at the local level, particularly the local environmental and social benefits. The chapter illuminates on future perspectives of integrating the climate co-benefits approach in urban development especially in the light of the newly instituted SCM.

9.3 Co-benefits in Development Sectors

The climate co-benefit theory mandates cross-linkages between different functional and administrative domains and development sectors to create synergy and positive benefits. An in-depth study of co-benefits accruing from energy generation, industries, transport, land use, buildings and construction, waste and biodiversity sectors in urban areas is paramount to understand the opportunities for co-benefits in each of the sectors. Dedicated sectoral chapters address the role of the respective sector in normatively and empirically contributing to India's climate change, the policy landscape, best practices if any and the potential co-benefits arising from technological, economic, social or regulatory mechanisms. For instance, Chap. 5 deals with the energy sector within the inter-linkages of economy, energy and the environment (commonly known as E3 or Nexus) in human settlements. In the absence of any previous investigations on E3 linkages in India, this study proceeds on the research premise that E3 challenges are most crucially played out in urban areas that possibly seek concerted technological and policy-oriented interventions to produce climate co-benefits. The chapter reviews global literature to theoretically comprehend E3 linkages in cities, followed by a detailed study on present conditions, issues and challenges of India's economy, energy and environment (emissions). The research underscores the role of urban India in these three sectors and generates, for the first time, super-imposed scenarios of the Nexus. The study culminates with a discussion on appropriate technology and policy/governance instruments that could possibly mitigate impacts and produce co-benefits in urban areas.

Climate co-benefits are often not the primary drivers for choosing policies and investments in the transportation sector. However, the growing awareness on the costs of climate change, the importance of the sector in the global greenhouse gas emissions profile and the growing opportunities provided by the emerging climate instruments have opened up a more holistic paradigm in assessing transportation investments. With growing negative externalities from the transport sector, limited availability of funds and long-term impacts of transport investments, there is an urgent need to maximize benefits through integration of multiple objectives including climate concerns in policies and projects. It is often assumed that applying the co-benefits approach to the transport sector is difficult, needs lot of resources and often is not straightforward. Chapter 6 tries to break this myth, as it rethinks co-benefits from urban transport. It establishes a case for quantifying co-benefits and presents a specific case study on the Chennai Metro Project. It shows how quantifying multiple co-benefits from transport projects could be simple yet effective in understanding the economic viability and long-term health impacts from the project. The application of the co-benefits approach in the assessment of transport investments aims to reveal the costs and benefits of the alternatives which are currently not captured in existing assessment approaches. In doing so, a clearer picture of the impacts of the different alternatives is generated and thus a better investment decisions can be made.

Chapter 7 assesses the role of the building and construction sector in India's climate scenario and the impact of climate friendly policies in achieving co-benefits in cities. The sector is known to play a significant role in creating employment and income and at the same time contributes significantly to greenhouse gas emissions. Hence it offers opportunities for climate co-benefits to reduce GHG emissions, mitigate climate change impact and promote economic growth and employment. The slums and informal housing sector requires special attention as it is more vulnerable to climate change impact. The chapter reviews Indian policies in the building sector which include the Energy Conservation Act 2001, Energy Conservation Building Code 2007, National Building Code, National Mission for Sustainable Habitat, rating systems for green buildings and so on. The chapter provides a brief assessment of the impact of these policies and makes suggestions for strengthening the same.

Rapidly increasing urbanization and economic development in India has impacted the quantity of waste generation. Presently, only 68% of Municipal Solid Waste (MSW) generated in the country is collected, of which 28% is treated by the municipal authorities. In the case of waste water, about 30% of the waste water generated from major cities of India is treated. Untreated waste leads to an adverse impact on public health and also creates various environmental problems including pollution of air, water and land resources. There have been efforts by the respective governments for initiating action on converting waste to energy but these initiatives are fragmented and have not been integrated into national policy frameworks. Further, waste to energy involves higher costs and a relatively higher degree of expertise amongst the governments about integration of various technologies. Chapter 8 aims to document and analyse waste to energy initiatives both in solid waste and waste water so as to identify and disseminate innovative urban practices. Among the various initiatives taken for 'waste to energy' in India, three cases have been selected, namely: Biogas-Fertilizer Plants (BGFPs) for generation, purification/enrichment, bottling and piped distribution of biogas in Talwade, Nasik district of Maharashtra; organic MSW based decentralized biomethanation plant at Pune city of Maharashtra and methane recovery and power generation from sewage treatment plants (STPs) in Surat city of Gujarat. This chapter makes an attempt to analyse and document their best practices for replication and thus find an implementable solution to the problem faced by many Indian cities.

Growth and development of a city modifies its biodiversity, which has a vital role in maintaining and improving the quality of urban environment. Chapter 9 in the book captures co-benefits associated with urban biodiversity by studying the growth and development of Delhi, the capital city of India. The research outlines an approach to assess co-benefits from the biodiversity pattern of the metropolis. The co-relation between habitat scales and levels of planning to arrive at a biodiversity profile of an urban area, are explained. The Global Biodiversity Assessment and Convention on Biological Diversity has brought forth the need to conserve biodiversity at global and local levels. Because of their diminishing numbers, the role of species in urban areas has become critical. So far, in cities, open spaces are supposed to mainly cater to aesthetic demands and recreation needs. But if one examines the environmental role they play in pollution abatement, water recharge, indicator of pollution, climatic amelioration, flood control and so on, there are many co-benefits. The strategy for conservation is related to the scale of habitat, thus a multi-scale strategy for conservation is explained in detail. The legal, governance and policy tools in India relating to national, sub-national and local levels are put forth to give a holistic picture of various aspects to be considered for mainstreaming biodiversity conservation in the urban planning process.

9.4 Promoting Co-benefits in the Urban Context: Innovations and Reforms

Cities are centres of knowledge and innovation, both technological and institutional, can make tackling the sector-based challenges noted above viable. Partnerships among different actors at multiple levels happen all the time in cities, and need to be understood and promoted by public policies from global to local level. Cities are also hubs for social and political movements that catalyse societal transformations. The sheer concentration of people, resources, knowledge, political power and economic activities in urban areas, if properly nurtured, can provide the political institutions, economies of scale and efficiency gains that lower the use of resources and energy, and thereby promote doing more with less, while offering fair outcomes to the most vulnerable people and the environment. Creating mechanisms facilitates development of local capacities in cities in order to scale up innovations. As many of the solutions to global concerns emerge at the local level, we require local and global efforts to create the capacity to innovate locally and spread those innovations globally to those who need them. Local groups must be able to adopt smart technologies for their immediate needs, absorb new practices and create the institutional mechanisms to increase their benefits.

These days, smart cities are creating a new buzzword across the world. Several examples boom in Japan, Europe, UAE and Singapore, while many others are shaping up on the drawing board. With the recently embarked Smart Cities Mission, the Government of India has strategically responded to both the international call for transformative sustainability, as well as growing domestic pressure in cities to innovate. Interestingly, there is neither an internationally accepted definition of a Smart City, nor does India have any national policy on urbanization. Within this science-policy vacuum, there is a fair degree of consensus on what a smart city looks like, but no understanding on what are the inputs and strategies to achieve one. With numerous expectations, inhibitions and euphoria around this theme, Chap. 10: 'Smart Cities: Opportunities to Enhance Quality of Life and Realize Multiple Co-benefits', attempts to systematically investigate what is a smart city, how it is different from similar prototypes like a sustainable, green and low-carbon city and what are the global best practices. The research addresses some important ideological, technical, societal, governance and financial challenges that India faces in attaining the '100 Smart Cities' goal, and explores how Smart Cities Mission could become a useful policy instrument in achieving multiple long-term co-benefits and enhancing quality of life.

But at the same time, India hosts the largest population in the world that lives without any access to a clean energy source like electricity and cooking gas. People with lack of energy access have been largely stuck up in vicious cycle of exclusion which may be a result or a by-product of their lower incomes. It is to be believed that energy, particularly electricity, has become the central focus of human existence and its inaccessibility is an indication of lack of access to many basic needs. The rural versus urban divide in access to energy is quite disturbing. While a majority of urban residents have privileged access to electricity due to its ready and economic availability in their vicinity, a large number of people from rural areas are poor, living in disadvantaged areas that continue to experience energy poverty. Given the constraints in the capacities of developing countries, it may be difficult to cater for the escalating energy needs in urban areas unless innovative methods are explored. However, what becomes very significant in the developing countries is the fact that the growing third sector is emerging very strong; in particular, social enterprises with their innovative strategies, processes and methods to provide access to energy and protect the environment from over usage of natural resources. With this background, Chap. 11, 'Social Entrepreneurship, energy and Urban Innovations', through a case study approach, attempts to explore how social entrepreneurs in India adopt better ways of providing energy to the excluded and marginalized social sections.

Other than innovations, promoting co-benefits in the urban context is also possible through policy reforms. The Government of India's policies like NMSH and the National Urban Transport Policy (NUTP) advocate integrated land use and transportation planning as a tool to reduce GHG emissions; there exists no policy to guide integrated land use and transportation in the cities or incorporate them in future decision making in similar matters. Chapter 12 devises a framework for such a state level integrated land use and transportation policy aimed at reducing GHGs and improving accessibility. The framework draws policy pointers from similar efforts globally to explore the barriers in implementing such a policy in the Indian context. The research also identifies existing schemes/programmes that can support the implementation for such a policy. The chapter concludes with a list of actions that would facilitate the implementation of an integrated land use and transportation policy in India. As noted above, mitigation and adaptation policies in an urban setting need to be complementary. Co-benefits that have been, in the past, essentially associated with the climate mitigation paradigm need to be integrated into climate adaptation of cities through improved urban planning. There is a need to address current risks and to begin building climate resilience into urban fabric and systems to tackle the likely future risks. Chapter 13 on promoting climate resilience in urban planning, offers a comprehensive discussion on one such initiative to integrate urban climate resilience accomplished in Gorakhpur, India under the ACCCRN programme.

An important actor in the urban reform agenda in India has been an active judiciary. Through public interest litigation, the Supreme Court (SC) and the High Courts have received claims from citizens, directly related to city development. Although the basis of the claims in front of the SC were the violation of a fundamental right, the judiciary was able to step into sectors such as land use, transport and crosscutting co-benefits such as public health by limiting air pollution. Now the case law of the SC has been particularly important for the capital city of New Delhi. Based on this example, Chap. 14 demonstrates how the SC, starting from the violation of the fundamental right to life, ended up making detailed policy and technological orders to local public authorities to tackle the problem of air pollution. The chapter explains the outcomes of these orders and co-benefits. Finally, it discusses the limits and opportunities of this rights-based approach for city development.

The cities have a key role to accrue co-benefits as they have jurisdictional and functional responsibility for planning and provision of services. The five-year plans, sectoral policies and projects comprise the urban policy framework in India and the past efforts have not paid much attention to climate co-benefits. It is only in the last decade that the programmes and policies have started paying attention to the climate co-benefits in their development agenda. There is a greater scope for mainstreaming a climate co-benefits approach into these policies and ensuring their effective implementation at the level of state and urban local bodies (ULBs). The governance and financing capacities of implementing agencies can play a key role in effective implementation of urban sector policies. Chapter 15, the concluding chapter, examines the means for mainstreaming climate co-benefits approach in urban policy, governance and finance to address climate impacts and reduce GHGs in India. It also sets up a way forward through a series of suggestions and recommendations that could possibly influence research, practice, institutions, finance and partnerships necessary to effectively actualize climate co-benefits in cities.

References

Ahmed, M., & Suphachalasai, S. (2014). *Assessing the costs of climate change and adaptation in South Asia*. Mandaluyong, Philippines: Asian Development Bank.

Angel, S., Sheppard, S. C., & Civco, D. L. (2005). *The dynamics of global urban expansion. Transport and urban development*. Washington, DC: World Bank.

Betsill, M., & Bulkeley, H. (2007). Looking back and thinking ahead: A decade of cities and climate change research. *Local Environment, 12*(5),447–456 (2007), doi:10.1080/13549830701659683.

Bulkeley, H., & Betsill, M. (2003). *Cities and climate change: Urban sustainability and global environmental governance. Studies in physical geography and environment*. Oxon: Routledge.

Census of India. (2011). *Provisional population totals-2011, Paper–II, Vol–II* (p. 1). New Delhi: Census of India.

CSE. (2014). Review of national mission on sustainable habitat. http://cseindia.org/content/national-mission-sustainable-habitat. Accessed on November 14, 2015.

Dhakal, S. (2009). Urban energy use and carbon emissions from cities in China and policy implications. *Energy Policy, 37*(11), 4208–4219 (2009). doi:10.1016/j.enpol.2009.05.020.

Dirgahayani, P. (2013). Environmental co-benefits of public transportation improvement initiative: The case of trans-Jogja bus system in Yogyakarta. *Indonesia. Journal of Cleaner Production, 58*(1), 61–73.

Dodman, D. (2009). Blaming cities for climate change? An analysis of urban greenhouse gas emission inventories. *Environ Urban, 21*(1), 185–201.

Doll, C. N. H., & Puppim de Oliveira, J. A. (2017). Climate co-benefits in urban areas. In C.N.H. Doll & Jose A. Puppim de Oliveira (Eds.), *Urbanization and climate co-benefits: Implementation of win-win interventions in cities* (pp. 1–23). Routledge.

Dubash, N.K. (2012). *Working paper 2012/1*: *The politics of climate change in India: Narratives of equity and co-benefits*. New Delhi: Centre for Policy Research.

Geng, Y., Ma, Z., Xue, B., Ren, W., Liu, Z., & Fujita, T. (2013). Co-benefit evaluation for urban public transportation sector—A case of Shenyang. *China. Journal of Cleaner Production, 58* (1), 82–91.

IEA. (2008). World energy outlook 2008 edition. international energy agency. http://www.worldenergyoutlook.org/media/weowebsite/2008-1994/weo2008.pdf. Accessed on February 12, 2015.

IGES. (2010). Fact sheet No. 1 What are co-benefits? The Institute for Global Environmental Strategies. http://pub.iges.or.jp/modules/envirolib/upload/3378/attach/acp_factsheet_1_what_co-benefits.pdf. Accessed on March 5, 2015.

Intergovernmental Panel on Climate Change (IPCC). (2007a). *Climate change 2007: Synthesis report. Climate change–Summary for policy makers*. Cambridge: Cambridge University Press.

Intergovernmental panel on climate change (IPCC). (2007b). In B. Metz, O. Davidson, R. Swart. and J. Pan (Eds.), *Fourth Assessment Report. Climate Change Mitigation*. Cambridge: Cambridge University Press.

Intergovernmental Panel on Climate Change (IPCC). (2014a). Contribution of Working Groups I, II and III to the Fifth Assessment Report of the Intergovernmental Panel on Climate Change. In R.K. Pachauri & L.A. Meyer (Eds.) [Core Writing Team] *Climate change 2014: Synthesis report*. Geneva, Switzerland: IPCC.

Intergovernmental Panel on Climate Change (IPCC). [Seto K. C., S. Dhakal, A. Bigio, H. Blanco, G. C. Delgado, D. Dewar, L. Huang, A. Inaba, A. Kansal, S. Lwasa, J. E. McMahon, D. B. Müller, J. Murakami, H. Nagendra, & A. Ramaswami]. (2014b). Human settlements, infrastructure and spatial planning. In Edenhofer, O., R. Pichs-Madruga, Y. Sokona, E.

Farahani, S. Kadner, K. Seyboth, A. Adler, I. Baum, S. Brunner, P. Eickemeier, B. Kriemann, J. Savolainen, S. Schlömer, C. von Stechow, T. Zwickel & J.C. Minx (Eds.), *Climate change 2014: Mitigation of climate change. Contribution of Working Group III to the Fifth Assessment Report of the Intergovernmental Panel on Climate Change*. Cambridge, United Kingdom and New York, NY, USA: Cambridge University Press.

Jiang, P., Dong, W., Kung, Y., & Geng, Y. (2013a). Analysing co-benefits of the energy conservation and carbon reduction in China's large commercial buildings. *Journal of Cleaner Production, 58*(1), 112–120.

Jiang, P., Yihui, C., Geng, Y., Dong, W., Xu, B., Xue, B., et al. (2013b). Analysis of the co-benefits of climate change mitigation and air pollution reduction in China. *Journal of Cleaner Production, 58*(1), 130–137.

Kurniawan, T. A., Puppim de Oliveira, J. A., Premakumara, D. G. J., & Nagaishi, M. (2013). City-to-city level cooperation for generating urban co-benefits: The case of technological cooperation in the waste sector between Surabaya (Indonesia) and Kitakyushu (Japan). *Journal of Cleaner Production, 58*(1), 43–50.

Martine, G. (2009). Population dynamics and policies in the context of global climate change. In J. M. Guzman, G. Martine, G. McGranahan, D. Schensul & C. Tacoli (Eds), *Population dynamics and climate change* (pp. 9–30). New York: UNFPA & London: IIED.

Mckinsey. (2010). *India's urban awakening: Building inclusive cities sustaining economic growth*. New Delhi: Mckinsey Global Institute.

MoEF. (2006). *National environment policy*. New Delhi: Ministry of Environment and Forests, Government of India.

MoEF. (2009). *India's GHG emissions profile*. New Delhi: Climate Modeling Forum, Ministry of Environment and Forests, Government of India.

MoEF. (2010a). *Climate change and India: A 4 × 4 assessment—A sectoral and regional analysis for 2030s*. New Delhi: Ministry of Environment and Forests, Government of India.

MoEF. (2010b). *India: Greenhouse gas emissions 2007*. New Delhi: Indian Network for Climate Change Assessment, Ministry of Environment & Forests, Government of India. pp. 1–54.

MoEF. (2015). *India's intended nationally determined contribution: Working towards climate justice*. New Delhi: Ministry of Environment & Forests, Government of India.

MoUD. (2011). *India's urban demographic transition: The 2011 Census results-provisional*. New Delhi: JNNURM Directorate and National Institute of Urban Affairs.

NAPCC. (2008). *National action plan for climate change 2008* (pp. 4–7). New Delhi: Ministry of Environment and Forests, Government of India.

OECD. (2009). *Cities, climate change and multilevel governance*. OECD Environmental Working Papers No 14. (pp. 30–44). Paris: OECD Publishing.

Oliver, J. G. J., Janssens-Maenhout, G., Muntean, M. & Peters, J. A. H. W. (2016). *Trends in global CO_2 emissions; 2014 report*. The Hague: PBL Netherlands Environmental Assessment Agency; Ispra: Joint Research Centre.

Puppim de Oliveira, J. A. (2009). The implementation of climate change related policies at the subnational level: An analysis of three countries. *Habitat International, 33*(3), 253–259.

Puppim de Oliveira, J. A. (2013). Learning how to align climate, environmental and development objectives: Lessons from the implementation of climate cobenefits initiatives in urban Asia. *Journal of Cleaner Production, 58*(1), 7–14.

Puppim de Oliveira, J. A., Doll, C. N. H., Kurniawan, T. A., Geng, Y., Kapshe, M., & Huisingh, D. (2013). Promoting win–win situations in climate change mitigation, local environmental quality and development in Asian cities through co-benefits. *Journal of Cleaner Production, 58* (1), 1–6.

Satterthwaite, D. (2009). The implications of population growth and urbanization for climate change. *Environment and Urbanization, 21*(2), 545–567.

Schroeder, H., & Bulkeley, H. (2009). Global cities and the governance of climate change: What is the role of law in cities?'. *Fordham Urban Law Journal, 36*(2), 313–359.
Sengupta, S. (2012). International climate negotiations and India's role. In K. Dubash Navroz (Ed.), *Handbook of climate change and India: Development, politics and governance* (pp. 101–117). New Delhi: Oxford University Press.
Sethi, M. (2015a). Location of greenhouse gases (GHG) emissions from thermal power plants in India along the urban-rural continuum. *Journal of Cleaner Production, 103,* 586–600. doi:10.1016/j.jclepro.2014.10.067.
Sethi, M. (2015b). Mapping global to local carbon inequities in climate governance—a conceptual framework. *International Journal of Development Research, 5*(8), 5387–5397.
Sethi, M., & Mohapatra, S. (2013). Governance framework to mitigate climate change: Challenges in urbanising India. In Huong Ha and Tek Nath Dhakal (Eds.), *Governance approaches to mitigation of and adaptation to climate change in Asia* (pp. 200–230). Hampshire, UK: Palgrave Macmillan.
Sethi, M., & Puppim de Oliveira, J. (2015). From global 'North–South' to local 'urban–rural': A shifting paradigm in climate governance? *Urban Climate, 14*(4), 529–543.
Seto, K. C., Güneralp, B. & Hutyra, L. R. (2012). Global forecasts of urban expansion to 2030 and direct impacts on biodiversity and carbon pools. In *Proceedings of the National Academy of Sciences*, 109, 16083–16088. doi:10.1073/pnas.1211658109, ISSN: 0027-8424, 1091–6490.
UN—United Nations. (1992). *United nations framework convention on climate change.* New York: United Nations.
UN—United Nations. (2007). *City planning will determine pace of global warming.* New York: United Nations.
United Nations (UN). 2016, *Draft outcome document of the united nations conference on housing and sustainable urban development (Habitat III),* Document A/CONF.226/4, e-book, United Nations. http://www.habitat3.org/the-new-urban-agenda. Accessed on November 10, 2016.
UN DESA. (2007). World urbanization prospects: The 2007 revision population database. United Nations, Department of Economic and Social Affairs, Population Division. http://www.google.co.in/url?sa=t&rct=j&q=&esrc=s&source=web&cd=1&ved=0ahUKEwiTp4Km_bnRAhXKOI8KHQN-BbgQFggeMAA&url=http%3A%2F%2Fwww.un.org%2Fesa%2Fpopulation%2Fpublications%2Fwup2007%2F2007WUP_Highlights_web.pdf&usg=AFQjCNEVD-nU6hxWN87kWBXwOSEX3_bfhQ&bvm=bv.143423383,d.c2I. Accessed on November 27, 2015.
UN DESA. (2011). Urban population development and the environment wallchart. United Nations, Department of Economic and Social Affairs, Population Division. http://www.un.org/en/development/desa/population/publications/urbanization/urbanPopDevEnv_2011.shtml. Accessed on May 7, 2015.
UN DESA. (2012). World urbanization prospects: The 2011 revision. United Nations, Department of Economic and Social Affairs, Population Division. http://www.google.co.in/url?sa=t&rct=j&q=&esrc=s&source=web&cd=1&cad=rja&uact=8&ved=0ahUKEwjLjtOI_rnRAhUFrY8K-HZ4bBVoQFggZMAA&url=http%3A%2F%2Fwww.un.org%2Fen%2Fdevelopment%2Fdesa%2Fpopulation%2Fpublications%2Fpdf%2Furbanization%2FWUP2011_Report.pdf&usg=AFQjCNF_Uxn6vzYP5TSK1M3to2WAkOZCiQ&bvm=bv.143423383,d.c2I. Accessed on December 5, 2015.
UNFCCC. (2015). Conference of parties, December 2015, FCCC/CP/2015/L.9. United nations framework convention on climate change. http://unfccc.int/resource/docs/2015/cop21/eng/l09.pdf. Accessed on January 24, 2016.
UN Habitat. (2006/2007). *State of the world cities report.* United Nations Human Settlements Programme, London/Washington DC: Earthscan.
UN Habitat. (2011). *Global report on human settlements 2011: Cities and climate change.* London/Washington DC: United Nations Human Settlements Programme & Earthscan.

UNU-IAS. (2013). *Urban development with climate co-benefits: Aligning climate, environmental and other development goals in cities*. Yokohama: United Nations University Institute of Advanced Studies.
United States Environmental Protection Agency. (2005). *Integrated Environmental Strategies (IES) Program*. Seoul, Korea: International Conference on Atmosphere Protection. CGE Training Workshop on Mitigation Assessments.
World Bank. (2010). *Cities and climate change: An urgent agenda*. New York: The World Bank.

Chapter 2
Co-benefits Assessment Tools and Research Gaps

Mahendra Sethi

Abstract Climate change and urban areas bear a symbiotic relationship. Research in the causes and impact of climate change has witnessed widespread growth since the 1980s and 90s, while cities have recently started taking interest in acknowledging their role towards climate. The co-benefits approach helps to guide interventions that are at the crossroads of climate action, development needs and improvement of local environment. There are numerous tools or models that are available to estimate climate, local environment and development co-benefits in urban areas. The aim of this chapter is to have a fair understanding of theoretical literature on generating and estimating co-benefits in this inter-disciplinary area by reviewing available assessment tools, and in the process, identify the grey zones or research gaps. The study reports in detail on 44 tools of different kinds—database, evaluation and simulation, across both mitigation and adaptation areas. It identifies the conceptual, methodological, empirical and policy-governance gaps in systematic evaluation of urban co-benefits. The chapter recommends how future research tools could lead to effective estimation of co-benefits through assimilation of concepts, methodical analysis, inclusion of available, reliable and comparable data, and science-policy applications. The findings can find pertinence to governments, policy-makers, researchers and practitioners alike as they help frame a new paradigm to ascertain co-benefits in the inter-disciplines of climate change and urban studies.

Keywords Co-benefits · Assessment tools · Mitigation · Adaptation
Evaluation · Simulation · Research gaps

M. Sethi (✉)
National Institute of Urban Affairs, New Delhi, India
e-mail: mahendrasethi@hotmail.com

M. Sethi and J.A. Puppim de Oliveira (eds.), *Mainstreaming Climate Co-Benefits in Indian Cities*, Exploring Urban Change in South Asia,
https://doi.org/10.1007/978-981-10-5816-5_2

1 Introduction: Climate Change and Urbanization

Climate change and urbanization—the two pose to be the greatest challenges currently facing humanity in the twenty-first century, and their effects are converging in dangerous ways (UN-Habitat 2011). This global trend prompts a comprehensive enquiry to systematically and strategically guide urbanization, urban policy and governance, with India as the case in point. India is undergoing similar patterns of rapid urbanization at 2.3% annually (UN DESA 2011) and as per estimates, significantly in non-megacities now with the prospect of its urban population breaking even with its rural population around 2039 (MoUD 2011). The evolution of policy and research discourse in the area of climate change, particularly GHG emissions and that of cities assessing their responsibilities towards it, is documented in Table 1. It is noteworthy that while research on the causes and impacts of climate change has witnessed widespread growth since the 1980s and 90s, cities have recently started taking interest in acknowledging their role towards climate, evaluating impacts, estimating footprints, performing participatory research, framing strategies, documenting and learning from best-practices, and so on, to chart out a coherent discourse at the global level.

Meanwhile the co-benefits approach helps to guide interventions that are at a crossroads of climate action, development needs and improvement of local environment. The co-benefits approach in climate change has surfaced as a means to achieve more than one outcome with a single policy (IPCC 2007, 2014). Climate co-benefits have been identified by various organizations as win–win opportunities to tackle climate change with other positive outcomes (OECD 2003; MOEJ 2008; Netherlands Environmental Assessment Agency 2009). The distinct feature about generating co-benefits is that it pro-actively envisages for positive externalities in climate friendly projects, cutting through multi-sectoral and multi-level challenges. There are numerous tools or models that are available to estimate climate, local environment and development co-benefits in urban areas.

The aim of this chapter is to have a fair understanding of theoretical literature on generating and estimating co-benefits in this inter-disciplinary area of cities and climate change by reviewing available assessment tools, and in the process, identify the grey zones or research gaps. Accordingly, the chapter reports on the normative understanding of estimating urban co-benefits in Sect. 2. This is supported by an exhaustive study of various tools that assist in estimating urban co-benefits across different sectors in Sect. 3, both climate mitigation and adaptation respectively. Section 3 discusses unresolved issues and knowledge gaps in systematic assessment of co-benefits, and the chapter concludes with possible research and practical applications in Sect. 4.

Table 1 Timeline showing developments of research and policy in disciplines of climate change and cities

Climate change	Cities
1976: Studies reveal CFCs (1975), methane and ozone (1976) seriously contribute to the greenhouse gas (GHG) effect	1992: City of Toronto in Canada becomes the first city in the world to inventorize its carbon emissions and formulate strategies to curb them
1987: Montreal protocol of the Vienna convention imposes international restrictions on emission of oxygen depleting substances	1996: Baldasano et al. publish the emission inventory for GHGs in the city of Barcelona
1988: WMO and UNEP establish the Intergovernmental Panel on Climate Change (IPCC)	2004: Dhakal accounts for energy emissions and carbon emissions for Asian megacities in China and Japan
1990: IPCC's 1st Assessment Report (AR): World has been warming and future warming likely. Industry lobbyists and some scientists refute the findings	2009: Satterthwaite theoretically investigates the contribution of urban areas across the globe to cause GHG emissions
1992: Conference in Rio de Janeiro produces UNFCCC, but US blocks call for serious action	2009: The world becomes majorly urbanized for the first time in human history
1995: IPCC 2nd AR detects 'signature' of human-caused GHG effect, declares serious warming likely in coming century	2009: Kennedy et al. publish emissions of 44 global cities and metropolitan regions
2001: IPCC 3rd AR finds 'new and stronger' evidence that humanity's emissions of GHGs are the main cause of warming seen in the second half of the 20th century	2009: Carney et al. assess emissions of 17 European cities, while ICLEI releases GHG data for 53 South Asian cities
2005: Kyoto Treaty goes into effect signed by major industrial nations except US, Japan, W. Europe. CO_2 levels in atmosphere reach 380 ppm	2010: Cities and climate change—An urgent agenda released by the World Bank
2007: IPCC's 4th AR concludes it is more than 90% likely that humanity's emissions of GHGs are responsible for modern day climate change	2011: Cities and climate change—Global Report on Human Settlements released by UN Habitat
2009: China overtakes the US as the world's biggest GHG emitter, although the US remains well ahead on a per capita basis	2009–12: Release of three international standards/protocol for city based GHG/carbon footprint analysis
2014: IPCC's 5th AR underpins the roles and responsibilities of settlements towards climate change	2014: Urban sustainability recognized as one of the aims of Sustainable Development Goals replacing MDGs post 2015
2015: The world concurs to the Paris Agreement, making Intended Nationally Determined Contributions as an instrument to curb GHGs	2015: New Urban Agenda adopted at the UN Conference on Housing and Sustainable Urban Development (Habitat III) in Quito, Ecuador emphasizes the need to address climate change in cities

2 Estimating Urban Co-benefits

The impacts of pursuing climate actions have been studied through traditional cost-benefit approaches (IPCC 2014). Internationally, mechanisms based on measurement, reporting and verification (MRV) have been accepted to guide climate action. These take into account multi-criteria analysis of risks, socio-economic and monetary costs and benefits. Measuring mitigation benefits within a particular sector is relatively easy based on activity data and level of technology. The assessment typically follows Kaya Identity taking into account emission factors, energy intensity, carbon intensity, and so on (Kaya 1990). Accordingly, UNFCCC has standard IPAT-based methodologies (Chertow 2000) for various GHG producing processes adhering to transparent, consistent, comparable, complete and accurate (TCCCA) principles, that while selecting an activity-technology mix, could be used to measure carbon saved, as a measure of co-benefits. However, as one moves from process-based activities (that typically follow life-cycle analysis or input-output analysis) to system-based ones relevant to settlements (which involve parallel activities generating GHG emissions from different functional sectors), it becomes a complicated procedure to measure their throughput. Estimation of co-benefits is impaired by inconsistencies in accounting for a city's GHG emissions. Similarly, there is apparently no single methodology to account for urban-level risks and vulnerabilities from diverse natural processes like floods, heat waves, sea-level rise, and so on. In order to simplify this rather analytical and calculation-based process, and assess its possible impacts and ramifications while selecting a policy alternative, there are several models or assessment tools being used worldwide that estimate actual or relative co-benefits and facilitate decision making (see Fig. 1 for clarity).

As a recent study by United Nations University—Institute of Advanced Studies (UNU-IAS) explains, any assessment tool or methodology has a trade-off between technical sophistication and its general application. Precise quantification requires a tool that will be difficult to use for all policy-makers from differing levels of technical expertise. This would limit its applicability and consequently its use.

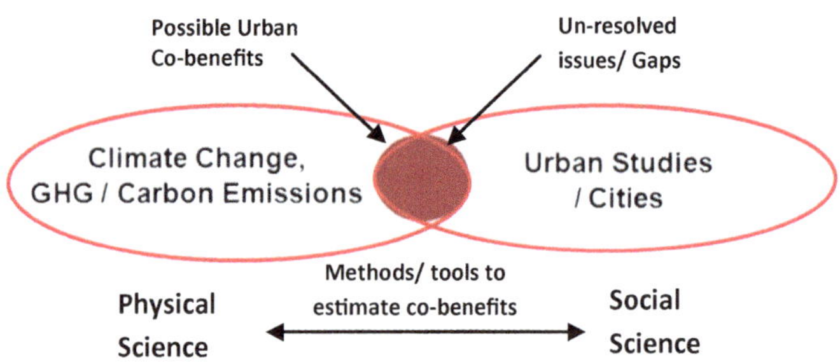

Fig. 1 Conceptual diagram showing research of co-benefits in the inter-disciplines of climate change and cities

More quantification also means more data requirements. This would be costly and time consuming to apply, too, because large amounts of data have to be collected and input into the system before the quantification tool can be used. Such a tool may not be viable in the context of many developing countries or in the case of small initiatives with limited resources (UNU-IAS 2013). Applying co-benefit tools in urban areas faces a similar challenge. Considering the need to promote a co-benefits approach in policies and projects, easy to use tools are required to explore, screen and forecast co-benefits associated with development alternatives and empirically facilitate decision making. Based on the above theory, this section makes an exhaustive review of 44 urban-co-benefit assessment tools, 27 in mitigation (Table 2) and 17 in adaptation (Table 3). The features of assessment tools are studied by accessing tools, reviewing their manuals, product literature, and so on, made freely available by their proponents on the internet.

3 Unresolved Issues and Knowledge Gaps

In the inter-disciplines of cities and climate change (mitigation and adaptation streams), the 44 co-benefit assessment tools could be classified into three broad categories:

(1) Informative or database tools: These tools offer a collection of data or information on certain urban, environment or development indicators to facilitate government and policy-makers. The main intent of these tools is to create awareness and provide fundamental information to stakeholders by showcasing best practices from other cities.
(2) Evaluation tools: These tools allow users to input their city data, and help them assess the existing situation, identify specific problems, evaluate the most suitable policy alternative from a series of choices. The scenarios are generated from standard programming within the tool and could not be significantly altered or customized by the users. The objective of these tools is to facilitate optimization and negotiation in current decision-making process.
(3) Simulation tools: These tools are advanced modelling tools, that could be simulated by the users to prepare alternative or future scenarios, create visualizations or generate impact assessments, forecast and modulate data figures or policy alternatives to create a range of real-like situations for enhanced decision-making.

Out of a sample of 44 co-benefit tools, both mitigation and adaptation studied here, about 23 (52%) are evaluation tools, 17 (39%) are simulation tools and only 4 (4%) are database tools. It needs to be mentioned that stand-alone information or database tools may seem scanty, but most of the evaluation and simulation tools do carry an in-built database for users to formulate and choose alternatives. Urban co-benefit tools in the mitigation area (27) exceed that of adaptation (17), the ones

Table 2 Tools used in assessing urban co-benefits in climate mitigation sectors

S. No.	Name	Description	Co-benefit sectors	Purpose
1	UNU-IAS transport tool (UNU-IAS 2013)	A quantitative spreadsheet of the transport sector with an institutional evaluation and emission reductions from local air pollution. The tool assesses implementation of policies in four areas: activity, shift, improve and fuel	Transport, local environment, GHG emissions	Ev
2	UNU-IAS energy system tool (UNU-IAS 2013)	A simulation model designed for evaluating the climate co-benefits of an urban energy system in the short term. The tool relates systematically the climate change based on the specific energy demand in different sectors in cities to the corresponding social, economic and technological factors that affect this demand from three main 'energy consumer' sectors: residential, commercial and service	Energy, residential/commercial, service sector, GHG emissions	Sm
3	UNU-IAS waste management tool (UNU-IAS 2013)	Evaluates the environmental impacts including GHG emissions and air pollutants accompanied by energy implications and cost-benefit analysis of the various waste management strategies by means of a life-cycle assessment. Considers a number of actions to recover material, energy and environmental impact through an integrated waste management system	Waste, energy, GHG emissions	Ev
4	HEAT+ (ICLEI 2014a)	International Council for Local Environmental Initiatives' online GHG emissions inventory tool, HEAT+ helps local governments account for GHG emissions, common air pollutants and volatile organic compounds. It helps in decision making by formulating a cost-effective manner in which to prepare inventory and forecast emissions, prepare action plans, track commitments, measure progress against targets, inform policy decisions, determine priorities, quantify progress, and report 'Scope' differentiated results	Energy, transport, local environment, waste, GHGs	Sm

(continued)

Table 2 (continued)

S. No.	Name	Description	Co-benefit sectors	Purpose
5	SynCity (2014)	The 'Synthetic city' tool kit facilitates the integrated modelling of urban energy systems including planning and layout of a new city, through to the socio-economic structure of the city and its activities, and the choice of energy carriers and technologies used to meet these demands	Urban planning (LU, spatial, socioeconomics), energy, GHG emissions	Sm
6	INDEX (Allen 2009)	INDEX is an integrated suite of interactive GIS planning support tools for: assessing community conditions, designing future scenarios in real-time, measuring scenarios with performance indicators, ranking scenarios by goal achievement and monitoring implementation. The tool designs and visualizes alternative planning scenarios, analyses and score their performance, and compares and ranks alternatives based on goal achievement plans	90 indicators that address land use, urban design, transportation, energy, local environment, urban planning (LU), GHG emissions	Db, Sm
7	I-PLACE^{3}S (SACOG 2014)	A web-based, publicly available modelling platform for measuring the climate impacts of the built environment developed by the State of California, has been tested to analyse the transportation and public health impacts of land development alternatives in Greater Seattle	Transport, public health, urban planning	Ev
8	Envision Tomorrow (2014)	The tool includes analysis of the physical and financial feasibility of development, and provides data on the carbon footprint of different scenarios-development using green building techniques, for example, or with greater density, walkable neighbourhood design, or multimodal transportation options	Residential/commercial buildings, urban planning, finance, transport, GHG emissions	Ev
9	Development pattern approach	A database of parcel-scale examples of streets, open space, and buildings across a range of densities, used in North Vancouver's sustainability master plan aimed at reducing GHGs by 80% by 2050. The spatial modelling incorporates GIS and Google SketchUp to produce dramatic visual and quantitative results when changes are made—replacing a single-family house with a duplex, for example	Urban planning (density, urban form), LULC, GHG emissions	Sm

(continued)

Table 2 (continued)

S. No.	Name	Description	Co-benefit sectors	Purpose
10	Practical evaluation tools for urban sustainability (PETUS 2014)	PETUS is an online tool to consider the impacts of buildings and infrastructure on the environment, society and the economy. The website includes information that can be used to analyse and improve the sustainability of urban infrastructure, of different sizes or types. It consists of alternative scenario building from energy, waste, water and sewage, transport, building and land use, etc.	Urban, planning (LU), buildings, energy, water, transport, waste, economy, LULC, GHG emissions	Sm
11	The BRTS evaluation and design (BEAD) tool (Shakti 2012)	Provide Bus Rapid Transit System (BRTS) planners and designers with the means to predict the performance of a designed/planned system and compare the same against alternate options based on passenger centric parameters such as passenger speeds and door-to-door travel times. This allows appropriate performance evaluation of systems based on its benefits in terms of attracting use and effecting long-term modal shift in favour of public transport	Transport (BRTS), local environment, GHG emissions	Ev
12	SMILE (EU 2014)	The SMILE database is a tool for local authorities to: find in-depth information and to exchange their experience and transfer their know-how, in the field of Sustainable Urban Transport Policies and Initiatives. It includes a compilation of 170 successful and replicable practices for sustainable mobility, designing specific measures and guidance for noise abatement planning	Transport, local-environment, GHG emissions	Db
13	Transport Emissions Evaluation Models for Projects (TEEMP) CAI Asia, together with ITDP, ADB, Cambridge Systematics and UNEP-GEF (2014)	Transport projects can either lead to net increase in greenhouse gas (GHG) and air pollutant emissions or they can result in emissions savings. Clean Air Initiative—Asia, has developed an Excel-based, free-of-charge spreadsheet model. The TEEMP tools were initially developed for evaluating the emissions impacts of ADB's transport project and have been modified and extended for GEF projects. These enable estimation of emissions in both 'project' and 'no-project' scenarios, using standard ASIF methodology and can be used for evaluating short- to long-term impacts of projects. TEEMP evaluates the impacts of transport projects on CO_2 emissions and to some extent	Transport, urban planning, (LU), GHG emissions, air-pollution	Ev

(continued)

Table 2 (continued)

S. No.	Name	Description	Co-benefit sectors	Purpose
		air pollutant emissions using data gathered during project feasibility and actual operations. The TEEMP enables rapid assessment of diverse transport project's (BRT, bikeway, MRTS, railway, etc.) CO_2 impacts and gives reasonable direction for action and alternate options evaluation		
14	CATCH project tool (EU 2013)	Aims to increase awareness of the impacts of travel behaviours, suggest ways that they can be changed, and indicate the impact that these changes could have. It is focused on climate change mitigation in the urban transport sector, but will also communicate other social, economic and environmental impacts of low carbon intensity travel behaviours	Transport, local-environment, socioeconomic, GHG emissions	Ev
15	Decision support system (DSS) (Carvalho n.d.)	Integrated in GIS for analysis of different transport policies, the tool assists transport administrators to enhance efficiency of the transportation supply while improving environmental and energy indicators. DSS works on three levels: the first performs the transport network analysis, the second assesses the energy consumption and pollutant emissions and the third evaluates the several policies selected. The evaluation of each policy scenario is based on a number of traffic, environmental and energy indicators. A multi-criteria analysis, where decision is based upon judging over appropriate weighted criteria, is adopted. Models are integrated in a GIS environment, which serves as the repository of the data as well as the user interface of the tool	Urban planning, energy, transport, local-environment GHG emissions	Db, Ev
16	Ecosmart landscape tool (University of California 2012)	Helps estimate the carbon and energy impacts of trees. The online tools provide quantitative data on carbon dioxide sequestration and building heating/cooling energy savings afforded by individual trees. Results can be used to estimate the GHGs benefits of existing trees, forecast future benefits, and facilitate planning and management of carbon offset projects. Carbon calculations are based on methodology	Residential/commercial, buildings, energy, LULC, GHG emissions	

(continued)

Table 2 (continued)

S. No.	Name	Description	Co-benefit sectors	Purpose
		approved by the Climate Action Reserve's Urban Forest Project Protocol. This software allows organizations to integrate public tree planting programmes into their carbon offset or urban forestry programmes		
17	The vulcan project (Arizona State University 2014)	The Vulcan project has achieved the quantification of the 2002 US fossil fuel CO_2 emissions at the scale of individual factories, power plants, roadways and neighbourhoods on an hourly basis. It has built the entire inventory on a common 10 km × 10 km grid to facilitate atmospheric modelling. In addition to improvement in space and time resolution, Vulcan is quantified at the level of fuel type, economic sub-sector, and county/state identification	Residential, commercial, industrial energy, urban planning (spatial), economy, GHG emissions	Db
18	Energy demand characterization (Webster et al. 2009)	Characterizing residential energy use involves quantifying total annual energy consumed for space heating and air conditioning, hot water, lighting and appliance use in a house or apartment unit and tested in Canada. It also involves describing features of the dwelling that impact on household energy consumption, such as geometry (size, number of storeys and other characteristics), and mechanical systems (such as furnace and hot water heater); occupancy is also factored into the characterization of residential energy use. By integrating residential energy figures, building characteristics and occupancy information in GIS, a place-based, scalable characterization of residential energy use can be achieved. Using this method, urban form variables contributing to energy use can also be analysed, yielding information to support a variety of urban planning decisions	Energy, urban planning (LU), residential, built-environment/urban form, GHG emissions	Sm
19	Neighborhood explorations: this view of density (SFLCV 2014)	The tool has a calculator that explores how neighbourhood density impacts the environment (land, materials, energy and driving). Wander around San Francisco. It shows photographs of differently dense neighbourhoods and reports their land use intensity, road area, water use, vehicles, parking status, average vehicle miles travelled or fuel use, air pollution and GHGs from driving	Transport, urban planning, GHG emissions, air-pollution, water	Db

(continued)

Table 2 (continued)

S. No.	Name	Description	Co-benefit sectors	Purpose
20	Tool for evaluating neighbourhood sustainability (CMHC 2015)	Develops a model of GHGs from personal urban transportation given variations in neighbourhood characteristics, including community and housing design, socioeconomic make-up, and locational factors. The main purpose of the study is to develop a spreadsheet tool that enables the user to estimate and compare various neighbourhood scenarios for GHGs from urban travel. The results provide valuable insight into how communities can be designed and planned to reduce GHG emissions from passenger travel in urban areas	Transport, urban planning, GHG emissions	Sm
21	Cool spots (Allen 2009)	An integrated land-use/transportation planning technique called Cool Spots is able to quantify GHG emissions of alternative community scenarios in real-time at public meetings, and in doing so, assist stakeholders in assembling climate-friendly urban development plans. Cool Spots are neighbourhoods where per capita GHG emissions are reduced below current community rates by co-locating energy-efficient buildings with multimodal travel environments and clean energy supplies. In a case study, the technique has been applied to a 300-acre transit station area in Elburn, Illinois using INDEX planning support software and a series of 'digital charrettes'. Stakeholders iterated to a preferred station area plan that reduces GHG emissions 32% below the community's current emission rate	LULC, urban planning, transportation, built-environment, energy, GHG emissions	Sm
22	Community energy and emissions inventory (CEEI 2014)	CEEI collects GHG data to calculate the size of each sector's carbon footprint in each local government jurisdiction across British Columbia. Additionally, it monitors supporting indicators from core sectors and other sources to help track the progress of local government efforts to reduce GHG emissions across their communities. It estimates GHG emissions from on-road transportation, buildings and solid waste. LULUC from deforestation and enteric fermentation from livestock available at District level	Energy, waste, buildings, transportation, LULCF	

(continued)

Table 2 (continued)

S. No.	Name	Description	Co-benefit sectors	Purpose
23	Community viz (APA Planners 2011)	CommunityViz provides options for 3D visualization in the Scenario 3D and Scenario 360 plugins. CommunityViz also allows users to export and view their work in ArcGIS Online, Google Earth and other KML/KMZ viewers such as ArcGIS Explorer. In CommunityViz Scenario 360, users can create their own analyses across multiple scenarios using custom formulas, indicators, and charts which all update dynamically in real time as the user makes changes on the map or to the calculations. CommunityViz contains 360 Indicators Wizard which can produce up to 101 indicators, the Custom Impacts Wizard to aid in designing selected indicators	Urban planning (LU), transportation and resource management	Sm
24	I-PLACE^{3}S (Urban Sim 2014a)	I-PLACE^{3}S is a software tool that facilitates a kind of planning known as scenario planning. It provides a web-based platform from which to communicate ideas, store data, and analyse potential outcomes. In this tool, several options are considered and objectively evaluated against quantifiable criteria	Transportation, LU, environment, health, economic development, demographics, etc.	Sm
25	MetroQuest (Planning Tool Exchange 2014a)	MetroQuest is an integrated 3D scenario planning software platform, designed to produce powerful visual displays as well as substantive information on scenarios, proposals, and tradeoffs. It is designed to substantively engage people in planning. Municipalities and planning agencies use MetroQuest for data analysis, decision-making, facilitation and dialogue, public participation and outreach, scenario planning using maps, charts and graphs, pictures, and text	Inter-sectoral	Sm
26	UPlan (ICE 2014)	Developed by Information Center for the Environment, UPlan is a simple rule-based urban growth model intended for regional or county level modelling. It estimates space needed for each land use type calculated from simple demographics and assigned based on the net attractiveness of locations to that land use (based on user input), locations unsuitable for any development and a general plan that determines where specific types of development are permitted	Urban planning (LU, demography), transport	Sm

(continued)

Table 2 (continued)

S. No.	Name	Description	Co-benefit sectors	Purpose
27	CITYgreen (Planning Tool Exchange 2014b)	A GIS software program for mapping, measuring, and analysing the storm water, summer energy savings, carbon storage and sequestration, air quality, and wildlife of urban ecosystems. CITYgreen software can provide complex ecological analysis for presentation use in the planning process and help communities determine steps needed to be taken to reach sustainable status	LULCF, urban planning, water, ecology	Ev
28	Athena—impact estimator for buildings (Planning Tool Exchange 2014c)	The Impact Estimator for Buildings gives architects, engineers and analysts access to advanced life-cycle inventory data without requiring advanced skills. The Impact Estimator provides a cradle-to-grave life-cycle inventory profile for a whole building. The inventory results comprise the flows from and to nature: energy and raw material flows plus emissions to air, water and land. The software reports footprint data for the following environmental impact measures consistent with the latest US EPA methodology: global warming potential, acidification potential, human health respiratory effects potential, ozone depletion potential, smog potential, and eutrophication potential and reports fossil fuel consumption	Residential/commercial buildings, energy, water, GHG emissions	Ev

Key Db database tools, *Ev* evaluation tools, *Sm* simulation tools, *GIS* geographical information system

Table 3 Tools used in assessing urban co-benefits in climate adaptation sectors

S. No.	Name	Description	Co-benefits sectors	Purpose
1	Urban CLIM (APN 2015)	An urban climate change decision support tool that considers major sectors: climate related hazards resilience, water, transport, and health. It follows a participatory assessment approach through working with urban policy-makers and planners from targeted Asian cities	Hazards resilience, water, health, transport	Ev
2	Cascadia's climate impact decision support tool (CIMPACT-DST) (USAID 2014)	Part of a pilot project in the city of Hue, Vietnam to develop and deploy a customized version of Cascadia's CIMPACT-DST for country's 750 cities that will be used by local city staff to improve the city's resilience to climate change impacts		Ev
3	Climate app (Deltares 2014)	Assessment of safety against flooding, heavy rainfall, drought and heat. It has been developed for worldwide application and has been tested in Ho Chi Minh City, Copenhagen and New Orleans. In the Netherlands, the app has been tested in Rotterdam, Delft and the Province of Utrecht. The app offers a wealth of information relevant for any new building, restructuring or renovation project. Based on simple criteria such as scale, land use and product type, the app selects and ranks possible climate adaption measures	Hazards resilience, buildings, urban planning (LU)	Ev
4	Sustainable urban development planner for climate change adaptation (SUDPLAN 2014)	The EU-funded project has an innovative web-based tool for urban planners that takes into account future climate change forecasts. It ensures that buildings and infrastructure stand the test of time and changing weather patterns. SUDPLAN provides urban planners an easy-to-use 3D modelling, simulation that combines historical climate data with future climate and emissions forecasts, including gridded information on present and future extreme rainfall, temperature, river run-off and air pollution. Instead of using global forecasts, SUDPLAN scales the data first to the European level and then to the urban level to generate best possible, localized precipitation, temperature, hydrological and air quality data from 1960 to 2050 and beyond. The system was field tested in four pilot trials in Stockholm, Wuppertal, Linz and a 150 km^2 area around Prague	Buildings, infrastructure, weather, hydrology, air quality	Sm

(continued)

Table 3 (continued)

S. No.	Name	Description	Co-benefits sectors	Purpose
5	Stadtklimalotse (Urban climate pilot) (Urban Nexus n.d.)	An online decision support tool for urban climate change adaptation, aimed at planners and policy-makers from small and medium-sized towns and cities, who have a need for quick and easy access to information. It is mainly focused on Germany and contains 138 measures in 10 fields of action (energy, health, tourism, water, infrastructure, transportation, green spaces, air quality, agriculture, forestry), 330 links to legislative texts and 61 examples for planning and implementation of measures; further background information on climate change research and an integrated tool for a quick urban vulnerability assessment. This has been applied in Stuttgart as an integrated regional approach to adaptation, encompassing the city and surrounding towns	Energy, health, tourism, water, infrastructure, transportation, green spaces, air quality, agriculture, forestry	Ev, Db
6	Green clime adapt (EU n.d.)	Green tools for urban climate adaptation. Demonstrates climate adaptation in urban areas using innovative green tools such as open stormwater systems, green facades and green roofs	Urban planning, built-environment, hydrology, local-temperatures/heat islands	Ev
7	Smart urban adapt (UN-Habitat 2009)	A cloud-based, user friendly hub, developed by ETH Zurich that integrates all city planning applications and services. Smart Urban Adapt's revolutionary tool integrates available data and models with visualization tools and an interactive front end to provide city planners and public authorities with the user-friendly software they need	Urban planning, infrastructure, hazard resilience	Ev
8	Impacts of climate change on urban infrastructure and the built environment (NIWA, MWH, GNS and BRANZ 2012)	It is a resource to help planners, engineers, asset managers, and hazard analysts in New Zealand urban councils understand and evaluate the potential impacts of climate change in their city. The Toolbox is designed with an overall 5-step evaluation framework represented by the 'trays' in the Toolbox. Within each tray are downloadable reports (or 'tools'), each with a specific purpose	Infrastructure, built-environment, hazard resilience	

(continued)

Table 3 (continued)

S. No.	Name	Description	Co-benefits sectors	Purpose
9	UN Habitat's Planning for climate change: a strategic, values-based approach for urban planners—toolkit (UN-Habitat 2014)	A resource and planning guide developed for city planners and other professionals to better understand, assess and take action on climate change at the local level. The manual is organized around a four-module strategic planning approach that correspond to four key strategic planning questions: What is happening? What matters most? What can we do about it? Are we doing it? It also plots risks and priority areas, capacity assessment of institutions, etc. Option identification, selection and implementation, monitoring and evaluation	Vulnerability assessment of weather and climate events, impacts on people (health, lives), places, institutions and sectors, hazard mapping, socio-economic and community-based sensitivity assessment, infrastructure (water, power), agriculture	
10	Urban adaptation support tool (EU and EEA 2014)	The Urban adaptation Support tool has been developed as an offshoot of the general Adaptation Support Tool on Climate-ADAPT platform, recognizing the need of European urban adaptation decision-makers and practitioners in cities to be guided through the main steps of the adaptation process. It gives them easy access to relevant adaptation information, data, tools and guidance specifically tailored for urban settings. The Urban Adaptation Support Tool consists of six steps: (a) explore risks and vulnerabilities to current and future climate (b) identify and assess adaptation options (c) develop and implement a climate change adaptation strategy and/or action plan, and (d) monitor its results		
11	Adapting to urban heat: a tool kit for local governments (Georgetown Climate Center 2012)	This Urban Heat Tool Kit is designed to help local governments reduce the effects of increased heat on their communities and citizens. It provides an analytic tool for policy-makers to consider a combination of four built-environment changes (cool roofs, green roofs, cool pavements, and urban forestry), providing clear criteria for selecting among these approaches. It also examines the roles government can play in pursuing these changes: shaping government's	Built-environment, LULUCF, urban heat	Ev

(continued)

Table 3 (continued)

S. No.	Name	Description	Co-benefits sectors	Purpose
		own operations, mandating or providing incentives for private choices, and engaging in public education. It offers a complete list of options and the means to select among them to fit particular circumstances		
12	Cities for climate protection ICLEI's adaptation framework (ICLEI 2014c)	This toolkit outlines an adaptive management process and provides a set of tools and exercises meant to assist councils as they work through the process, to: (a) establish an interdisciplinary approach to information gathering for the development of climate change scenarios; (b) identify their current risk management systems in light of how appropriate they are to guide decisions in an environment with complex, changing inputs; (c) examine climate change scenarios and projections to understand the potential impacts climate change may have on councils and their communities, and the associated risks/opportunities; (d) analyse, evaluate and prioritize the risks/opportunities identified by council according to its local and regional context, adaptive capacity and the likelihood and consequence of each risk/opportunity; I explore treatment options for the prioritized risks/opportunities to develop an adaptation action plan; (f) establish strategies for monitoring the implementation of the adaptation action plan and reviewing its outcomes; (g) build the personal capacity of participants to deal with complexity and uncertainty	Climate vulnerability, human health, biodiversity, water, agriculture, infrastructure and the built environment, public communications, natural resources, sports grounds, economic development and emergency services, financial-legal implications	Ev
13	UK local climate impacts profile (LCIP) (UKCIP 2014)	LCIP is a simple tool designed to help organizations to assess their exposure to the weather. It can be used as a standalone tool, or as a step in a risk-based framework such as the Adaptation Wizard. It assesses current and future climate vulnerability, explored adaptation options and then helps to monitor and review	Impacts of filter by climate risk: coastal impacts, cold weather, drought, flooding, heatwave, wind damage, agriculture, built environment, business, leisure	Ev

(continued)

Table 3 (continued)

S. No.	Name	Description	Co-benefits sectors	Purpose
14	Disaster resilience scorecard for cities (UNISDR 2014)	Developed by IBM and AECOM for UNISDR, based on its 'Ten Essentials', the scorecard provides a set of assessments that will allow cities to understand how resilient they are to natural disasters. It is intended to enable cities to establish a baseline measurement of their current level of disaster resilience, to identify priorities for investment and action, and to track their progress in improving their disaster resilience over time. It consists of 85 disaster resilience evaluation criteria and focuses on the following aspects: research, organization, infrastructure, response capability, environment and recovery. Each evaluation criterion is broken down to set out the aspect of disaster resilience being measured, an indicative measurement and the measurement scale (from 0 to 5, where 5 is best practice)	Vulnerability and resilience of ecosystem services, food, shelter, fuel, infrastructure (water, energy, health, gas, telecom, transport), administration, computer systems, LU, buildings, law and order institutions, planning, etc.	Ev
15	GeoNode (World Bank 2012)	A platform for the analysis, management, and the web-based publication of geospatial data. It brings together a mature and stable open source software projects under a consistent and easy-to-use interface allowing users, with little training, to quickly and easily share data and create interactive maps. A new approach to spatial data infrastructure focuses on users and collaboration with simple web-based tools, allowing them to: (i) edit metadata in the same place; (ii) set up privacy controls to restrict access as needed; (iii) download data in a variety of formats; (iv) use data in the system to create maps; and (v) export maps to other websites or PDF	Spatial planning, sea-level rise, inundation hazard, demography, infrastructure investment decisions	Sm
16	Indonesia scenario assessment for emergencies (World Bank 2012)	A suite of tools that close the loop between sharing data and actionable information to support resilient decision-making. InaSAFE is designed to work as a web-based tool on top of the GeoNode open source geospatial data management platform or a desktop system using the QuantumGIS open source software. It combines the critical elements of GIS analysis with the ability to quantify impact metrics that can be used for informed decision-making. Through the pilot engagement of	Spatial planning, flood hazards, demography, infrastructure (dredging) investment decisions	Sm

(continued)

Table 3 (continued)

S. No.	Name	Description	Co-benefits sectors	Purpose
		the Building Urban Resilience Initiative, key stakeholders were exposed to beta functionality of InaSAFE and future work will further develop applications in infrastructure investment decisions		
17	Toolkit for resilient cities (ICLEI 2014b)	A research project by Arup, RPA and Siemens, it is a manual that provides an evaluation framework to assess impacts from risks, hazards and vulnerabilities for city stakeholders on five principles (resilience performance indicators) of robustness, redundancy, diversity and flexibility, responsiveness and coordination	Water, energy, transport, building systems	Ev

Key Db database tools, *Ev* evaluation tools, *Sm* simulation tools, *GIS* geographical information system dealing with transportation and energy sector being the most common ones. About 18 out of 27 (i.e. two-thirds) of the mitigation tools consider transport co-benefits, either individually or as a part of other sectors; while 14 out of 27 (around half of them) consider energy co-benefits, demand side, supply side or both. Another two-thirds (i.e. 17) of the mitigation tools consider spatial parameters like urban planning, density, urban form, built-environment, land use, and so on, though only seven of them (Heat+, Syncity, DSS, Vulcan, CEEI, Community Viz and MetroQuest) show some semblance towards following an integrated approach. Meanwhile, only six co-benefit tools (excluding I3 place and Metroquest) actually use a GIS-based interface to evaluate or simulate city emissions. It is notable that while tools dealing with mitigation co-benefits are more GHG quantification-based and abundant in practice, co-benefit tools in adaptation are comparatively scarce, deal with non-quantifiable data, and rather end up recreating a normative but meaningful decision-making process. They typically follow the following four basic steps: (a) identification of risks, hazards and vulnerabilities; (b) framing possible policy alternatives; (c) selecting the most optimal adaptation/resilience pathway; and (d) implementation and monitoring of results. At best, it is the integrated mitigation tools and some of the adaptation tools that attempt to bring together the assessment of climate action, development needs and local environment.

But producing comprehensive, three-dimensional, empirical and accessible tools for assessing climate change in urban areas is a daunting task. As noted above, most of them lack a systems approach, with no single tool that is common for mitigation and adaptation. Even within mitigation tools, there is hardly any tool that accounts for sequesterization of GHG emissions. A crucial need is for tools that can be applied to a range of scales—from an individual building, major project, or new neighborhood, right up to a broader region, where there are many variables. City managers should be able to predict definitively whether any planning alternative for future, greenfield or redevelopment, offsets emissions or contributes to a resilient society. The above analysis shows that while researchers and software developers have begun to create meticulous co-benefit tools, it needs to be seen how often and seriously practitioners and decision-makers apply them to report alternatives, and appraise new city developments. At the same time, like the inconsistent emission inventories that account for GHGs, there are several research gaps in estimating intermediate urban co-benefits.

A *conceptual gap* arises from lack of sufficient scientific understanding in prevailing knowledge or theories about urban causations to climate change (carbon footprints) or climate impacts on cities (risks, vulnerabilities, hazards, etc.). There are several tools that deal with micro-level assessments, such as impacts of an individual or a building project, but are not relevant to discern the implications of larger phenomenon like national or regional urbanization. Conceptual gaps are a consequence of our limitation in perceiving cities as a system. It has been acknowledged that an urban system is essentially an open system with extensive cross-boundary interactions for food, water, energy, mobility, material and services. Therefore, in urban regions, a clear understanding of the concept of urban carbon footprints is needed (Dhakal 2010). Revi (2008) proposes one of the first climate agendas for cities in India. It considers the likely changes that climate change will bring in temperature, precipitation and extreme rainfall, drought, river and inland flooding, storms/storm surges/coastal flooding, sea-level rise and environmental health risks, and who within urban populations are most at risk. It describes a possible urban climate change adaptation framework, including changes needed at the national, state, city and neighbourhood levels, and linkages to mitigation, but falls short of presenting a similar mitigation-centric research for Indian cities. On the other hand, Sethi and Mohapatra (2013) put forth a mitigation inclusive urban governance framework for Indian cities. In the process, they acknowledge that it is imperative to further research on how urban emissions vary with size, structure and form of the urban centre and the corresponding activity/land use mix. As such it is pertaining to follow a holistic approach rather than a piecemeal or silos/sectoral conceptualization of causations and impacts of climate change.

A *methodological gap* is caused by a gamut of reasons that inhibit proper assessment of a city's GHG emissions or climate vulnerability. There is no standard tool to account for climate needs-mitigation and adaptation, arising from urbanization/urban issues at different scales. There is no universally accepted methodology amongst cities to inventorize their GHG throughput, risks and vulnerabilities, thereby stretching inconsistencies in assessing climate co-benefits across different sectors, making it difficult to compare cities and draft climate strategies. The study of assessment tools corroborates some of the existing literature which reckons that a key gap remains on analyses involving the opportunity to optimize the urban system as a whole in an integrated fashion. Going beyond the conventional carbon management opportunities in cities, few studies have explored, separately, the opportunities to sequester carbon in urban systems too—integrated analysis, not only limiting to the efficiency gains in traditional sectors but also covering all infrastructure, technology, urban design and sinks (noting their other ecosystem services as co-benefits too) can provide a comprehensive picture of carbon management opportunities (Dhakal 2010). As pointed out, this is because of pre-occupancy of emission-based methodologies on Kaya Identity, activity data and impact factors. It has been stressed that city systems would require developing a new type of methodologies going beyond product-based life-cycle assessments, input–output analysis and household income–expenditure surveys. The case is even more pressing for developing countries that do not have an indigenous tool and rely on imported models. Further there is also an issue of comparability here as well, in terms of methodologies followed, sectors covered, causations and impacts considered, different urban definitions and spatial boundaries, and so on. With the multitude of agencies and institutions pushing their own assessment tools, it seems quite unlikely to have a single common tool to assess urban co-benefits in the near future. At the same time there is plenty of scope for other methods and metrics, like the ones associated with spatial or physical parameters of the urban-metabolism kind as demonstrated in Sethi (2017), to be employed for expansively estimating co-benefits.

An *empirical gap* crops up from lack of sufficient scientific or evidential data on climate impacts or carbon footprints to support the prevailing theoretical knowledge. This study of 44 assessment tools demonstrates little application in the Indian case, and validates prevailing literature. Gillenwater (2011) while studying inter-linkages between climate science and research on one hand and sustainability discourse on the other (to which urban studies is also a part of), identifies a large gap in GHG measurement and management. A large volume of literature on embodied energy has been published using input–output analysis, material flows, and life-cycle assessment at national, sectoral and product levels, but their application to study the carbon footprint of cities is less—more so while understanding correlations between urban development pathways and GHG emissions (Dhakal 2010). In the case of India, it has been established that certain spatial determinants of the city like form, density and public transport, are responsible for carbon emissions, but there is hardly any evidence on the nature of this causation, particularly for India (Sethi 2013). It is only possible with availability of reliable and ready-to-use empirical data on cities. A lot of relevant data is generated by the public sector. Master plans that carry urban baselines exist only for 23% of Indian cities (The Hindu 2012), data for GHG emissions is available for less than 50 cities, while resilience and adaptation plans are underway for about a handful of them.

A *policy-governance gap* is a consequence of limited knowledge with cities about their roles and responsibilities to effectively respond and take climate action. In developing countries, the capacity, resources and jurisdiction of city governments are limited. Thus, the role of municipal government is absolutely necessary but not sufficient for urban carbon management. In spite of best intent, owing to major conceptual, methodological and empirical gaps, one rarely sees a case where an assessment tool has been used fittingly to influence policy discourse. Research advances are needed to fill the science-policy vacuum, in determining rational and innovative ways of reducing urban vulnerabilities and GHGs alike. City authorities can effectively act as a facilitator if not actor in many cases and such roles need a better understanding (Dhakal 2010). This resonates with the existing literature in urban mitigation that points out that the local government can govern in four ways: self-governing (reducing GHG from municipal actions and activities), governing through legislating, governing by provisioning and governing by enabling (Alber and Kern 2008). Therefore, future research ought to focus on who can influence how much of cities' emissions and climate impacts within major urban governance stakeholders, and to understand the opportunities and challenges to streamline existing modes of governance to match with such influence or the intended one. It is high time that carbon footprint analysis, co-benefit tools and climate resilience strategies should be made integral to urban policy framework like regional plans, master plans and local area plans using regional, city and local information, norms and standards, from public environmental agencies, planning authorities, and so on.

4 Conclusion

In order to realize urban co-benefits in climate research, policy and local action, it is imperative to rationally evaluate causations and impacts of climate change against their development implications. Urban co-benefit assessment tools facilitate informing best practices, evaluating alternative paradigms of development and at times simulating prospective impacts. In this process there are still several grey zones, or research gaps of the conceptual, methodical, empirical and governance kind, that pose to conceal some of the co-benefits. In order to deal with the unresolved issues in effective application of co-benefit tools, they should further focus on: (a) assimilation of mitigation sectors and adaptation risks, vulnerabilities and hazards; (b) greater application of spatial-temporal and real-time analysis; (c) utilization of available national/regional datasets and city indicators; (d) outputs that could directly orient governance strategies, through greater science-policy convergence; and (e) mandatory application of assessment tools by city authorities.

The analysis of co-benefit tools used worldwide (in the inter-disciplines of climate change and cities) made here does not critically prioritize or differentiate tools in terms of their preferred place of application—developing versus developed cities. It needs to be put into perspective that assessment tools are only a medium to achieve the desired co-benefits. Normatively it is contingent upon the intended policy objective a city adopts (to mitigate, adapt or generate co-benefits depending upon a variety of local factors like its institutional or financial capacity, socio-economic circumstances and political pressures, etc.), which may or may not be necessarily adhering to a development (developed versus developing) schism, given that cities globally show lesser variability in development indicators than their respective nations. Also, given the paucity of data for environmental indicators in cities of the developing world, it would be virtually impossible and probably too early to corroborate such assertions. Nevertheless, these arguments are in principle valid and prospective in understanding cities of the developing world, thus they serve as significant research questions that would influence further investigations in this area.

Hence, there should be assessment tools to guide policy at every scale of intervention. For instance, at the national level, models that evaluate national economic, energy and emission pathways to suggest policy alternatives and urbanization strategies. Co-benefit models need to back-cast capped emission levels and accordingly establish co-relation with urbanization, development thresholds, mid-term and long-term goals of urban planning, within the country or region. Then at the city-region scale, there are technologies that measure spatial causation and impacts real time, with accurate ground positioning. In this regard, a spatial and temporal dataset for cities and city-regions could be very helpful to assess their climate co-benefits. In India, the Ministry of Urban Development launched the National Urban Information System (NUIS) Scheme in March 2006 to develop GIS databases for 137 towns/cities in the country in two scales, that is, 1:10,000 and 1:2000 (MoUD 2014). Till September 2017, the total number of towns in the NUIS

Scheme whose data could be readily used is 152. At the micro-level within cities, models should focus on how co-benefits could be generated from efficient flows within the urban metabolism, namely natural resources, energy processes, exchange of goods and services, land or space utilizations and mobility. This should help to better understand how to plan and design human settlements in a climate responsive manner, bearing lower impacts and smaller carbon footprints. It would also illuminate the existing knowledge on what should be an appropriate settlement strategy in various geographical regions, notwithstanding their individual capacities to develop.

An integrated urban tool is highly needful that takes into account risks, vulnerabilities and hazards on cities and also impacts of city-based GHG emissions from various sectors on the climate. The tool should be technically competent to downscale national scenarios on urbanization, GHG emissions, climate variability, and so on, as regional level inputs, and at the same time upscale local urban data, user and their behavioural data collected through surveys at locality to city scale. The tool should also be able to quantify co-benefits based on simulation of different factors, particularly user choices, spatial planning, density, type of built-form and their energy needs, city structure, land use mix and transportation/mobility. The tool ought to perform evaluation of scenarios, real time simulation and periodic monitoring for policy guidance, decision-making, regulations and mid-term course corrections. The outcomes can find pertinence to governments, policy-makers, researchers and practitioners alike as it shall help framing a new paradigm to ascertain co-benefits in the inter-disciplines of climate change and urban studies.

References

Alber, G., & Kern, K. (2008). Governing climate change in cities: Modes of urban climate governance in multi-level systems. OECD international conference, competitive cities and climate change, 2nd annual meeting of the OECD roundtable strategy for urban development, Milan. http://www.oecd.org/document/32/0,3343,en_21571361_41059646_41440096_1_1_1_1,00.html. October 9–10.

Allen, E. (2009). Designing a cool spot neighborhood: An urban planning technique to reduce GHG emissions. In Condon, et al. (Eds.), *Policy, urban form, and tools for measuring and managing GHG emissions*. Lincoln: Lincoln Institute of Land Policy.

APA Planners. (2011). In: D. Walker & T. Daniels (Eds.), *The planners guide to community viz: The essential tool for a new generation of planning*. Chicago & Washington, DC: APA Planners Press.

APN. (2015). UrbanCLIM. Asia Pacific Network. http://www.apn-gcr.org/resources/items/show/1749. Accessed on September 5, 2015.

Arizona State University. (2014). The vulcan project. http://vulcan.project.asu.edu/. Accessed on September 17, 2015.

CAI Asia. (2014). Transport emissions evaluation models for projects (TEEMP). CAI, ITDP, ADB, Cambridge Systematics & UNEP-GEF. http://cleanairinitiative.org/portal/TEEMPTool. Accessed on September 22, 2015.

Carvalho. (n.d.). *Decision support system*. http://www.eolss.net/Sample-Chapters/C09/E4–20-04-09.pdf. Accessed on June 5, 2015.

CCP/ICLEI. (2014). Adaptation toolkit. ICLEI. http://archive.iclei.org/fileadmin/user_upload/documents/ANZ/CCP/CCP-AU/Projects/AI/AdaptationToolkit/Toolkit_CCPAdaptation_Final.pdf. Accessed on August 11, 2015.

CEEI. (2014). Community energy & emissions inventory. Province of British Columbia. http://www2.gov.bc.ca/gov/topic.page?id=9CD252EC63C84868AC2325F70E21683C. Accessed on August 18, 2015.

Chertow, M.R. (2000). The IPAT equation and its variants. *Journal of Industrial Ecology 4*, 13–29 (2000). doi:10.1162/10881980052541927.

CMHC. (2015). Greenhouse gas emissions from urban travel: Tool for evaluating neighbourhood sustainability. Canada Mortgage & Housing Corporation. http://www.cmhc-schl.gc.ca/odpub/pdf/62142.pdf?fr=1420702400777 and https://www03.cmhc-schl.gc.ca/catalog/productDetail.cfm?cat=123&itm=62&lang=en&fr=1420702327433. Accessed on August 24, 2015.

Deltares. (2014). Climate app. http://www.deltares.nl/en/news/news-item/item/17651/climate-app-the-new-tool-for-global-urban-climate-adaption-measures. Accessed on July 07, 2015.

Dhakal, S. (2010). GHG emissions from urbanization and opportunities for urban carbon mitigation. *Current Opinion in Environmental Sustainability, 2,* 277–283.

Envision Tomorrow. (2014). Envision tomorrow: A suite of urban and regional planning tools. *Fregonese Associates.* http://www.envisiontomorrow.org/about-envision-tomorrow/. Accessed on July 24, 2015.

EU. (2013). Carbon aware travel choice (CATCH) project tool. http://www.carbonaware.eu/. Accessed on August 2, 2015.

EU. (2014). Sustainable urban transport policies and initiatives (SMILE). http://ec.europa.eu/environment/urban/urban_transport.htm. Accessed on August 11, 2015.

EU. (n.d.). GreenClimeAdapt—Green tools for urban climate adaptation. http://ec.europa.eu/environment/life/project/Projects/index.cfm?fuseaction=search.dspPage&n_proj_id=3263&docType=pdf. Accessed on July 18, 2015.

EU & EEA. (2014). Urban adaptation support tool. http://climate-adapt.eea.europa.eu/tools/urban-ast/step-0–0. Accessed on August 5, 2015.

Georgetown Climate Center. (2012). *Adapting to urban heat: A toolkit for local governments.* http://www.sustainablecitiesinstitute.org/topics/climate-adaptation-and-resilience/urban-heat/adapting-to-urban-heat-a-toolkit-for-local-governments. Accessed on July 9, 2015.

Gillenwater, M. (2011). Filling a gap in climate change education and scholarship. *Greenhouse Gas Measurement & Management 1*, 11–16. doi:10.3763/ghgmm.2010.0012.

ICE. (2014). UPlan tool. http://ice.ucdavis.edu/project/uplan. Accessed on July 22, 2015.

ICLEI. (2014a). HEAT+ tool. http://heat.iclei.org/heatplusgpc/indexnew.aspx. Accessed on October 12, 2015.

ICLEI. (2014b). *Toolkit for resilient cities.* http://resilient-cities.iclei.org/fileadmin/sites/resilient-cities/files/Images_and_logos/Resilience_Resource_Point/SiemensResilience_InteractPDF_2013–09-25.pdf. Accessed on October 6, 2015.

ICLEI. (2014c). *Local government climate change adaptation toolkit.* http://archive.iclei.org/index.php?id=adaptation-toolkit0. Accessed on October 19, 2015.

IPCC. (2007). Intergovernmental panel on climate change. In B. Metz, O. Davidson, R. Swart, & J. Pan (Eds.), *Fourth assessment report. Climate change mitigation.* Cambridge: Cambridge University Press.

IPCC. (2014). Intergovernmental panel on climate change [Seto K. C., Dhakal S., Bigio A., Blanco H., Delgado G. C., Dewar D., Huang L., Inaba A., Kansal A., Lwasa S., McMahon J. E., Müller D. B., Murakami J., Nagendra H., & Ramaswami, A. (2014). Human settlements, infrastructure and spatial planning]. In O. Edenhofer, R. Pichs-Madruga, Y. Sokona, E. Farahani, S. Kadner, K. Seyboth, A. Adler, I. Baum, S. Brunner, P. Eickemeier, B. Kriemann, J. Savolainen, S. Schlömer, C. von Stechow, T. Zwickel, & J. C. Minx (Eds.), *Climate change 2014: Mitigation of climate change. Contribution of working group III to the fifth assessment report of the intergovernmental panel on climate change.* Cambridge, United Kingdom and New York, NY, USA: Cambridge University Press.

Kaya, Y. (1990). *Impact of carbon dioxide emission control on GNP growth: Interpretation of proposed scenarios*. Paris: Response Strategies Working Group.
MOEJ—Ministry of the Environment Japan. (2008). *The co-benefits approach for GHG emission reduction projects*. Tokyo: MOEJ Project.
MoUD. (2011). *India's urban demographic transition: The 2011 census results-provisional*. New Delhi: JNNURM Directorate and National Institute of Urban Affairs.
MoUD. (2014). National urban information system. http://moud.gov.in/nuis. Accessed on November, 28 2014.
Netherlands Environmental Assessment Agency. (2009). *Co-benefits of climate policy*. Netherlands Environmental Assessment Agency. http://www.unep.org/transport/gfei/autotool/understanding_the_problem/Netherlands%20Environment%20Agency.pdf. Accessed on February 12, 2015.
NIWA, MWH, GNS & BRANZ. (2012). *Impacts of climate change on urban infrastructure and the built environment: Toolbox handbook*. http://www.niwa.co.nz/climate/urban-impacts-toolbox. Accessed on October 23, 2015.
OECD. (2003). *The forgotten benefits of climate change mitigation: Innovation, technological leapfrogging, employment, and sustainable development*. Paris: Organisation for Economic Co-operation and Development.
PETUS. (2014). *Practical evaluation tools for urban sustainability*. http://www.petus.eu.com/. Accessed on October 6, 2015.
Planning Tool Exchange. (2014a). *MetroQuest tool*. http://www.planningtoolexchange.org/tool/metroquest. Accessed on October 1, 2015.
Planning Tool Exchange. (2014b). CITYgreen tool. http://www.planningtoolexchange.org/tool/citygreen-0. Accessed on October 1, 2015.
Planning Tool Exchange. (2014c). Athena tool. http://www.athenasmi.org/our-software-data/impact-estimator/. Accessed on October 4, 2015.
Revi, A. (2008). Climate change risk: A mitigation and adaptation agenda for indian cities. *Environment and Urbanization, 20*(1), 207–230.
SACOG. (2014). *The sacramento area council of governments*. http://www.sacog.org/scenario-planning-tools. Accessed on September 17, 2015.
Sethi, M. (2013). Research perspective for low carbon cities-decoding urbanisation and GHG emission in India. *Urban India, 33*(1), 18–31.
Sethi, M., & Mohapatra, S. (2013). Governance framework to mitigate climate change: Challenges in urbanising India. In H. Ha, & T. N. Dhakal (Eds.), *Governance approaches to mitigation of and adaptation to climate change in Asia* (pp. 200–230). Hampshire, UK: Palgrave Macmillan.
Sethi, M. (2017). Climate Change and Urban Settlements—A spatial perspective of carbon footprint and beyond. Routledge, London, U.K: Taylor & Francis.
SFLCV. (2014). *San Francisco league of conservation voters*. http://www.sflcv.org/density/. Accessed on September 11, 2015.
Shakti. (2012). *The BRTS evaluation and design (BEAD) tool*. http://shaktifoundation.in/report/bead-toolkit-project-report/. Accessed on September 19, 2015.
SUDPLAN. (2014). *SUDPLAN project—Sustainable urban development planner for climate change adaptation*. http://sudplan.eu/. Accessed on October 7, 2015.
SynCity. (2014). Synthetic city for modeling urban energy systems. http://secondlife.com/destination/syn-city. Accessed on October 21, 2015.
The Hindu. (2012). *Three-fourths of Indian cities functioning without a master plan*. http://www.thehindu.com/todays-paper/tp-national/tp-newdelhi/threefourths-of-indian-cities-functioning-without-a-master-plan/article3838187.ece. Accessed on March 22, 2016.
UKCIP. (2014). UK local climate impacts profile. http://www.ukcip.org.uk/wizard/current-climate-vulnerability/lclip/. Accessed on October 29, 2015.
UN DESA. (2011). *Urban population development and the environment wallchart*. United Nations, Department of Economic and Social Affairs, Population Division. http://www.un.org/en/development/desa/population/publications/urbanization/urbanPopDevEnv_2011.shtml. Accessed on August 5, 2015.

UN-Habitat. (2009). Smart urban adapt toolkit. http://www.cepal.org/dmaah/noticias/paginas/3/35703/session-7-un-habitat.pdf. Accessed on August 14, 2015.

UN-Habitat. (2011). *Global report on human settlements 2011: Cities and climate change*. London, Washington, DC: United Nations Human Settlements Programme & Earthscan.

UN-Habitat. (2014). UN habitat's planning for climate change: A strategic, values-based approach for urban planners—toolkit. Nairobi, Kenya: United Nations Human Settlements Programme (UN-Habitat).

UNISDR. (2014). *Disaster resilience scorecard for cities* (*Working Document*), *Version* 1.5. The United Nations Office for Disaster Risk Reduction. http://www.unisdr.org/2014/campaign-cities/Resilience%20Scorecard%20V1.5.pdf. Accessed on August 21, 2015.

University of California. (2012). *EcoSmart landscape tool*. http://www.ecosmartlandscapes.org/. Accessed on November 17, 2015.

UNU-IAS. (2013). *Urban development with climate co-benefits: Aligning climate, environmental and other development goals in cities*. Yokohama: United Nations University Institute of Advanced Studies.

Urban Nexus. (n.d.). Synthesis report: Annex of case studies in urban climate resilience. http://www.urban-nexus.eu/www.urban-nexus.eu/images/Final%20version%20Synthesis%20Paper%20WP2%20Annex.pdf. Accessed on December 15, 2015.

Urban Sim. (2015). *I-PLACE^{3}S tool*. http://www.sacog.org/services/scenario-planning/pdf/I-PLACE3S%20User%20Guide%20for%20Internet_Sept2010.pdf. Accessed on January 9, 2016.

USAID. (2014). Vietnam climate impacts decision-support tool. http://www.ccrdproject.com/adaptation-partnership/cris/vietnam-climate-impacts-planning-tool-workshop. Accessed on January 5, 2016.

Webster, J., Tomalty, R., & Korteling, B. (2009). Residential energy use characterization and mapping: Research and development of information for community energy and greenhouse gas planning in Canada. In: *45th ISOCARP congress 2009*. http://www.isocarp.net/Data/case_studies/1604.pdf. Accessed on January 8, 2016.

World Bank. (2012). *Building Urban resilience: Tools for building urban resilience: Integrating risk information into investment decisions*. http://www-wds.worldbank.org/external/default/WDSContentServer/WDSP/IB/2012/08/01/000386194_20120801040625/Rendered/PDF/714870WP0P124400JAKARTA0CAN0THO0WEB.pdf. Accessed on January 22, 2016.

Part II
Contextualizing Co-benefit Issues: Across Spatial Scales and Sectors

Chapter 3
South Asian Perspective: A Case of Urban Air Pollution and Potential for Climate Co-benefits in India

Bhola Ram Gurjar, Toshimasa Ohara, Mukesh Khare, Priyanka Kulshrestha, Vandana Tyagi and Ajay Singh Nagpure

Abstract This chapter provides an analysis of the urban pollution and potential for co-benefits in India, making a comparison with other South Asian countries. Most of the South Asian countries are at the stage of developing or emerging economies, and India is the fastest growing economy among them. The Indian subcontinent is comprised of hundreds of densely populated large and medium size cities including five megacities (each with a population of more than 10 million). Constantly increasing energy-intensive urban activities in burgeoning cities and biomass burning in rural areas of India are responsible for a large share in the unacceptably high emissions of health-endangering and environment-polluting gaseous and particulate pollutants. The air pollution problem in India and the rest of the South Asia came into the spotlight by the South Asian haze called the Atmospheric Brown Cloud (ABC), which is said to be responsible for annually 100,000s premature deaths in the South Asian region (Lelieveld et al. 2001). Given the increasing trend of energy use and emissions in Indian cities, the present study is a step to make governments and people aware of the extent and intensity of the ambient as well as

B.R. Gurjar (✉) · V. Tyagi
Department of Civil Engineering and Centre for Excellence in Transportation Systems, Indian Institute of Technology Roorkee (IIT-R), Roorkee 247667, India
e-mail: indiabrgurjar@gmail.com

T. Ohara
Regional Atmospheric Modeling Section, Asian Environment Research Group, National Institute for Environmental Studies, 16-2 Onogawa, Tsukuba 305-8506, Japan

M. Khare
Department of Civil Engineering, Indian Institute of Technology Delhi, New Delhi 110016, India

P. Kulshrestha
Lady Irwin College, University of Delhi, New Delhi 110001, India

A.S. Nagpure
Center for Science, Technology, and Environmental Policy, Hubert H. Humphrey School of Public Affairs, University of Minnesota, Twin Cities, USA

M. Sethi and J.A. Puppim de Oliveira (eds.), *Mainstreaming Climate Co-Benefits in Indian Cities*, Exploring Urban Change in South Asia,
https://doi.org/10.1007/978-981-10-5816-5_3

the indoor air pollution problem. The analysis presented in this study is expected to help initiate appropriate policy measures and suitable action plans to limit emissions and adopt ways based on the co-benefits approach that promote sustainable development.

Keywords Air pollution · GHG emissions · Co-benefits · South Asia · Cities

1 Introduction

Air pollution has been a matter of concern for a long time, following industrialization and urbanization arising from the fossil fuel-based economy. In developing countries, air pollution has become more aggravating due to increased developmental activities—industrialization, growing cities, increasing traffic, rapid economic development, and higher energy consumption levels (Shukla et al. 2008, 2013; Gurjar et al. 2010; Nagpure et al. 2013). South Asia contains only about 3% of the world's land mass but has 22% of the world's population. As a result, this is one of the most densely populated regions in the world, with a population more than one-fifth of the world total (EIA 2004). South Asian countries have shown an average annual growth rate of 5% (Devarajan and Nabi 2006). Nevertheless, it is home to 43.4% of the world's poor living on less than $1.25 a day in 2005 prices. India and other South Asian countries are in a stage of rapid economic growth, hence industrialization and developmental activities are leading to a huge amount of air pollution emissions. With a very large size and density of population, inefficient technologies and feeble control measures, energy consumption in India and South Asia is discharging a huge amount of pollutants into the air. Unfortunately, most of the South Asian countries lag behind in air quality management, resulting in detrimental effects on human health and the environment. For example, in South Asia, the large north–south gradient in the Atmospheric Brown Cloud (ABC) dimming has altered both the north–south gradients in sea surface temperatures and land–ocean contrast in surface temperatures. This in turn has slowed down the monsoon circulation and decreased rainfall over the continent. On the other hand, heating by black carbon has warmed the atmosphere at elevated levels from 2 to 6 km, where most tropical glaciers are located. This has strengthened the effect of GHGs on retreat of snow packs and glaciers in the Hindu Kush-Himalaya-Tibetan glaciers (Ramanathan and Feng 2009).

Rising air pollution in the Indian subcontinent is continuously degrading the ambient air quality (Lelieveld et al. 2001; Gurjar et al. 2008; Nagpure et al. 2011, 2013, 2016) which is responsible for constantly rising numbers of death cases and respiratory illness (Gurjar et al. 2010; Kumar et al. 2011). According to the UNEP Assessment Report (2002), about two million total mortality and 1.4 million acute respiratory illness (ARI) deaths occur annually only due to atmospheric pollution in South Asia (UNEP Assessment Report 2002). The recent study published by the World Health Organization (WHO 2014) ranked South Asian megacity Delhi as the most polluted city of the world. In Delhi, at least one schoolchild out of ten was found to be suffering from asthma (South Asia Brief 2010). Similarly according to Nagpure et al. (2014) the number of premature deaths in megacity Delhi has risen by 100% from 1991 to 2010. Silva et al. (2013) have estimated ozone-related mortality on a global scale and found that it is widespread globally, as ozone has increased essentially everywhere from anthropogenic activities, but is greatest in highly populated and highly polluted areas of India and East Asia, which account for 68% of the global total. Regional premature annual deaths in India from anthropogenic outdoor air pollution (2000–1850), for ozone (respiratory) and $PM_{2.5}$, have been estimated equal to 118,000 and 397,000, respectively (Silva et al. 2013). Recent estimates by Likhvar et al. (2015) have reported that due to ozone and $PM_{2.5}$ the largest increase in the number of cardiovascular (CV) deaths occurred in South Asia (85% of total Asia increases), the largest part of which belong in India (over 0.15 million). In Kathmandu, the metropolitan city of Nepal, annually 1600 premature deaths only occur due to PM_{10} exposure (Schwela 2009), and around 2500 people die prematurely annually in Pakistan urban areas due to air pollution exposure (Munir 2009).

2 Anthropogenic Emissions from South Asian Countries

The sources of pollution are common to almost all the South Asian countries, but emission contributions by the sources vary with area. The major energy sources causing air pollution are fossil fuels and biomass along with dominating sectors like power plants and motor vehicle activities (Gurjar et al. 2004; Dutkiewicz et al. 2009; Nagpure et al. 2016). In addition, many other sources/sectors like chemical plants, oil refineries, incinerators, agriculture and livestock, metal production and plastic factories contribute to the ever-growing problem (Gurjar et al. 2015).

2.1 Short-Lived Pollutants (NO_x, SO_x, CO, BC, OC and PM) Emissions and Source Contribution

Among the criteria, air pollutants there are certain pollutants whose life span is much shorter. Such kind of short-lived pollutants, which have been discussed here in this section, include nitrogen oxides (NO_x), sulphur oxides (SO_x), carbon monoxide (CO), black carbon (BC), organic carbon (OC), and particulate matter (PM). Some of them are also known as short-lived climate pollutants (SLCPs), such as BC particles, methane and tropospheric ozone. The SLCPs are not only harmful to human health but also contribute significantly to climate change.

2.1.1 Nitrogen Oxides (NO_x) Emissions

According to Aardenne et al. (1999), NO_x emissions from South Asia were 3924 Gg during 1990, while WRI-EarthTrends (2010) estimated 5011 Gg. Dentener et al. (2004) estimated NO_x emissions from industries and power plants as 1300 and 2000 Gg for 1990 and 2000, whereas these were 1000 and 2000 Gg from the transport sector in 1990 and 2000. As per Lelieveld et al.'s (2001) study, out of the total NO_x emitted in South Asia (6000 Gg) during 1999, fossil fuels combustion from transport and the industrial sector accounted for the highest share (2600 Gg), followed by agriculture (2000 Gg). During 1991, biomass combustion contributed 802 Gg of NO_x in India, 115 Gg in Pakistan, 21 Gg in Nepal and 16 Gg in Sri Lanka (Bhattacharya et al. 2000). The sources of nitrogen oxides vary from country to country in South Asia. For example, road transport is the main source of NO_x emission in India as compared to industry and the power sector. Another important source of NO_x emission is the industrial process, especially the production of nitric acid, used in fertilizer manufacturing. In Bhutan and Maldives, however, the power plants dominate in NO_x emissions. In Bangladesh, on the other hand, NO_x is mainly emitted during energy consumption including energy transformation, industries (e.g., processing of iron and steel industries), transport and biomass burning. Another major source of NO_x emissions is burning of agricultural residues. In Nepal, major sources of NO_x are associated with the combustion of fossil fuels and from fuel combustion in industries, especially the cement industry (Khwaja et al. 2012).

Figure 1 summarizes the recent estimations and projections of NO_x emissions based on emission inventory developed by Kurokawa et al. (2013). In 2000, South Asia contributed about 6.73 Tg of NO_x emissions, that is, around 25% of total NO_x emissions (27.33 Tg) from Asia, in which India contributed the major share of 5.83 Tg (87%) followed by Pakistan's 0.52 Tg (8%). Although NO_x emissions in South Asia increased from 6.73 Tg in 2000 to 11.1 Tg in 2008, its share in Asia

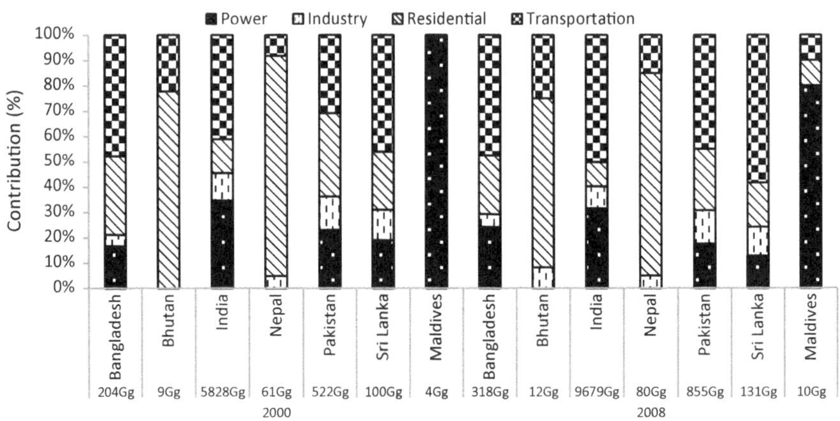

Fig. 1 Country and sector wise NO_x (Gg) emission trends (2000–2008) in South Asia

was almost constant (~24–25%). During 2000–2008, India has been the dominating contributor (83–87%) of NO_x in South Asia. From a sector-based activity perspective, until 2000 the power sector and transportation sector were almost equal contributors of NO_x in South Asia, but in 2008 the transport sector dominated the scene (Kurokawa et al. 2013). Kurokawa et al. (2013) have further stated that total NO_x emissions as NO_2 in 2008 (growth rate between 2000 and 2008) have been 53.9 Tg (+54%) for Asia and 11.1 Tg (+56%) for India. Road transport emissions in India were the largest contributor to NO_x emissions and doubled from ~ 2.4 Tg in 2000 to 4.8 Tg in 2008.

With respect to citywide emissions, a study conducted by Guttikunda and Calori (2013) found that megacity Delhi is contributing 376 Gg/year NO_x annually and the transportation sector shares the highest percentage (53%) of total NO_x followed by diesel generator (25%) and industrial (11%) emissions. For Kolkata city, an Asian Development Bank (ADB 2005) study estimated that the NO_x emissions from all sectors were 131 Gg/year in 2003 and raised to173 Gg/year in 2008. The study projected that annual emissions will rise to 241 Gg/year in the year 2014. The study, then projected that transport sectors contributed about 96 Gg/year NO_x emissions in 2003 and is expected to be 174 Gg in 2014.

2.1.2 Sulphur Dioxide (SO_2) Emissions

According to WRI-Earth Trends (2010), SO_2 emissions in South Asia ranged from 5719 to 9050 Gg during 1990–2005, with India as the major contributor. Urban areas including megacities are the largest contributor of SO_2 in India. The estimated SO_2 emission from megacity Delhi in 2010 was about 37 Gg/year (Guttikunda and Calori 2013). Streets et al. (2000) estimated that during 1997, excluding the Maldives, all the South Asian countries accounted for 19% of the total Asian SO_2 emissions and 16% was from India. SO_2 emissions in India are relatively less, but the rate of increase almost doubled from 1985 to 2005. In 2000, Indian SO_2 emissions were estimated to be 4.26 Mt (Lu et al. 2010). Karim et al. (1997) evaluated emissions from different energy consumption sectors in Bangladesh. It was found that 47% of SO_2 was emitted from the transport sector due to high sulphur content in petroleum products and extensive use of diesel fuel (Karim 1999). Bhattacharya et al. (2000) estimated SO_x emissions during 1991 from biomass combustion and found 863 and 162 Gg of it from India and Pakistan; while Nepal and Sri Lanka contributed 24 and 9 Gg during 1993.

Figure 2 presents country and sector wise SO_2 emission trends (2000–08) in South Asia. South Asia's share in Asian SO_2 emission is about 21–23% during 2000–08, in which India's contribution in South Asian SO_2 emissions is the largest (85–88%). Maximum emission of SO_2 in South Asia is from power and industrial sectors, which collectively are responsible for around 87% of SO_2 emission in South Asia. The projections by Ohara et al. (2007) until 2020 exemplify the SO_2 emissions from India and other South Asian countries up to 10.2 Tg (66%) and 2.5 Tg (91%), respectively, presenting a hasty emission growth. Nevertheless, the

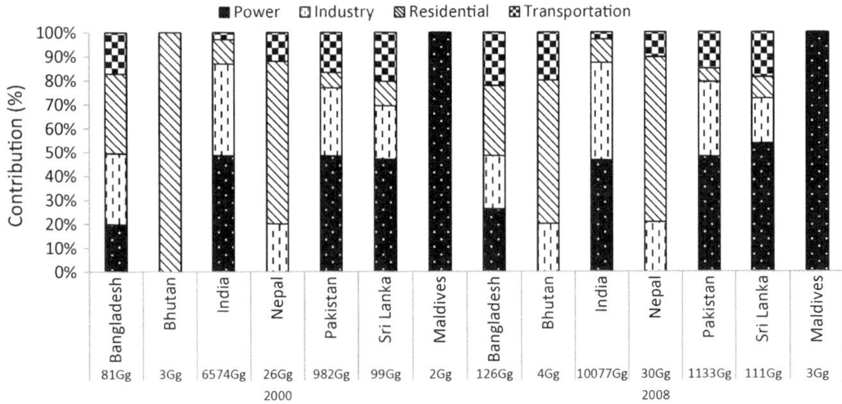

Fig. 2 Country and sector wise SO_2 emission trends (2000–08) in South Asia

major sources of SO_2 in South Asia differ from country to country. In Bangladesh, the main sources of emissions of SO_2 are vehicles, brick kilns, paper and pulp industries, oil refineries and sulphuric acid production plants. The high emission factors of diesel-powered trucks (1.13 g sulphur dioxide/km) and minibuses indicate that substantial SO_2 emissions come from these sources.

South Asian megacity Delhi was the second largest SO_2 contributor among the megacities in India during 1990 to 2000. SO_2 emissions from all sectors in Delhi ranged between 90 and 113 Gg/year during 1991–2000 (Gurjar et al. 2004). Similar to Delhi, megacity Mumbai had significantly enhanced SO_2 emissions in comparison to other South and South-East Asian megacities with annual emissions in excess of 200 Gg during the year 2000 (Guttikunda et al. 2001). According to Bhanarkar et al. (2005) annual SO_2 emissions from all sources in Mumbai were about 55 Gg during 2010. For another important South Asian megacity Kolkata, annual SO_2 emissions from all sources were 200.7 Gg/year in 2010 and are expected to rise to 310.6 Gg/year in 2020.

2.1.3 Carbon Monoxide (CO) Emissions

According to Dickerson et al. (2002), residential biofuel combustion during 2000 resulted in 37,700 Gg of CO from South Asia, with India contributing 29 Tg. During 1990–2005, India ranks first in CO emissions (51,000–63,000 Gg) followed by Bangladesh (7000–9000 Gg) and Pakistan (7000–8000 Gg). CO emissions from India were 79,000 Gg (26% of total Asia emissions) in 2000 with biofuel combustion as the principal contributor (71% of total emissions from India) followed by industrial biofuel and transport oil use (each 9% of total emissions from India) (Ohara et al. 2007). Streets et al. (2003a) and Barletta et al. (2002) also observed similar findings. According to Bhattacharya et al. (2000), during 1991, the CO emissions from biomass combustion were 9941 and 2761 Gg from India and Pakistan, respectively; while 894 and 740 Gg in Nepal and Sri Lanka during 1993. Shrestha and Malla (1996) stated that in Kathmandu (Nepal) highest CO emissions during 1993 were from transport (23 Gg) followed by residential (9.8 Gg) and industrial (5 Gg) sectors. According to United States Energy Information Administration (USEIA), vehicles in Pakistan emit almost 25 times more CO in comparison to an average US vehicle, probably due to old age and poorly maintained vehicles with almost negligible pollution control devices (Barletta et al. 2002).

India is the largest contributor (>80%) of CO in South Asia, and the residential/domestic sector is the chief emitter of CO in South Asia followed by industry. In India, CO emissions from the domestic, industrial and road transport sectors increased monotonically between 2000 and 2008. However, Guttikunda and Calori (2013) have estimated that in megacity Delhi the power plants' share is the highest percentage (31%), followed by the transport (18%) and domestic sector (14%) in CO emissions. The domestic sector has been the largest contributor during 2000–2008, but the contribution fraction is decreasing (63 to 58%). Growth rates of emissions in the road transport sector have increased and reached nearly 20% of total emissions in 2008.

2.1.4 Black Carbon (BC) and Organic Carbon (OC) Emissions

Ohara et al. (2007) found that during 2000, South Asia contributed about 47% of total BC emissions from Asia (2.73 Tg) in which India's share was 29% (0.80 Tg) of the total emissions. The residential sector has dominated (92%) in South Asia in case of BC emissions. According to Dickerson et al. (2002), the BC emissions during 2000 from residential, transport and power sectors in South Asia were 551, 43 and 2.7 Gg, respectively (Dickerson et al. 2002). Venkataraman et al. (2005) estimated that total BC emissions in India were 410 Gg/year, with 170 Gg/year from biofuel burning (particularly fuel wood used in cook-stoves), 140 Gg/year from open biomass burning, and 100 Gg/year from fossil fuel burning in the residential sector. Parashar et al. (2005) have stated that BC emissions in India from bio-fuels (including forest biomass) and fossil fuels (residential, power, transport and industrial sector) ranged from 399 to 1430 Gg for the year 1995. In a recent study conducted by Gadhavi et al. (2015), it has been found that northern India is a major source of anthropogenic BC particles, but southern India also has significant BC emissions.

In addition to BC, emissions of organic carbon (OC) are also important for air pollution studies. OC emissions in India for 1996–1997 were 715 Gg, with 231 Gg from fossil fuel (power, industrial, residential and transport) and 484 Gg from biofuel combustion (Reddy and Venketraman 2002a, b). Parashar et al. (2005) estimated that OC emissions from bio-fuels (including forest biomass) and fossil fuels ranged between 1193 and 3335 Gg for the year 1995. Total OC emissions in Asia were estimated to be 8.88 Tg in 2000, with 3.27 Tg from India, which was higher than China (2.56 Tg). The country and sector wise OC emission trends (2000–2008) in South Asia are shown in Fig. 3.

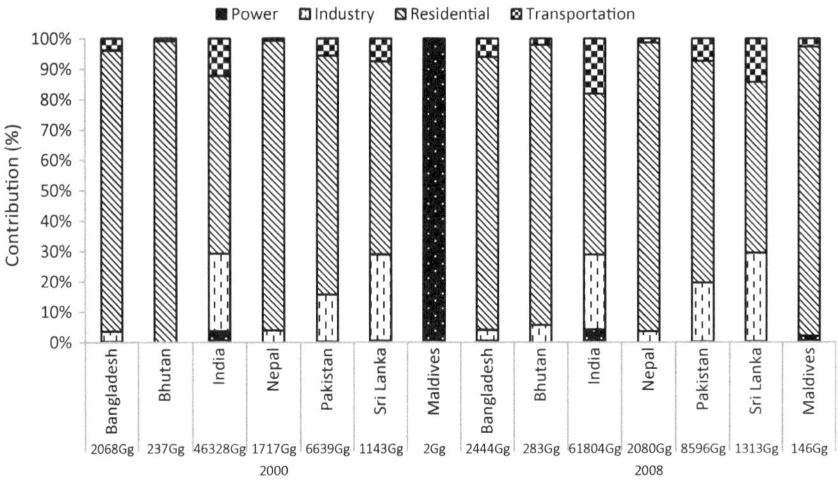

Fig. 3 Country and sector wise OC emission trends (2000–2008) in South Asia

Domestic biofuel combustion accounted for the major share (92%) of OC in India (Ohara et al. 2007). The OC emissions estimated by Streets et al. (2003a) (TRACE-P) for 2000 were 2800 Gg for India, 300 Gg for Pakistan, 200 Gg for Bangladesh, and 100 Gg for Nepal. According to Streets et al. (2003b), during 2000 the OC emissions from biomass burning were 650 Gg in India followed by 81 Gg in Bangladesh, 52 Gg from Pakistan and Nepal each. This illustrates that India is the chief contributor of BC and OC in South Asian region, which mainly come from burning of biofuels. As Fig. 4a, b show, the BC and OC contributions of South Asia and the total BC and OC emissions of Asia during 2000–08 increased from 28 to 31% and 37 to 39%, respectively. The residential/domestic sector is the biggest contributor of both BC and OC emissions in South Asia and the transport sector is second. On a country scale, in both cases India's contribution in South Asia was 76–79% followed by Pakistan with 12–14%.

2.1.5 Particulate Matter (PM) Emissions

Particulate emissions in India during 2000 and 2005 were 5160 and 4170 Gg, respectively (Garg et al. 2003). Bhattacharya et al. (2000) estimated that during 1991–1993 the total suspended particulate (TSP) emissions from biomass and fossil fuel combustion were 2794 Gg from India (72%) followed by 725 Gg from Pakistan (18%), 219 Gg from Nepal (6%) and 103 Gg from Sri Lanka (3%). During the period 1992–93, annual TSP and PM_{10} emissions in Mumbai were 32 and 16 Gg (Kumar and Joseph 2006). Total PM emissions from the Indian transport sector during 2003–04 were 153 Gg (Ramachandra and Shwetmala 2009). In Kathmandu (Nepal), vehicles (38%) and re-suspended dust from poorly maintained

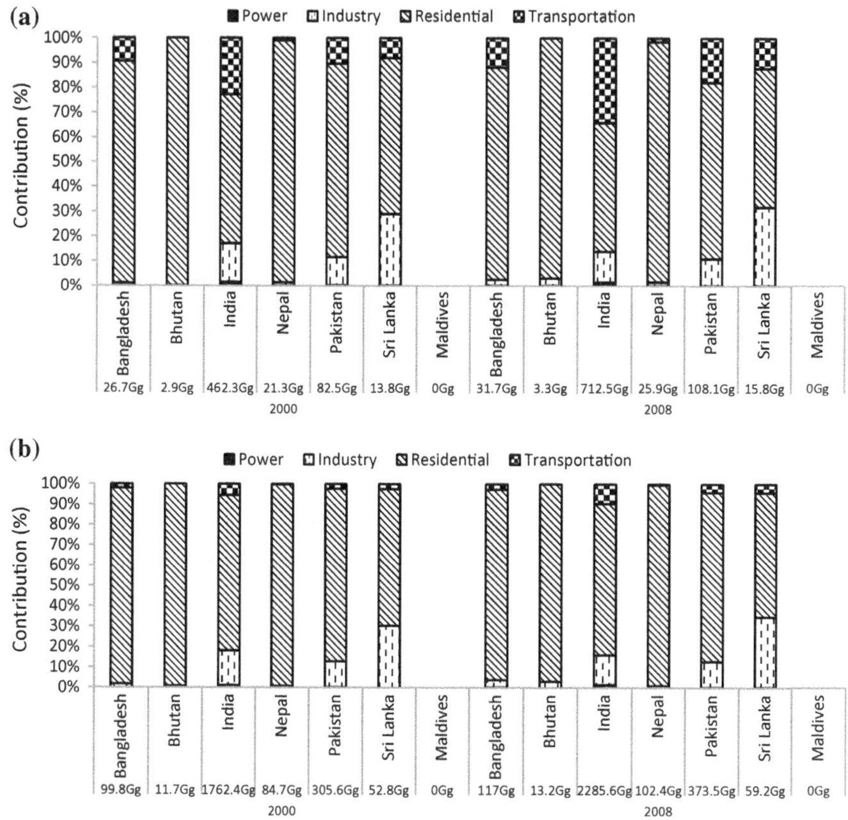

Fig. 4 **a** Country and sector wise BC emission trends (2000–08) in South Asia, **b** country and sector-wise OC emission trends (2000–08) in South Asia

and unclean roads (25%), along with agriculture (18%) and brick kilns (11%) followed by minor emissions from other industries were identified as key PM_{10} emission sources. PM_{10} emissions from road traffic in Kathmandu are approximately 16 times higher than CO and 26 times than NO_x (Schwela 2009).

According to Greenstone (2015), half of the Indian population (660 million) are living in the areas where air pollution levels exceed the National Ambient Air Quality Standard (NAAQS) for fine particulate pollution. Greenstone (2015) observed that Indian megacity Delhi is 45% more polluted then rapidly developing city Beijing, which scientific community finds hard to believe because in the past there have been studies that rated Beijing more polluted than Delhi (e.g., Gurjar et al. 2008, 2010). According to recent media reports (e.g., The Economist 2015), a WHO study reveals that India is home to 13 of the top 20 most polluted urban areas in the world with worst fine particulate ($PM_{2.5}$) air pollution. Both urban and rural inhabitants are exposed to serious respiratory and cardiovascular health hazards due to high $PM_{2.5}$ pollution. Weak implementation of emission norms might be

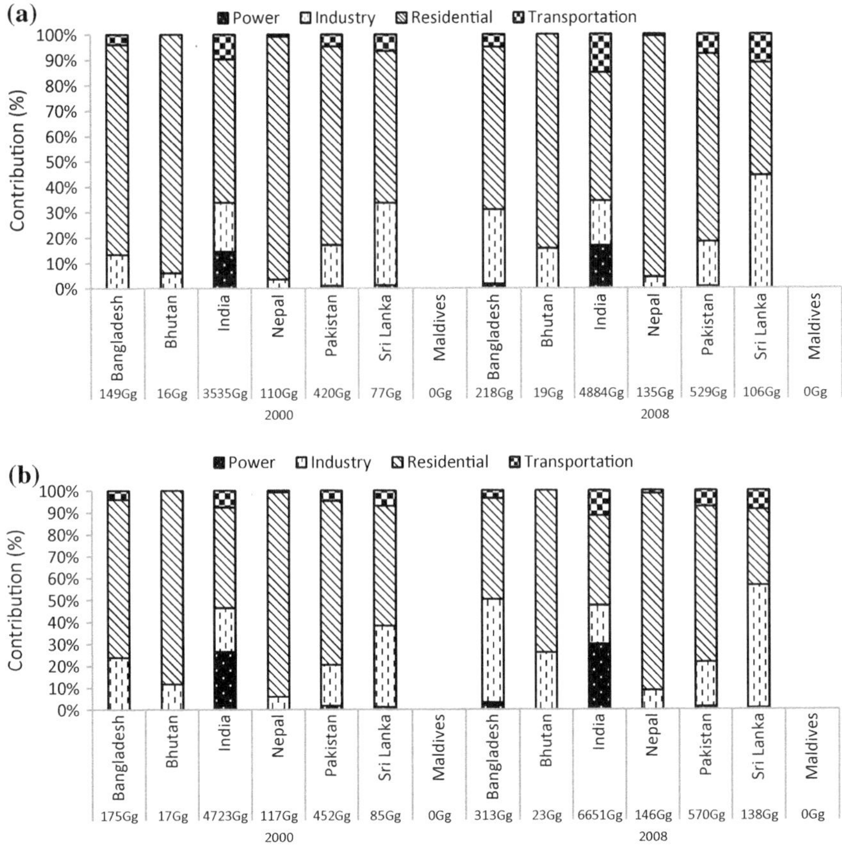

Fig. 5 **a** Country and sector wise $PM_{2.5}$ emission trends (2000–08) in South Asia, **b** country and sector wise PM_{10} emission trends (2000–08) in South Asia

responsible for increasing air pollution in Indian cities. However, in rural areas, too, the air quality is poor due to high air pollution emissions from burning of solid fuels for cooking.

About 24–26% of total Asian emissions of $PM_{2.5}$ and PM_{10} are contributed by South Asia. India is the largest emitter (82–85%) of total $PM_{2.5}$ and PM_{10} emissions in South Asia. On sector-based emission evaluation, it is observed that the residential sector emits the largest share of these particulates. This could be because of inefficient cook-stoves widely used in rural areas of South Asia (see Fig. 5a–b). Not surprisingly, Munir (2009) has estimated that in the absence of proper control measures, the residential sector would contribute 84% of total PM emissions in Pakistan until 2020.

For PM emission South Asian megacity Delhi ranked top among most polluted cities of the world. The important sources of PM in South Asian cities are small-scale industries, domestic coal burning, thermal power plants, transportation, and biomass burning (cow dung, crop residue) (Gurjar et al. 2004; Reddy and Venkataraman 2002a). The recent study conducted by Guttikunda et al. (2011) indicated that among all anthropogenic sources, road dust, power plants, transport and municipal solid waste (MSW) burning shared the highest percentage—59% (35 Gg), 12% (7.33 Gg), 8% (4.99 Gg) and 8% (4.90 Gg) respectively—in 2010 in megacity Delhi. It is interesting to note that open burning of MSW emerges as one of the major contributors of $PM_{2.5}$ and PM_{10} emissions, especially in cities of developing countries (Guttikunda et al. 2014; Wiedinmyer et al. 2014; Nagpure et al. 2015). Similar to Delhi, Mumbai was amongst the three cities in the world having the highest level of PM (World Bank 1997) during the last decade. In 1992–93 it was estimated that PM_{10} emissions in Mumbai were about 16 Gg (URBAIR 1997). Like Delhi, the study done by CPCB (2010) suggests that road dust suspension due to vehicular activities (8 Gg, 30%), power plants (6 Gg, 21%), solid waste burning (4 Gg, 14%) and construction activities (2 Gg, 9%) are the biggest contributors for most of the PM emissions in Mumbai. The condition of PM_{10} in Kolkata is almost similar to other Indian megacities like Delhi and Mumbai (Chakraborty and Bhattacharya 2004). The Asian Development Bank (ADB 2005) study found that the total annual PM_{10} emission in Kolkata was 76 Gg during 2003 with major contributions by road dust (61%) followed by vehicles (21%), industries and power plants (9%) and other sources (9%) such as biomass burning, railway, domestic cooking light distillate oil burning, coal fuel kilns in the brick industry.

2.2 *Long-Lived Greenhouse Gases (GHGs)*

During 1990, CO_2 emissions from fossil fuel combustion (power, transport, industry and residential) were highest (89%) in India (159,000 Gg), followed by Pakistan (18,000 Gg), Sri Lanka (1100 Gg), Nepal (300 Gg) and Bhutan (30 Gg) (Azad et al. 2006). As per Oliver et al. (2014), India's share in total GHGs emissions of the world in 2000—and about 2000 million metric tons GHGs of the world was only contributed by the Indian subcontinent in 2012—was about 6%. GHG emissions from India are expected to increase up to 4–7.3 billion metric tonnes in 2030–31 which aggregates to 2.77–5 metric tonnes/capita (MoEF 2009). Bhattacharya et al. (2000) estimated that of the total emissions from biomass combustion in 1991, 77% CO_2 was from India, while it was 16, 4 and 3% from Pakistan, Nepal and Sri Lanka, respectively. In Pakistan, among different sectors—such as energy, industrial processes, agriculture, land-use change and forestry, and waste sectors—energy corresponds to the highest share in anthropogenic emissions (Khan and Baig 2003).

According to WRI-Earth Trends (2010), CH_4 emissions from biomass combustion were highest (76%) from India, followed by Pakistan. South Asia accounts for approximately 50% of the total global anthropogenic CH_4 emissions (UNEP 2002). In South Asia, CH_4 emissions are estimated mainly from the agricultural sector (livestock, rice cultivation, manure management etc.), biomass burning and waste and wastewater treatment activities (Dentener et al. 2004). CH_4 emissions from the agriculture sector in India increased from 19,500 Gg in 2000 to 20,500 Gg in 2005 (Garg et al. 2003). In 1991, CH_4 emissions from biomass combustion in India were 1797 Gg, and 364 Gg in Pakistan. In 1993, it was 109 and 89 Gg for Nepal and Sri Lanka respectively (Bhattacharya et al. 2000). Bhatia et al. (2004) estimated that CH_4 emissions from Indian rice fields for 1994–1995 were 2904 Gg. Yamaji et al. (2003 and 2004) stated that during 1995 and 2000, CH_4 emissions from livestock were highest in India (~ 11,000–12,000 Gg), which was much more than the CH_4 emissions from Pakistan (1800 Gg) and Bangladesh (900 Gg). In Pakistan, the highest emissions of CH_4 were from stationary sectors (1323 Gg) with 98% from the residential sector (predominantly owing to biomass combustion) and the commercial sector emitted merely 16.84 Gg (Khan and Baig 2003).

Very limited work has been carried out on N_2O emissions and that has mainly focused on the agricultural sector, as it is the major source of N_2O. In India, N_2O emissions have increased from 0.31 to 0.41 Tg during 2000–2005 (Garg et al. 2003). Bhatia et al. (2004) stated that the Indian agriculture sector emitted 80 Gg of N_2O during 1994–1995. As stated by Khan and Baig (2003) the highest N_2O emissions in Pakistan during 1999–2000 were emitted from the agricultural and residential sector (95%), with minor contributions from the power, industrial and transport sectors.

2.3 *Air Pollution and Society*

Poor air quality is one of the most serious environmental problems in urban areas of most of the developing countries. The report published by WHO in December 2014 says that 13 of the top 20 most polluted cities are from India. Most recently, estimates from the Global Burden of Disease (GBD) study indicate that outdoor particulate pollution is the fifth major cause of premature death and disability life years lost in India after high blood pressure, indoor air pollution, tobacco smoking and poor nutrition, with about 695,000 premature deaths per year (IHME 2013). Further, for outdoor air pollution sources concentrated near residential areas, related emissions can infiltrate indoors, contributing to the indoor air quality problems.

The disparities in access of clean household energy presents a major challenge to improving health and protecting the environment—the WHO study indicated that indoor air pollution is the second largest cause of death in India. Moreover, improper management of the public transport system to cope with the rapid pace of urbanization, growing industrial production has heightened the burden of outdoor air pollution related mortalities. Recently Wiedinmyer et al. (2014) and Nagpure et al. (2015) have highlighted that open waste burning is an important contributor for increasing air pollution. Nagpure et al. (2015) suggested that most of the waste burning is happening in low socio-economic status areas because of low waste management efficiency. Because of higher exposure poor people, children and pregnant women are the most vulnerable populations with respect to indoor and outdoor air pollution.

3 Co-benefits of Air Pollution and GHGs Reduction Policies

3.1 India

The co-benefits approach includes the direct and indirect effects associated with various air pollution and climate change policy options that are favourable to human welfare. The impact on human health is the most significant and known co-benefit allied with air quality policy. Implementations of regulation and technology change to reduce air pollution have many co-benefits related to air quality. There is a tremendous potential to reduce GHG emissions and bring much cleaner air to Indian cities. The South Asian regional and local perspective on co-benefits mainly addresses the health, energy access, security, and other socioeconomic problems associated with development.

As the impacts of air pollution on climate and health came into sight through scientific studies and publications, the governments of South Asian countries started to take a keen interest to reduce the air pollution load with a co-benefit approach. In general, for the reduction of emissions, it is appropriate to promote switching to alternate and energy efficient fuels, which not only affects the regional air quality, but also protects the fuel economy. In South Asia, a noteworthy attainment has been the absolute removal of lead in gasoline by mid-2002, making it as the first developing region of the world to do so (Kojima et al. 2004). The co-benefits approach is relatively new, therefore this approach still needs conceptual and methodological clarifications. Current applications of the approach are mostly sector-based initiatives because city governments are largely administered and function in that way. With the transport sector accounting for 21% of the world's energy-related carbon dioxide emissions (Grazi et al. 2008), and given the critical role it plays in the effective functioning of a city, it is necessary to strengthen the links between the transport sector and the co-benefits approach via analytical tools and policy instruments (Doll and Balaban 2013). With this viewpoint, some of the major policy measures initiated in India for the reduction of emissions and concentration of ambient air pollutants and attainment of better air quality for health benefits are briefly described below; Table 1 provides a summary of relevant policy instruments and/or technological interventions for other countries of South Asia:

- In 1998, the Indian Supreme Court directed to use compressed natural gas (CNG) as public transport fuel to improve air quality of megacity Delhi, and this was further extended to other cities (Timilsina and Shrestha 2009). The sale of sub-standard lubricants is prohibited in India, as their addition to two-stroke engines of gasoline driven two- and three-wheelers increases particulate emissions (Kojima et al. 2004). The government has also published Gazette Notification vide G.S.R. 84(E) on 9 February 2009 for introducing Bharat Stage IV and III norms (equivalent to Euro IV and III norms), as applicable, for new vehicles in selected cities. The benefits and co-benefits objectives

Table 1 Summary of relevant policy instruments undertaken in other South Asian countries

Country	Policy measures for co-benefits
Bangladesh	• Government of Bangladesh announced ban on Fix Chimney Kiln (FCK) technology (effective from year 2010) to reduce emissions from brick kilns • Motor vehicle ordinance (1983) was re-examined as a measure to control the rising vehicular emissions. Import policy had provisions for introduction of minimum Euro II diesel vehicles • The use of CNG as auto fuel is promoted by reducing the import duty on CNG vehicles and engines and their spare parts • Introduction of Deliberate Transport Plan for Dhaka, for instance, environment-friendly mass transit introduction, which is like the Bus Rapid Transit (BRT), Light Rail Transit (LRT) and metro system • Phasing out of baby taxis since January 2003, contributed to a reduction of 40% in $PM_{2.5}$, 51% in HCs and 35% in CO (Clean Air Bangladesh 2008) • The Ministry of Power, Energy and Mineral Resources of the government of Bangladesh introduced its first Renewable Energy Policy. The policy endorses creating an enabling environment for renewable energy technologies and supports attempts to stimulate market development for improved cook stoves • The government led (Phase II) Improved Cook Stove Programme has widely disseminated improved cook stove models suitable for Bangladesh (implemented from 1988 to 2001) • The Sustainable Energy for Development Programme was implemented by the GTZ (2004–10) with the objective to promote improved cook stoves through a commercial, sustainable approach • USAID: Reduction of Exposure to Indoor Air Pollution through Household Energy and Behavioral Improvements: The programme disseminated 580 improved cook stoves and was implemented in the north-west of Bangladesh from 2005 to 2007. The main objectives were increasing awareness of indoor air pollution; changing household energy behaviour; and promoting and developing a commercial market for improved cook stoves and entrepreneurs • Biogas Programme: This programme was implemented from 1988 to 2003 with the main objective of research and demonstration of biogas technology, and constructed 22,100 biogas plants (ESMAP 2010)
Bhutan	• In 1996, the Ministry of Finance prohibited the import of second-hand vehicles and new two-stroke two-wheelers • In 1999, the Government established type approval standards, which suggested that all new vehicles should be Euro 1 type approval • Bhutan began to import unleaded petrol during 2001 and ultra-low sulphur diesel (ULSD) in 2003 (0.025% sulfur-content diesel) (CAI-Bhutan 2006) • A hydropower project supported by ADB and the governments of Austria and Japan has been registered under the Clean Development Mechanism (CDM). The 114-megawatt Dagachhu hydropower project in Bhutan will promote cross-border power trade and reduce GHG pollution in the South Asia region. The hydropower project is expected to reduce about 500,000 tonnes of CO_2 emission annually, especially through exports to India, which relies heavily on coal-fired power plants for its electricity generation (Environmental Protection 2010) • As per the constitution of Bhutan, at least 60% of the landmass must be under forest cover at all times, and the government has strictly followed its policy of environmental sustainability, with approximately 51% of Bhutan's land mass in the form of protected areas and forests (Koljonen and Lehtila 2012)

(continued)

Table 1 (continued)

Country	Policy measures for co-benefits
Sri Lanka	• Government of Sri Lanka banned the import of two-stroke engine three-wheelers from January 2008 and full engine, engine blocks and cylinder heads after 2011 in order to evade local assembly of two-stroke engines (Clean Air Sri Lanka 2008) • Sri Lanka launched the implementation of a vehicle emission testing programme on 15 July 2008 and is progressively decreasing sulphur in diesel (Kojima et al. 2004) • Supreme Court of Sri Lanka has ordered the preparation and implementation of a Strategic Plan for Traffic Management in Greater Colombo (2008–10) (Clean Air Sri Lanka 2008) • Around 800,000 improved cook stoves have been distributed (25% of the households in Sri Lanka) and US$1.5 million has been spent since 1971 (Amaresekara et al. 2004)
Nepal	• Government of Nepal decided to ban inefficient and more polluting 'Moving Chimney Bull's Trench Kilns' (MCBTK) from Kathmandu during the year 2004 • Interim National Plan (2008–11) was initiated for promotion of environment friendly transport and to give suitable consideration to vehicular emission control (Clean Air Nepal 2008) • Alternative Energy Promotion Center (AEPC)/Energy Sector Assistance Programme (ESAP) have launched Biomass Energy Component in ESAP phase II programme, to promote various biomass energy technologies such as Improved Cook Stove (ICS), biomass briquette, bio-fuel and gasification. This five-year (2007–11) programme was an extension of the ESAP phase-I which was focused on promotion of ICS in Nepal (Indoor Air Pollution and Health Forum Nepal 2008) • In the Tenth Five Year Plan (2002–07), it was decided to install 200,000 biogas plants and distribute 250,000 improved cooking stoves in about 45 districts of Nepal (Shakya and Shakya 2007)
Pakistan	• Government of Pakistan started a comprehensive biogas programme in 1974 and had commissioned 4550 plants by 1990 • Before 1988, only 2500 improved cook stoves were constructed in Pakistan. The Fuel Efficient Cooking Technologies project led to the production and dissemination of approximately 40,000 stoves in 1990 • Approximately 17,000 energy-efficient and living condition improvement products were installed in various households (under the Building and Construction Improvement Program (BACIP) established in 1997) benefiting nearly 70,000 people across 125 villages. This included fuel efficient 'smoke-free' cooking stoves with chimneys, as well as wall and floor insulation, and roof hatch windows to reduce dust particles and improve indoor heating (Colbeck et al. 2010) • Other policy measures include implementation of EURO-II vehicular emission standards, making vehicle fitness inspection compulsory for all private vehicles, phasing out two-stroke vehicles, and relocating brick kilns and providing alternate technology for manufacturing of brick (NEP 2005)

associated with these policies are: (i) ensuring sustainable, safe, affordable and uninterrupted supply of auto fuels; (ii) optimum utilization of infrastructure for importing crude and crude products, processing, storage and transportation; (iii) assessing the future trends in emissions and air quality requirements from the viewpoint of public health; and (iv) adopting such vehicular emission standards that will be able to make a decisive impact on air quality.

- The Indian government introduced the National Program for Improved Chulhas (NPIC) in 1983, providing a minimum subsidy of 50% to households that purchased the improved cook stove to reduce indoor air pollution. Renewable energy cooking options such as solar cookers were also introduced in some states, providing subsidies to the producers, which were to be passed on to the consumer (Bhattacharya and Cropper 2010).
- The Ministry of New and Renewable Energy (MNRE), Government of India, has initiated various schemes to harness different renewable sources of energy. As a result, a total grid-connected renewable power generation capacity of 15,694 MW was achieved by 31 December 2009, which is about 10% of the total installed power generating capacity in the country. It includes wind power of 10,925 MW, small hydropower of 2,559 MW, biomass power of around 2137 MW, and around 6 MW solar power. A capacity addition of 14,000 MW was targeted during the 11th Plan period to take the renewable power generating capacity to nearly 25,000 MW by 2012. This momentum is likely to be sustained and it is envisaged that the renewable power capacity in the country will cross 87,000 MW by 2022 (MNRE 2009). Such initiatives would result in significant reduction of air pollution, GHG emissions and associated health risks in India.
- The Government also published a notification on the Revised National Ambient Air Quality Standards, 2009 in the Official Gazette on 16 November 2009. These ambient air quality standards provide a legal framework for controlling air pollution and protecting public health (MOEF 2009).

4 Conclusion and Recommendations

Rising air pollution is a major environmental threat affecting human health on the Indian subcontinent. India is the major contributor for most of the pollutants and power plants are the largest contributor for most of the SO_2 emission in South Asian countries. However, for CO, BC, OC and PM the residential sector is the major source. For NO_x the accountability of residential and transportation sectors is highest among all other sectors in almost all countries of South Asia. Most of the governments have introduced several policies and programmes to mitigate the increasing air pollution, but the emissions of these pollutants have constantly increased during the last decade because of rising population, economic growth and development activities. Doll et al. (2013) have emphasized that the Government of

India's aim to achieve 100% sanitized cities has a particular relevance to the co-benefits approach for which a scoring system for 423 cities has been developed to assess their performance in 22 indicators in three key areas of sanitation (output, process and outcome). For example, current scores run from 73% in Chandigarh to 16% in Churu (Rajasthan). From the perspective of the co-benefits approach, this not only helps in identifying potential cities for co-benefits research but also provides a framework and baseline measures to assess co-benefits related improvements. Whilst a number of policy instruments are available to achieve co-benefits in Indian cities, no one measure is likely to succeed alone and therefore a mix of approaches is necessary. For example, various measures such as the development of the metro using overseas aid was complemented by implementing a partial Bus Rapid Transit System (BRTS) and also by making CNG the compulsory fuel for all public transport vehicles in megacity Delhi. Although Delhi's experience offers a means to understand how such a mix of policy measures might help develop the co-benefits approach, a city-specific perspective is necessary to highlight where local gaps exist that may offset the benefits of certain projects at city-scale (Doll et al. 2013).

Based on the past and recent trends, the following is the summary of key points for the improvement of air quality through various co-benefits programmes in cities of India and other South Asian countries:

- Develop and implement the clean or low emission technologies and more strict pollutant emission standards through:
  - Penetration of technology to reduce emissions of brick kilns
  - More strict emission control for vehicles, industries and power plants
  - Improvement of cook stoves and energy shift from biofuel to gas and electricity
- Accelerate the amendments of laws and regulations required for prevention and control of atmospheric pollution with the following perspectives:
  - Regional control of the total emissions of atmospheric pollutants in high polluted regions
  - Comprehensively improve the atmospheric environment in megacities
  - Prohibition against open burning of waste and crop residue
- Introduce the co-benefit approaches bearing on public health and climate policy, such as:
  - Traditional co-benefits due to energy saving and energy shift
  - Emerging co-benefits due to reduction of Short Lived Climate Pollutants (SLCP), such as BC, O_3, and CH_4
- Achieve better understanding of air pollution in South Asia based on the scientific approaches, for example:
  - Development of emission inventories for multi-species and multi-scale and reduction of their uncertainties

- Air quality modelling of urban/regional/continental scales
- Impact assessment of atmospheric pollution on human health, ecosystem, climate change, and others
- Future projection based emission scenarios and policy assessment of air pollution control options.

Acknowledgements The United Nations University Institute for the Advanced Study of Sustainability (UNU-IAS), Tokyo, Japan provided the major support for this study. Support received from the National Institute for Environmental Studies (NIES), Tsukuba, Japan, and the Max Planck Society (Germany) and Max Planck Institute for Chemistry, Mainz (Germany) through 'Max Planck Partner Group for Megacities and Global Change' at IIT Roorkee is also acknowledged. We extend special thanks to Dr. Jose A. Puppim de Oliveira for providing us the direction and motivation to conduct this study.

References

ADB-Asian Development Bank. (2005). *Strengthening environmental management at the state level cluster, Component e-strengthening environmental management at West Bengal Pollution Control Board*. Final Report, Volume V-Air Quality Management, November 2005.

Aardenne, J. A., Carmichael, G. R., Levy, H., II, Streets, D., & Hordijk, L. (1999). Anthropogenic NO_x emissions in Asia in the period 1990–2020. *Atmospheric Environment, 33*, 633–646.

Amaresekara, R. M., Gunawardena, A., Sugathapala, A. G. T., & Steele, P. (2004). Reducing Indoor air pollution in Sri Lanka—Is the situation improving? *First National Symposium on Air Resource Management in Sri Lanka*. www.cleanairnet.org/caiasia/1412/articles-60276_presentation.pdf. Accessed on August 6, 2010.

Azad, A. K., Nashreen, S. W., & Sultana, J. (2006). State of energy consumption and CO_2 emission in Bangladesh. *Ambio, 35*(2), 86–88.

Barletta, B., Meinardi, S., Simpson, I. J., Khwaja, H. A., Blake, D. R., & Rowland, F. S. (2002). Mixing ratios of volatile organic compounds (VOCs) in the atmosphere of Karachi, Pakistan. *Atmospheric Environment, 36*, 3429–3443.

Bhanarkar, A. D., Rao, P. S., Gajghate, D. G., & Nema, P. (2005). Inventory, of SO2, PM and toxic metals emissions from industrial sources in Greater Mumbai. *India. Atmospheric Environment, 39*, 3851–3864.

Bhatia, A., Pathak, H., & Aggarwal, P. K. (2004). Inventory of methane and nitrous oxide emissions from agricultural soils of India and their global warming potential. *Current Science*, 87(3).

Bhattacharya, S., & Cropper, M. L. (2010). *Options for energy efficiency in India and barriers to their adoption: A scoping study. Resources for the future*. Discussion paper, RFF-DP-10-20. 1616 P St. NW, Washington, DC.

Bhattacharya, S. C., Salam, P. A., & Sharma, M. (2000). Emissions from biomass energy use in some selected Asian countries. *Energy, 25*, 169–188.

CAI-Bhutan. (2006). *Country synthesis report on urban air quality management: Bhutan*. Discussion Draft, December, 2006. Asian Development Bank and the Clean Air Initiative for Asian Cities (CAI-Asia) Center.

Chakraborty, D., Bhattacharya, P. (2004). Air quality management Strategy for Kolkata city under India Canada Environment Facility Project – A Case Study. West Bengal Pollution Control Board. http://www.cleanairnet.org/baq2004/1527/articles-59332_bhattacharya.pdf.

CPCB. (2010). *Air quality monitoring, emission inventory and source apportionment study for Indian cities*. New Delhi, India: Central Pollution Control Board, Government of India.

Clean Air Bangladesh. (2008). *Summary of progress on improving air quality*. CAI-Asia, Clean air initiative for Asian cities.

Clean Air Nepal. (2008). *Summary of progress on improving air quality*. CAI-Asia, Clean Air Network-Nepal.

Clean Air Sri Lanka. (2008). *Summary of progress on improving air quality*. Country Network Sri Lanka, Clean Air Sri Lanka.

Colbeck, I., Nasir, Z. A., & Ali, Z. (2010). The state of indoor air quality in Pakistan—A review. *Environmental Science and Pollution Research, 17,* 1187–1196.

Dentener, F., Stevenson, D., Cofala, J., Mechler, R., Amann, M., Bergamaschi, P., et al. (2004). The impact of air pollutant and methane emission controls on tropospheric ozone and radiative forcing: CTM calculations for the period 1990–2030. *Atmospheric Chemistry and Physics, 5,* 1731–1755.

Devarajan, S., & Nabi, I. (2006). *Economic growth in South Asia: Promising, un-equalizing… Sustainable? South Asia region*. Washington DC: World Bank.

Dickerson, R. R., Andreae, M. O., Campos, T., Mayol-Bracero, O. L., Neusuess, C., & Streets, D. G. (2002). Analysis of black carbon and carbon monoxide observed over the Indian Ocean: Implications for emissions and photochemistry. *Journal of Geophysical Research, 107*(D19), 8017.

Doll, C. N. H., & Balaban, O. (2013). A methodology for evaluating environmental cobenefits in the transport sector: Application to the Delhi Metro. *Journal of Cleaner Production, 58*(1), 61–73.

Doll, C. N. H., Dreyfus, M., Ahmad, S., & Balaban, O. (2013). Institutional framework for urban development with co-benefits: The Indian experience. *Journal of Cleaner Production, 58*(1), 121–129.

Dutkiewicz, V. A., Alvi, S., Ghauri, B. M., Choudhary, M. I., & Husain, L. (2009). Black carbon aerosols in urban air in South Asia. *Atmospheric Environment, 43,* 1737–1744.

EIA. (2004). *EIA country analysis briefs. South Asia regional overview, October 2004*. http://www.eia.doe.gov/emeu/cabs/nepal.html. Accessed on: March 26, 2010.

Environmental Protection. (2010). *Bhutan hydropower project to lower GHG Emissions in South Asia. Environmental protection: The solution resource for managing air, water, energy and waste Issues*. April 2010. http://eponline.com/articles/2010/04/14/bhutan-hydropower-project-to-lower-ghg-emissions-in-south-asia.aspx. Accessed on March 11, 2010.

ESMAP. (2010). *Improved cookstoves and better health in Bangladesh: Lessons from household energy and sanitation programs*. Washington, D.C: World Bank.

Gadhavi, H. S., Renuka, K., Kiran, V. R., et al. (2015). Evaluation of black carbon emission inventories using a Lagrangian dispersion model—A case study over southern India. *Atmospheric Chemistry and Physics, 15,* 1447–1461.

Garg, A., Shukla, P. R., Ghosh, D., Kapshe, M., & Rajesh, N. (2003). Future greenhouse gas and local pollutant emissions for India: Policy links and disjoints. *Mitigation and Adaptation Strategies for Global Change, 8,* 71–92.

Grazi, F., van den Bergh, J. C., van Ommeren, J. N. (2008). An empirical analysis of urban form, transport, and global warming. *The Energy Journal, 29*(4), 97–122

Greenstone, M. (2015). Lower pollution, longer lives life expectancy gains if India reduced particulate matter pollution. *EPW Economic & Political Weekly, l*(8). http://www.epw.in/system/files/SA_L_8_210215_Michael_Greenstone_0.pdf. Accessed on March 09, 2015.

Gurjar, B. R., Butler, T. M., Lawrence, M. G., & Lelieveld, J. (2008). Evaluation of emissions and air quality in megacities. *Atmospheric Environment, 42*(7), 1593–1606.

Gurjar, B. R., Nagpure, A. S., & Kumar P. (2015). Gaseous emissions from agricultural activities in megacity Delhi. *Ecological Engineering, 75*(C), 123–127.

Gurjar, B. R., Nagpure, A. S., Kumar, P., & Sahni, N. (2010). Pollutant emissions from road vehicles in mega-city Kolkata, India: Past and present trends. *Indian Journal of Air Pollution Control, 10*(2), 18–30.

Gurjar, B. R., van Aardenne, J. A., Lelieveld, J., & Mohan, M. (2004). Emission estimates and trends (1990–2000) for megacity Delhi and implications. *Atmospheric Environment, 38*(33), 5663–5681.

Guttikunda, S. K., & Calori, G. (2013). A GIS based emissions inventory at 1 km × 1 km spatial resolution for air pollution analysis in Delhi, India. *Atmospheric Environment, 67,* 101–111.

Guttikunda, S., Calori, G., Velay-Lasry, F., & Ngo, R. (2011). *Air quality forecasting system for cities: Modeling architecture for Delhi, simple interactive models for better air Quality. SIM-Air Working Paper Series 36*. Delhi, Urban Emission Info.

Guttikunda, S. K., Goel, R., & Pant, P. (2014). Nature of air pollution, emission sources, and management in the Indian cities. *Atmospheric Environment, 95,* 501–510.

Guttikunda, S. K., Thongboonchoo, N., Arndt, R. L., Calori, G., Carmichael, G. R., & Treet, D. G. (2001). Sulphur deposition in Asia: Seasonal behaviour and contributions from various energy sectors. *Water, Air, and Soil pollution, 131,* 383–406.

IHME. (2013). *The global burden of disease 2010*: *Generating evidence and guiding*. Seattle: Institute for Health Metrics and Evaluation.

Indoor Air Pollution and Health Forum Nepal. (2008). Workshop on Clean Indoor Air: The right to a healthy life. *Quarterly Newsletter of Indoor Air Pollution and Health Forum Nepal, 1*(3).

Karim, M. (1999). Traffic pollution inventories and modeling in metropolitan Dhaka, Bangladesh. *Transportation Research Part D, 4,* 291–312.

Karim, M., Matsui, H., Ohno, T., & Hoque, S. (1997). *Current state of traffic pollution in Bangladesh and metropolitan Dhaka*. Presentation at the air & waste management association's 90th annual meeting & exhibition, Toronto, Canada, June 8–13, 1997. http://www.eng-consult.com/pub/awma97pap.htm. Accessed on: 28 April 2010.

Khan, B., & Baig, M. A. (2003). Pakistan: Preliminary national greenhouse gas inventory. *Journal of Applied Sciences and Environmental Management, 7*(2), 49–54.

Khwaja, M. A., Umer, F., Shaheen, N., Sherazi, A., & Shaheen, F. H. (2012). *Air pollution reduction and control in South Asia. Working paper Series 121.* Islamabad, Pakistan: Sustainable Development Policy Institute (SDPI).

Kojima, M., Akbar, S., Brandon, C., & Gwilliam, K. (2004). *Toward cleaner urban air in South Asia: Tackling transport pollution, understanding sources. ESM 281.* Washington, D.C: Joint UNDP/World Bank Energy Sector Management Assistance Programme (ESMAP).

Kumar, P., Gurjar, B. R., Nagpure, A., & Harrison, R. M. (2011). Preliminary estimates of particle number emissions from road vehicles in megacity Delhi and associated health impacts. *Environmental Science and Technology, 45,* 5514–5521.

Kumar, R., & Joseph, A. E. (2006). Air pollution concentrations of $PM_{2.5}$, PM_{10} and NO_2 at ambient and kerbsite and their correlation in metro city—Mumbai. *Environmental Monitoring and Assessment, 119,* 191–199.

Kurokawa, J., Ohara, T., Morikawa, T., Hanayama, S., Greet, J. M., Fukui, T., et al. (2013). Emissions of air pollutants and greenhouse gases over Asian regions during 2000–2008: Regional emission inventory in Asia (REAS) version 2. *Atmospheric Chemistry and Physics Discussion, 13,* 10049–10123.

Lelieveld, J., Crutzen, P. J., Ramanathan, V., Andreae, et al. (2001). The Indian Ocean experiment: Widespread air pollution from South and Southeast Asia. *Science, 291*(5506), 1031–1036.

Likhvar, V. N., Pascal, M., Markakis, K., et al. (2015). A multi-scale health impact assessment of air pollution over the 21st century. *Science of the Total Environment, 514,* 439–449.

Lu, Z., Streets, D. G., Zhang, Q., Wang, S., Carmichael, G. R., Cheng, Y. F., et al. (2010). Sulfur dioxide emissions in China and sulfur trends in East Asia since 2000. *Atmospheric Chemistry and Physics Discussion, 10,* 8657–8715.

MNRE. (2009). *Annual Report 2009-10*. New Delhi: Ministry of New and Renewable Energy, Government of India.

MOEF. (2009). *Annual Report 2009–10*. New Delhi: Ministry of Environment and Forests, Government. of India.

MoEF—Ministry of Environment and Forests. (2009). *India's GHG emissions profile results of five climate modeling studies*. http://www.moef.nic.in/downloads/home/GHG-report.pdf, Accessed on March 09, 2015.

Munir, T. (2009). *Assessment of air pollution and GHG mitigation strategies in Pakistan using the GAINS Model. Interim Report: IR-09-027*. Vienna: International Institute for Applied Systems Analysis.

Nagpure, A. S., Gurjar, B. R., & Kumar, P. (2011). Impact of altitude on emission rates of ozone precursors from gasoline-driven light-duty commercial vehicles. *Atmospheric Environment, 45,* 1413–1417.

Nagpure, A. S., Gurjar, B. R., Kumar, V., & Kumar, P. (2016). Estimation of exhaust and non-exhaust gaseous, particulate matter and air toxics emissions from on-road vehicles in Delhi. *Atmospheric Environment, 127,* 118–124.

Nagpure, A. S., Gurjar, B. R., & Martel, J. C. (2014). Human health risks in national capital territory of Delhi due to air pollution. *Atmospheric Pollution Research, 5.*

Nagpure, A. S., Ramaswami, A., & Russell, A. (2015). Characterizing the spatial and temporal patterns of open burning of municipal solid waste (MSW) in Indian Cities. *Environmental Science and Technology, 49*(21), 12904–12912.

Nagpure, A. S., Sharma, K., & Gurjar, B. R. (2013). Traffic induced emission estimates and trends (2000–2005) in megacity Delhi. *Urban Climate, 4,* 61–73.

NEP. (2005). *National Environment Policy*. Islamabad: Ministry of Environment, Government of Pakistan.

Ohara, T., Akimoto, H., Kurokawa, J., Horii, N., Yamaji, K., Yan, X., et al. (2007). An Asian emission inventory for the period 1980–2020. *Atmospheric Chemistry and Physics, 7,* 4419–4444.

Oliver, J. G. J., Janssens-Maenhout, G., Muntean, M., & Peters, J. A. H. W. (2014). *Trends in global CO2 emissions; 2014 report*. The Hague: PBL Netherlands Environmental Assessment Agency; Ispra: Joint Research Centre.

Parashar, D. C., Gadi, R., Mandal, T. K., & Mitra, A. P. (2005). Carbonaceous aerosol emissions from India. *Atmospheric Environment, 39,* 7861–7871.

Ramachandra, T.V. & Shwetmala (2009). Emissions from India's transport sector: Statewise synthesis. *Atmospheric Environment*, 43, 5510-5517.

Ramanathan, V., & Feng, Y. (2009). Air pollution, greenhouse gases and climate change: Global and regional perspectives. *Atmospheric Environment, 43*(1), 37–50.

Reddy, M. S., & Venkataraman, C. (2002a). Inventory of aerosol and sulphur dioxide emissions from India: I—fossil fuel combustion. *Atmospheric Environment, 36,* 677–697.

Reddy, M. S., & Venkataraman, C. (2002b). Inventory of aerosol and sulphur dioxide emissions from India. Part I—Biomass combustion. *Atmospheric Environment, 36,* 699–712.

Schwela, D. (2009). *Policy Brief: Strategic Approach for Air pollution Reduction in Kathmandu Valley*. Stockholm: Stockholm Environment Institute.

Shakya, G.R. & Shakya, I. (2007). *Final Report: Policy Gaps in household energy and indoor air pollution in Nepal*. Kathmandu: Practical Action Nepal.

Shrestha, R. M., & Malla, S. (1996). Air pollution from energy use in a developing country city: The case of Kathmandu. *Nepal. Energy, 21*(9), 194–785.

Shukla, S. K., Nagpure, A. S., Kumar, V., Baby, S., Shrivastava, P., Singh, D., et al. (2008). Impact of dust emission on plant vegetation of vicinity of Cement Plant. *Environmental Engineering and Management Journal, 7*(1), 31–35.

Shukla, S., Nagpure, A., Sharma, R., Sharma, D., & Shukla, R. N. (2013). Assessment of environmental profile in the vicinity of Indian cement industry. *International Journal of Environmental Technology and Management, 16*(4), 2013.

Silva, R. A., West, J. J., Zhang, Y., et al. (2013). Global premature mortality due to anthropogenic outdoor air pollution and the contribution of past climate change. *Environmental Research Letter, 8*, 034005 (11 pp).

South Asia Brief. (2010). *US Aid: From the American people. Making cities work*. http://www.makingcitieswork.org/www/urbanWorld/south-asia/index/content. Accessed on March 22, 2010.

Streets, D. G., Bond, T. C., Carmichael, G. R., Fernandes, S. D., Fu, Q., He, D., et al. (2003a). An inventory of gaseous and primary aerosol emissions in Asia in the year 2000. *Journal of Geophysical Research, 108*(D21), 1–30.

Streets, D. G., Tsai, N. Y., Akimoto, H., & Oka, K. (2000). Sulfur dioxide emissions in Asia in the period 1985–1997. *Atmospheric Environment, 34,* 4413–4424.

Streets, D. G., Yarber, K. F., Woo, J. H., & Carmichael, G. R. (2003b). Biomass burning in Asia: annual and seasonal estimates and atmospheric emissions. *Global Biogeochemical Cycles,* GB002040.

The Economist. (2015, February 7). Air pollution in India: Breathe uneasy. *The Economist.* http://www.economist.com/news/asia/21642224-air-indians-breathe-dangerously-toxic-breathe-uneasy. Accessed on March 09, 2015.

Timilsina, G. R., Shrestha A. (2009). *Why have CO_2 emissions increased in the transport sector in Asia? Underlying factors and policy options. Policy Research Working Paper 5098.* The World Bank Development Research Group.

UNEP Assessment Report. (2002). *The Asian Brown Cloud: Climate and other environmental impacts, 2002. South Asian haze: Air pollution, ozone and aerosols.* Centre for Clouds, Chemistry and Climate. ISBN: 92-807-2240-9.

UNEP. (2002). *Global Environment Outlook 3. Past, present and future perspectives. State of the environment and policy retrospective, 1972–2002, Atmosphere.* www.unep.org/Geo/geo3/english/pdfs/chapter2-7_atmosphere.pdf.

URBAIR. (1997). In J. J. Shah & T. Nagpal (Eds.), *Urban air quality management in Asia, Greater Mumbai report. World Bank Technical Paper No. 381.* Washington, DC.: World Bank.

Venkataraman, C., Habib, G., Eiguren-Fernandez, A., Miguel, A. H., & Friedlander, S. K. (2005). Residental biofuels in South Asia: Carbonaceous aerosol emissions and climate impacts. *Science, 25*(307), 1454–1456.

WHO-World Health Organization. (2014). *Public health, environmental and social determinants of health (PHE), Ambient (outdoor) air pollution in cities database 2014.* http://www.who.int/phe/health_topics/outdoorair/databases/cities/en/. Accessed on March 09, 2015.

Wiedinmyer, C., Yokelson, R. J., & Gullett, B. K. (2014). Global emissions of trace gases, particulate matter, and hazardous air pollutants from open burning of domestic waste. *Environmental Science and Technology, 48*(16), 9523–9530.

WRI-EarthTrends. (2010). *World Resource Institute, Climate and Atmosphere.* http://earthtrends.wri.org/searchable_db/index.php?action=select_variable&theme=3. Accessed on May 12, 2010.

World Bank. (1997). Urban air quality management strategy in Asia—Greater Mumbai report. In J. J. Shah & T. Nagpal (Eds.), *Urban air quality management in Asia, Greater Mumbai report. World Bank Technical Paper No. 381.* Washington, DC.: World Bank.

Yamaji, K., Ohara, T., & Akimoto, H. (2003). A country-specific, high-resolution emission inventory for methane from livestock in Asia in 2000. *Atmospheric Environment, 37,* 4393–4406.

Yamaji, K., Ohara, T., & Akimoto, H. (2004). Regional-specific emission inventory for NH_3, N_2O and CH_4 via animal farming in South, Southeast, & East Asia. *Atmospheric Environment, 38,* 7111–7121.

Chapter 4
Aligning Global Environmental and Local Urban Issues

Usha P. Raghupathi, Richa Sharma and Aastha Joshi

Abstract India's population of 1.21 billion (in 2011) makes it the second most populous country in the world. About 31% of India's population lives in urban areas and the urban population has exhibited a higher decadal growth rate (31.8%) than its rural counterpart (12.2%), indicating that the country is urbanizing at a rapid pace. With an increasing number of people living in urban areas, the GHG emissions and environmental problems have also increased, owing to lacunae in urban planning and management. These GHG emissions ultimately contribute to climate change at both local and global levels. Therefore, measures to deal with climate change need to focus on both mitigation and adaptation. This study examines the status of a climate co-benefits approach in Indian cities, illustrated through various *sectoral case studies*. The study also attempts to examine how environmental issues in Indian cities align with local development issues. It discusses the context of climate and environmental co-benefits in urban India and relates it to urban policies. Implementation of the ACCCRN programme in three cities—Indore, Surat and Gorakhpur—is discussed to highlight climate change resilience issues at the local level and their environmental and social benefits. The final section deals with the future perspectives of integrating a climate co-benefits approach in urban development especially in light of the '100 Smart City Mission' of India.

Keywords ACCCRN · Global · Local environment · Co-benefits · Policy

1 Introduction

Climate change is inevitably becoming central to all economic and development issues. The recently concluded Conference of the Parties (COP 21) in Paris (December 2015) has recognized that 'climate change represents an urgent and

U.P. Raghupathi (✉) · R. Sharma · A. Joshi
National Institute of Urban Affairs, New Delhi, India
e-mail: usha16ji@gmail.com; uraghupathi@niua.org

M. Sethi and J.A. Puppim de Oliveira (eds.), *Mainstreaming Climate Co-Benefits in Indian Cities*, Exploring Urban Change in South Asia,
https://doi.org/10.1007/978-981-10-5816-5_4

potentially irreversible threat to human societies and the planet' and the increase in global average temperature should be contained to well below 2 °C above the pre-industrial levels and efforts should also be made to limit the temperature increase to 1.5 °C (UNFCCC-COP21 Paris Agreement, p. 2). The Agreement recognizes that, 'the social, economic and environmental value of voluntary mitigation actions and their co-benefits for adaptation, health and sustainable development' (p. 15, point 109, Paris Agreement).

Co-benefits, also known as ancillary benefits, are defined as the secondary, collateral or indirect benefits arising from a climate change policy. The co-benefits approach addresses climate change concerns while helping improve the local environment (Puppim de Oliveira et al. 2013a). The local environmental problems in cities range from air pollution, water pollution and waste management to concretization of cities and reduction in green cover. Urbanization results in an increase in greenhouse gas (GHG) emissions, that mainly originate from activities such as construction, manufacturing, transport, energy consumption and waste generation. India is the fifth largest emitter of carbon dioxide (CO_2) in the world after the United States, China, the European Union and Russia. Studies conducted in 2007 estimate that India emitted almost 1.72 billion tonnes of CO_2 equivalents. By 2030 India is expected to emit over 5.22 billion tonnes of CO_2 (Sethi and Mohapatra 2013). Meanwhile, India's urban population, which stood at 377 million in 2011, is projected to increase to 600 million by 2030. A co-benefits approach will help India align the goal of sustainable development with urbanization.

Planning for development often neglects the problem of GHG emissions as several compelling local development needs tend to get priority over local environment and global climate change concerns. The existing policies are not fully equipped to deal with the issues of urbanization and environment in an integrated manner (Puppim de Oliveira et al. 2013b). The co-benefits approach can help achieve more than one outcome in a single policy and can help developing countries implement climate friendly policies to deal with both global and local environmental issues while still catering to local development needs. This chapter focuses on the co-benefits approach and how environmental issues align with local development issues. It identifies co-benefits in the context of nationwide urban policies in India. The chapter explores how the opportunity to integrate co-benefits in different sectors exists in India today by reviewing projects executed on the ground. The chapter then discusses the Asian Cities Climate Change Resilience (ACCCRN) programme and the projects undertaken in three ACCCRN cities in India that deal with climate change resilience issues at the local level having environmental and social benefits. Finally, the chapter concludes with a way forward.

2 Environmental Concerns and Co-benefits

The global GHG emissions have been rising at an average of 2.2% per year between 2000 and 2010, compared to 1.3% per year between 1970 and 2000 (IPCC WG III 2014). Eighteen per cent of the world's population living in developed countries emit 47% of global emissions, whereas 82% residing in developing countries emit the remaining 53% (UN Habitat 2011). The IPCC AR5 indicates that of the last 1400 years, the period 1983–2012 is likely to have been the warmest 30 years. The CO_2 emissions from fossil fuel combustion and industrial processes have contributed a major portion of the total GHG emissions during the period 1970–2010.

With increasing population, land requirements for housing, agriculture and industrial use has increased. To meet the demands of urban growth, land conversion is growing. Agricultural land and green spaces are being converted to urban use (buildings, infrastructure etc.) which have led to degradation of many terrestrial ecosystems, as non-economic ecosystem functions are rarely given their due credit. Deforestation and concretization have further affected natural systems in cities. Changing climate patterns, economic globalization, population growth, increasing use of natural resources and rapid urbanization are putting pressure on terrestrial ecosystems as never before, and virtually all of them are under stress. The losses from deforestation and forest degradation are likely to cost the global economy huge sums of money. The pressure on protected areas (for biodiversity conservation) is also increasing (Hunsberger and Evans 2012). Policies that cater to the demands of society while maintaining environmental sustainability are needed. Some steps have been taken for reforestation and greening of cities and it suggests that all is not lost, and recovery may be possible.

The recently concurred Paris Agreement (COP 21) states that sustainable development requires adaptation measures that strengthen resilience and reduce vulnerability to climate change. It recognizes that multiple actors are involved in adaptation—local, sub-national, national, regional and international—and that mitigation and adaptation are complementary by stating that 'greater levels of mitigation can reduce the need for additional adaptation efforts'. Adaptation measures are specific to each country and, therefore, adaptation should be integrated into the 'relevant socio-economic and environmental policies and action' (p. 25, Article 7, Paris Agreement). In this aspect, solutions to global environmental challenges and concerns will have to be addressed at the local level and local development needs will have to accommodate global concerns. For instance, optimizing energy consumption at local levels can have local as well as global benefits. As the world, and India, urbanize, cities will be key to reducing emission of GHGs. Mitigation and adaptation measures would be required to build city resilience and need implementation at the local level. Partnership amongst different actors at all levels (global to local) will be able to promote policies that lead to sustainable development. For this, a study of national policy landscape and development sectors becomes necessary to evaluate potential global climate and local environmental co-benefits therein.

3 Policy Perspective

India is the world's second most populous country with a population of 1.21 billion (Census of India, 2011) of which approximately 31% reside in urban areas. With a growing population in Indian cities, GHG emissions are increasing and so are the environmental problems (Doll et al. 2013). Recognizing the problem of global climate change, India voluntarily declared a goal of reducing the emissions intensity of its GDP by 20–25%, over 2005 levels, by 2020. A number of policy measures were launched to achieve this goal. As a result, the emission intensity of India's GDP has decreased by 12% between 2005 and 2010 (MoEF 2015).

Climate change and energy are now a focus of local, state and national attention around the world. India has begun implementing a diverse portfolio of policies nationally and within individual states to improve energy efficiency, develop clean sources of energy, and prepare for the impacts of a changing climate (Leiserowitz and Thaker 2012). In this regard, some of the policies of the Government of India that relate to climate change and have an impact on local environment and development, particularly in urban domain are discussed below.

3.1 India's National Action Plan on Climate Change (NAPCC), 2008

Recognizing the enormous threat posed by climate change, the prime minister launched the National Action Plan on Climate Change (NAPCC) in June 2008. The NAPCC 'identifies measures that promote (India's) development objectives while also yielding co-benefits for addressing climate change effectively'. One of the guiding principles of NAPCC is: 'Protecting the poor and vulnerable sections of society through an inclusive and sustainable development strategy, sensitive to climate change' (NAPCC, Government of India 2008). It is, thus, clear that environmental, climate and social co-benefits are built into the approach of NAPCC and the poorest sections of the society have been kept in focus. The NAPCC focuses on both mitigation and adaptation. The eight sub-missions under NAPCC are:

1. *National Solar Mission*: Aims to increase the share of solar energy in the total energy mix in the country. As solar energy allows decentralized distribution, remote and less accessible areas can be provided energy. The mission aims to empower people at the grassroots level through the provision of energy.
2. *National Mission for Enhanced Energy Efficiency*: Aims to decrease energy consumption in large energy-consuming industries, with a system for companies to trade energy-savings certificates. The mission also aims to make energy-efficient appliances affordable and reduce energy consumption through demand-side management. By 2014–15 the mission's goal is to have an annual

fuel savings in excess of 23 million tonnes of oil equivalent; cumulative avoided electricity capacity addition of 19,000 MW; and CO_2 emission mitigation of 98 million tonnes per year.

3. *National Mission on Sustainable Habitat*: The mission focuses on promoting energy efficiency in buildings, urban waste management and recycling, including generation of energy from waste, and modal shift to public modes of transport leading to reduction in GHGs.
4. *National Water Mission*: Aims at conserving, minimizing wastage of water, and equitable water distribution. It also focuses on improving water use efficiency by 20% and on recycling of wastewater. The Mission calls for strategies to accommodate fluctuations in rainfall and river flows by enhancing water storage methods and rain water harvesting.
5. *National Mission for Sustaining the Himalayan Ecosystem*: Aims to conserve biodiversity, forest cover, and other ecological values in the Himalayan region. The Mission also calls for empowering local communities to play a greater role in managing ecological resources.
6. *National Mission for a Green India*: Aims at enhancing ecosystem services such as carbon sinks. It calls for afforestation and increasing land area under forest cover from 23 to 33%.
7. *National Mission for Sustainable Agriculture*: Aims to make agriculture more resilient to climate change by identifying new varieties of crops and alternative cropping patterns. It suggests expansion of a weather insurance mechanism to safeguard farmers from climate change.
8. *National Mission on Strategic Knowledge for Climate Change*: Aims to improve research and technology development through increased international collaboration. It also encourages private sector initiatives to develop innovative technologies for mitigation and adaptation.

All the above sub-missions, will have economic and social co-benefits along with climate change benefits. The two energy related sub-missions will help provide electricity to remote and inaccessible areas and make energy consumption efficient. They would also enhance productivity, provide better livelihood opportunities and improve the quality of life of people. Improved power supply will also lead to improved health and education levels. The co-benefits of sub-missions on Sustainable Habitat, and Water will help conserve natural resources and reduce GHG emissions. Improving water use efficiency, recycling waste water, and rain-water harvesting will result in greater water availability, which can help better health and productivity, leading to improved incomes. By safeguarding thc Himalayan Ecosystem (sub-mission 5), the source of water for rivers along the plains will be safeguarded and flooding will be controlled. The sub-mission would also help in providing sustainable livelihoods for local people. Under the Green India sub-mission, planting trees in degraded areas will increases livelihood opportunities for local people by promoting trade in NTFPs (Non-timber Forest Products) and also preserving biodiversity.

3.2 National Urban Sanitation Policy (NUSP), 2008

The National Urban Sanitation Policy's vision is to make urban India clean, healthy and livable. It lays special emphasis on making sanitation facilities affordable for urban poor and women. The policy goals of NUSP include awareness generation and behaviour change, making cities open defecation free, and providing integrated city-wide sanitation by re-orienting institutions and mainstreaming sanitation, safe disposal, and proper operation and maintenance of sanitary facilities (NUSP 2008). The co-benefits of this policy include improved public health (and therefore personal health) and improved productivity. It will have significant environmental benefits and by dealing with waste in a proper manner it will help reduce GHG emissions.

3.3 National Water Policy, 2012

Water is a scarce resource. The National Water Policy 2012 recognizes climate change as a factor that could impact the availability of water in the future. With a growing population the demand for water will increase and with wide variation in temporal and spatial availability of water due to climate change, and inequitable distribution of water, there could be water conflicts in future. The adaptation strategies include increasing water storage in ponds, reservoirs, increasing ground water availability through rainwater harvesting, recycling of wastewater, demand management and improving water use efficiency. The policy emphasizes conservation and efficient utilization of water resources. These measures will yield co-benefits by reducing pumping requirements, the use of fossil fuels and hence GHG emissions. The benefits of improved availability of water in urban areas will be on the health and quality of life of citizens, especially the poor. It will also improve productivity and income of the poor.

3.4 National Urban Transport Policy (NUTP), 2006

India's rapid economic growth, rise in income levels, and increasing levels of urbanization combined with lack of adequate public transport have led to immense mobility problems in cities. Higher incomes have also led to rising social aspirations which often directly results in the purchase of personal motor vehicles. The consequence of rapid increase in urban population combined with lack of or inadequate public transport has been congestion and chaos on roads. This causes loss of productivity, hence GDP, deterioration of air quality in cities and increase in GHG emissions contributing to global warming. To sustain economic growth and competitiveness of cities, and provide citizens with good quality of life, provision of an efficient system of transportation in urban areas is essential. As a response to these concerns, the Government of India formulated the National Urban Transport Policy (NUTP) 2006. The main objective of NUTP is 'to ensure safe, affordable, quick, comfortable, reliable and sustainable access (and mobility) for the growing number of city residents to jobs, education, recreation and such other needs within (Indian) cities'. The initiatives that can yield environmental, climate and social co-benefits to achieve the objectives of NUTP are:

- *Integrating land use and transport planning*: Mobility is the basis of the physical and economic growth of cities. The urban form of a city is shaped by the land use plan or the master plan of the city. Transport plan on the other hand is based on the mobility needs of the city. Integrating land use planning with transport planning will help the city grow in a sustainable manner. It can prevent urban sprawl and, therefore, reduce travel distances, reduce fuel consumption and reduce air pollution. The transit oriented development approach, which is being implemented in some cities (e.g. Delhi), integrates land use planning with transport planning.
- *Equitable allocation of road space*: Road space in cities is planned with the objective of moving vehicles efficiently and not people. Private motor vehicles get priority over public transport. Public transport (buses) that moves large numbers of people does not get the road space it deserves and non-motorized vehicles and pedestrians get squeezed in the road space. Equitable allocation of road space focusing on moving people, giving priority to public transport, non-motorized vehicles and pedestrians will have environmental and social co-benefits. Making space for non-motorized vehicles and pedestrians will especially help the poor travel or walk safely on roads.
- *Priority to the use of public transport*: According to NUTP 'public transport occupies less road space and causes less pollution per passenger-km than personal vehicles'. Public transport is a more sustainable form of transport than personal vehicles. People should be encouraged to switch from personal modes of transport to public modes and from motorized modes to non-motorized modes of transport. This will yield positive co-benefits for environment and the society. Giving priority to public transport is beneficial to both the environment and the poor. This will reduce congestion on roads and reduce air pollution.
- *Priority to non-motorized transport*: Bicycles are generally used by the poorer sections of society for transport to work. They are not considered as an option for travel by most people for a number of reasons that include: no separate road space, large distances of travel, weather conditions, safety and, most importantly, a bicycle is associated with the poor. To encourage non-motorized transport and address safety issues, road design should be changed to have special lanes for cyclists. Providing proper footpaths in road design will encourage people to walk and will also provide safety to pedestrians. The environment and health co-benefits of this will be encouraging.

- *Use of cleaner technologies*: Most of the motorized vehicles today either use diesel or petrol, contributing significantly to air pollution. CNG is now being widely used in cities for buses and also in cars and auto-rickshaws. Electric buses have been tried but have their limitations. Renewable sources of fuel need to be considered in future for sustainable development and to reduce the dependence on fossil fuel. This will improve air quality in cities and help reduce GHG emissions. It will also help India reduce its import bill (for importing fossil fuel) and save valuable foreign exchange.

3.5 URDPFI Guidelines, 2014

The Draft Urban and Regional Development Plans Formulation and Implementation (URDPFI) Guidelines 2014, brought out by the Ministry of Urban Development, Government of India, is the revised version of the UDPFI Guidelines of 1996. The 2014 revised guidelines include the climate change aspects in planning, specifying the benefits derived from such planning. These guidelines provide a framework for plan preparation and implementation. The design of urban centres, the built environment, transportation system and so on, determine the environment of cities and towns. Planning is required to manage the negative impacts of urbanization on the environment and make urban areas sustainable. A high-density built environment and lack of green spaces create urban heat islands. Providing green spaces and trees can cool the environment and reduce the energy requirements for cooling. Building compact and green cities will help in reducing pollution levels and improving the quality of life in urban areas. The URDPFI guidelines discuss Green Cities and their benefits. It states that Green Cities are those where the economic growth and development does not negatively impact the environment and natural resources, and the pressure on ecosystem services is minimized. The co-benefits of the planning approach given in URDPFI will lead to better environment in cities, improved productivity and economic growth, and improved health of citizens, due to lower pollution levels, and greater availability of green spaces, amongst others.

3.6 Recent Urban Sector Programmes in India

Since the latter half of 2014, the Government of India has launched a number of missions such as the Smart Cities Mission, Atal Mission for Rejuvenation and Urban Transformation (AMRUT), National Heritage City Development and

Augmentation Yojana (HRIDAY), and Swachh Bharat Mission. These missions aim to provide/upgrade infrastructure, make cities liveable, encourage use of technology and use of renewable energy, and focus on sustainable development. The Smart Cities Mission and AMRUT also focus on protection of natural environment, and adoption of climate resilient and energy efficient policies and regulations. India has initiated a plan of action for clean energy, energy efficiency in various sectors to achieve lower emission intensity in the automobile and transport sector, a major thrust to non-fossil based electricity generation and a building sector based on energy conservation (MoEF 2015). Apart from the Government initiatives, there are other examples across the country that have helped reduce emissions and create co-benefits in different sectors. Some initiatives taken across sectors include:

- Transport—provision of efficient public transport and switching to cleaner fuels such as compressed natural gas (CNG)
- Energy—demand side management, energy efficiency in buildings
- Solid waste management—segregation of waste at source, efficiency improvement in collection and transport, methane avoidance and waste to energy and
- Liquid waste management—low energy sewage treatment and methane recovery (Kapshe et al. 2013).

4 Sectoral Examples of Co-benefits

This section presents examples of co-benefits at state and city levels covering different urban sectors (Fig. 1).

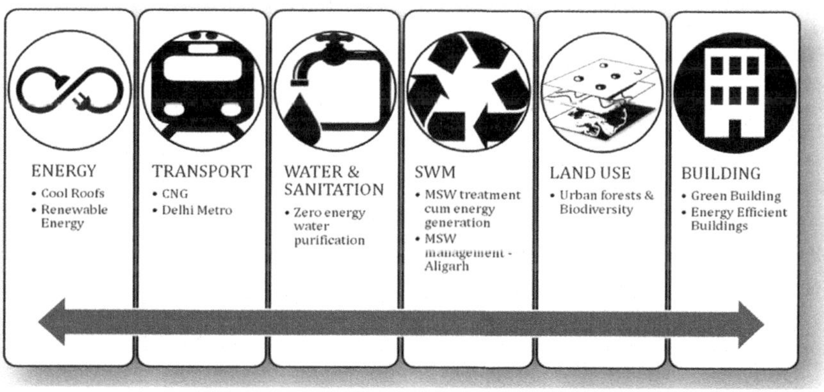

Fig. 1 Identifying co-benefits in different sectors

4.1 Energy

4.1.1 Cool Roofs, Hyderabad

Cool roof is a roofing technology based on solar reflectance (reflect solar radiations) and thermal emittance (instead of transferring heat to the building, they emit it back into the atmosphere) of the roof surface. Cool reflective roofs are usually light or white coloured and so they absorb less heat which lowers down the surface temperature of the roof. These could be made of reflective paints, coverings, tiles or shingles. A variety of cost effective options are available for cool roofs, such as simple lime wash with adhesive, and polyurethane based insulation materials.

The project: Cool roofs for reducing energy demand for cooling is increasingly becoming an important measure for urban heat island (UHI) mitigation in cities. The cool roof project in Hyderabad was an initiative of Lawrence Berkeley National Laboratory and IIIT, Hyderabad. It aimed at mitigation of UHI. The project applied various measures to achieve this objective, including field evaluations, model simulations, and knowledge transfer (Akbari 2011). The specific objective of this project was to conduct field studies (to quantify and demonstrate the effect of cool roofs in reducing energy use in buildings), perform meteorological simulations (to estimate the effect of UHI mitigation on urban climates such as air-temperature reduction), and carry out outreach activities through training programmes, conference sessions, websites, and kiosks (to disseminate knowledge and create awareness). The monitoring of the project was carried out in 2006.

Environmental Benefits: The project exhibited that in a tropical city like Hyderabad, whitening of previously black roofs could save 20–22 kWh of energy per square metre of roof area corresponding to 14–26% of energy use for air-conditioning in commercial buildings. Similarly, application of white coating on uncoated concrete roofs could save 10–19% (13–14 kWh/m^2). Assuming an annual increase of 100,000 m^2 of new roof construction in the next 10 years, the whitening of concrete roofs is estimated to save 13–14 GWh of electricity in the 10th year and 73–79 GWh would be the 10-year cumulative savings. This, when extrapolated to country level, would at least save 730–790 GWh. In terms of money, the savings would be approximately Rs. 93–101 per m^2 of roof on air conditioning and the 10-year cumulative savings for cooling in India would be over Rs. 5 billion. With meteorological simulations, the study also concluded that urban heat island control strategies (combination of increased surface albedo and vegetative cover) could help reduce the air temperature in Hyderabad by 2 °C. More intensive strategies could bring the temperature down by 2.5–3.5 °C.

Climate Co-benefits: If the projections of energy saved from this project are converted into CO_2 equivalents, it is found that approximately 11–12 kg CO_2 would be reduced per year per square metre of whitened flat concrete roof area in India. And the 10-year cumulative CO_2 reduction in India would be 0.6–0.65 MT. The study primarily aimed at mitigating the UHI in the city. In other words, the target is to address the local environmental issue at city scale. However, the project holds potential to generate global climate co-benefits in terms of reducing CO_2 emissions which could be easily amplified from city to country level. The outcomes from the study could thus be used in developing cool-roof building standards and framing related policies.

4.1.2 The 5 MW Solar PV Power Project, Patan District, Gujarat, India

The Project: The project is a part of 'National Solar Power Development Programme, India' for which the coordinating entity is Emergent Ventures India Pvt. Ltd (UNFCCC 2012a). The programme has seven more solar power projects; 5 MW Solar PV power project in Bikaner (Rajasthan), 5 MW Solar PV power project in Surendranagar (Gujarat), 5 MW Solar PV power project in Kutch (Gujarat), 5 MW Solar PV power project in Jodhpur (Rajasthan), 6 MW Solar PV power project in Patan (Gujarat), 5 MW Solar PV power project in Sabarkantha (Gujarat) and 5 MW Solar PV power project in Patan (Gujarat). The project was initiated in 2012. The implementing agency is M/s. Surana Telecom and Power Limited. The project proposes to develop a 5 MW grid connected solar PV power project in the Patan District of Gujarat State in India.

Environmental Benefits: The project will provide an average of 7045 MWh of renewable power annually to the NEWNE grid system in India. Since the project envisages utilizing solar energy, a clean fuel, for power generation, it is expected to have a positive impact on the environment. Additionally, solar energy has no associated GHG emissions. The power supplied from the project would replace the fossil fuel consumption which would contribute in overall reduction of emissions.

Climate Co-benefits: The electricity thus generated would contribute to GHG reductions estimated at 6683 tCO_2e/annum. The decrease in thermal electricity consumption reduces the amount of GHGs and other pollutants such as CO, SO_x, NO_x, produced by fossil fuel consumption.

4.2 Water—Zero Energy Water Purifier

The Project: The project envisages energy free water purifiers and cleaning kits in Kerala state. It aims at avoiding the use of conventional fossil fuel-based energy-intensive methods of water purification (e.g. boiling) while providing safe and clean drinking water to people at affordable prices (UNFCCC 2014). The technology was approved by Government of India in 2008, and a baseline survey for this was conducted in 2011.

Environmental Benefits: The project would help in resource conservation through cleaner means of water purification. It would also help in improving the health of the people by providing affordable access to safe drinking water.

Climate Co-benefits: The project would reduce CO_2 emission by 17,549 tCO_2e over a 10 year crediting period on an annual average.

4.3 Sanitation: Energy from Sewage—Anjana Sewage Treatment Plant (STP), Surat

The Project: In response to environmental concerns, the Surat Municipal Corporation (SMC) initiated a green energy STP in 1997 that productively utilized energy rich components of the sewage gas from the Anjana Sewage Treatment Plant (Kapshe et al. 2013). The project was funded by the Ministry of New and Renewable Energy (MNRE) under the national programme on Energy Recovery from Urban and Industrial Wastes. The plant consists of systems for gas collection, gas cleaning, gas storing, and excess gas flaring.

Environmental Benefits: The primary aim of the project was to address the local environmental concern of sewage disposal through treatment plant. The project was, however, designed in such a manner that it provided other benefits such as fuel conservation (the plant uses the energy it generates), revenue and employment generation.

Climate Co-benefits: Apart from other benefits, this sewage-based power plant, helps reduce GHG emissions in the atmosphere in several ways. It prevents methane emission from the digester as is the case with conventional STPs. In addition to this the generated electricity reduces consumption of grid power, thus saving the fuel and reducing the carbon emission associated with it. In terms of CO_2, the plant is able to save 22,844.06 tonnes. Other STPs installed based on similar technology have earned Certified Emission Reduction (CERs) units. For example, Singapore STP has earned 18,000 CERs per annum; the Karanj STP is earning 195,000 CERs per annum.

4.4 Solid Waste Management—Municipal Solid Waste Management, Aligarh

The Project: The Municipal Solid Waste Management project at Aligarh, India was completed in 2012, aimed at avoidance of methane production from decay of biomass through controlled combustion, gasification or mechanical/thermal treatment, and avoidance of methane emissions through composting (UNFCCC 2012b). The project consisted of an integrated processing complex involving collection and transportation of waste from the city, a composting plant, a plant for preparation of RDF along with recovery of recyclable material, and a brick manufacturing unit. The project has a capacity to process 220 TPD of MSW.

Environmental Benefits: The project achieved efficient collection and transportation of municipal solid waste from the city. The project also helped in avoiding leachate production. It reduced the requirement of land filling area for dumping, demand for chemical fertilizers and fossil fuels.

Climate Co-benefits: The project has contributed to reduction in GHG emission by avoiding methane production, which would have otherwise taken place due to dumping of waste in anaerobic conditions. An average annual emission reduction of 11,427 tCO_2 and total GHG emission reduction of 79,992 tCO_2e is estimated for the project.

4.5 Transport and Air Quality

4.5.1 Delhi Metro

The Project: The population of Delhi has been growing rapidly (it has over 14 million inhabitants) and this has resulted in traffic woes and high levels of air pollution in the city. To address these issues, the Delhi Metro Rail Corporation (DMRC) was set up by the Government of India and the Government of Delhi on 3 May 1995. The Delhi metro is being built in phases:

- Phase I—Completed 58 stations covering 65 km of route length, 13 km of this being underground and 52 km being on surface or elevated.
- Phase II—Comprises 85 stations covering a route length of 124.6 km
- Phase III—Comprises 69 stations covering a route length of 103 km
- Phase IV—To be completed by 2021, with the network spanning 413 km

The first stretch of metro rail corridor became operational in December 2002 though Phase I got completed only in 2005. Phase II of Delhi Metro has also been completed and work on Phase III is currently ongoing. According to a study conducted by the Central Road Research Institute (CRRI) in 2011, Delhi Metro has helped in removing about 1.17 lakh vehicles from the streets of Delhi. At present, the Delhi Metro has a ridership of about 2 million passengers a day.

Environmental Benefits: Delhi Metro has helped improve the city's air quality and eased commuting. Delhi Metro also contributes to rain water harvesting (along Blue Line). DMRC is also looking forward to harness solar energy.

Apart from reduction in air pollution the Delhi Metro provides multiple benefits such as time saving to passengers due to reduction in congestion on roads and due to lower travel time of the metro, reduction in accidents, reduction in traffic jams and congestion, fuel savings, cost saving to car owning passengers in terms of capital cost and operation and maintenance costs of cars (Murty et al. 2006).

Climate Co-benefits: Apart from a variety of environmental benefits, the Delhi Metro has proved its potential for climate co-benefits by earning carbon credits for mitigating GHG emissions (approximately 6.3 lakh tonnes each year). DMRC is receiving carbon credits worth Rs. 0.47 billion each year (DMRC, 2014a, 2014b). Each passenger, when switching to metro instead of personal car or public bus, reduces emissions (CO_2) by approximately 100 gm for every 10 km trip.

4.5.2 Use of Compressed Natural Gas (CNG), Delhi

The Project: In order to reduce air pollution in the city of Delhi, all public transport vehicles were made to switch their fuel to compressed natural gas (CNG) during 2001–03. This task was not easy and had to be done through a Supreme Court ruling in July 1998. At the beginning of 2005, 10,300 CNG buses, 55,000 CNG three-wheelers taxis, 5000 CNG minibuses, 10,000 CNG taxis and 10,000 CNG cars were running on Delhi's roads. As per the Press Information Bureau, there were 340,000 CNG vehicles in Delhi in 2010. This included 15,000 CNG buses, 240,000 CNG private cars, 45,000 light commercial vehicles, the remaining being auto-rickshaws and taxis. Findings from Goyal (2003), Kathuria (2004) and Ravindra et al. (2006) suggest the following co-benefits:

Environmental Benefits: CNG being the most environment friendly fuel helped improve the air quality of Delhi. CNG powered vehicles usually emit 85% less NO_X, 70% less reactive-hydrocarbons, and 74% less of CO. The CNG vehicles have also reduced ozone precursor emission. CNG has the added advantages of being low cost, providing longer service life, improving fuel consumption and engine efficiency, and reducing engine noise.

Climate Co-benefits: Methane and black carbon emissions are critical contributors to the change in carbon dioxide equivalent emissions. Switching to natural gas resulted in a 30% increase in CO_2 equivalents without incorporating the impact of aerosols. However, when aerosol emissions were considered, the net effect was estimated to be a 10–30% reduction in CO_2 equivalents. Hence, although switching to CNG was mainly aimed at reducing NO_X and SO_X emissions to improve air quality, CNG simultaneously helped in lowering CO_2 equivalent emissions considerably. Reduction in air pollution reduces respiratory illnesses and improves health of citizens.

4.6 Land Use—Urban Forest

The Project: 'Rehabilitation of Degraded Wastelands at Dera Mandi in Southern District of National Capital Territory of Delhi through Reforestation' is the Clean Development Mechanism (CDM) that is being implemented by the Department of Environment, Forests and Wildlife, Government of NCT of Delhi, for reforestation of the degraded grasslands in the project area of 365.7 ha (UNFCCC 2013). The project was initiated in 2008 and approved in 2013. It is a 30-year project. The project aims at carbon sequestration through reforestation. Other aims of the project are:

- Earning carbon credits by growing trees
- Providing climate change mitigation services by planting trees
- Eco-restoration of the degraded land and biodiversity conservation
- Improving soil and land productivity
- Providing critical environmental sustenance to the megapolis of Delhi
- Generating economic, social and educational options involving stakeholders.

Environmental Benefits: The project would help in conserving the urban biodiversity in Delhi along with improving the city's air quality. It will also help in restoration of degraded land in the Southern District of Delhi. The trees enhance microclimate by providing shade and acting as wind breakers. Urban forests also help in mitigating the impacts of locally developed heat islands.

Climate Co-benefits: It is estimated that the project would be able to save 12,138 tonnes CO_2 equivalent per annum. The co-benefits of the project would include improved health of people living in the project area and better air quality in surrounding regions.

4.7 Building Sector—Green Building, Dehradun and Kolkata

The Project: Oil and Natural Gas Corporation (ONGC) Limited has a green building office in Dehradun and Kolkata, with energy efficiency features (UNFCCC 2011a, b). The first ONGC green building was inaugurated in 2013 in Dehradun. The building has been constructed following the New Construction (NC) standards under Indian Green Building Council—Leadership in Energy and Environmental Design (IGBC-LEED) Green Building Rating System. The green buildings have active and passive intelligence features to maximize energy efficiency. The projects attempt to create an eco-friendly and energy efficient workspace. Various energy efficiency measures have been taken which aim at improving the efficiency of heating, ventilating, and air conditioning (HVAC) systems and improving the insulation of the building. The projects would reduce GHG emissions in three ways —decrease power consumption through features that reduce heating and cooling demands of the building; decrease power consumption through installation of efficient equipments/systems (HVAC, Lighting fixtures etc.); and decrease power consumption at variable loads though installation of systems such as Variable Frequency Drives (VFDs), Occupancy sensors etc.

Environmental Benefits: The Green Buildings envisage reducing net electricity consumption by implementing energy efficiency measures. These will reduce the electricity load on the grid, and help in saving equivalent amount of fossil fuel used for electricity generation, along with reduction in associated GHG emissions.

Climate Co-benefits: By various energy efficient measures adopted by the Green Building, the Dehradun project estimates a total reduction of 5,145 tCO_2 equivalents in GHG emissions while the Kolkata project aims to reduce GHG emission by 13,167 tCO_2 equivalents.

5 ACCCRN Case Studies and Best Practices in India

ACCCRN, the Asian Cities Climate Change Resilience Network, was launched in 2008 with the financial support of the Rockefeller Foundation to build climate change resilience in cities of Asia. According to ACCCRN, *climate change resilience is the ability to survive, recover from, and even thrive in changing climatic conditions* (ACCCRN 2014). Understanding the impacts that extreme events such as floods, prolonged drought, cyclones and so on, can have on cities and the ability to minimize the negative effects and respond to changing conditions are an integral part of the resilience of the city (ACCCRN 2014). Since its inception in Asia, ACCCRN has been active in ten small to medium-sized cities across four countries: India, Indonesia, Thailand and Vietnam. The network aims to improve the ability of cities to withstand, prepare and recover from the impacts of climate change (ACCCRN 2013).

Efforts to reduce greenhouse gas emissions is the central focus of climate change at the global level, but the emissions that are already present in the earth's atmosphere are leading to global warming. Possible impacts of climate change include rising sea levels, increase in incidences of diseases, loss of glaciers, coastal erosion, loss of biodiversity and many more. These impacts are a threat to urban communities, governments, industries and businesses around the world. ACCCRN seeks to address these challenges. In India, three cities were identified to implement the ACCCRN programme: Gorakhpur, Surat and Indore. The interventions in these cities include lake restoration, peri-urban agriculture, strengthening disease surveillance, establishment of urban health centres, end-to-end early warning system for flood management, and so on. While these projects are helping the cities cope with local climate change issues, there are co-benefits too, which are detailed in Table 1.

Mysore, an ACCCRN city in the second phase, is being assisted in the implementation of their city resilience strategy under the ACCCRN project. The assistance enables local governments to assess their climate risks in the context of urbanisation, poverty and vulnerability and formulate corresponding resilience strategies. The resilience interventions in the city include: encouraging the use of solar power for commercial and residential buildings; waste to energy conversion to meet the energy demand; use of sustainable building design in new constructions; awareness generation for energy efficiency; and encouraging the use of energy efficient products. The city resilience strategy includes identification of funding options for the strategies, project planning and project implementation (ICLEI 2015).

Table 1 Details of projects being implemented in Indian cities under the ACCCRN Programme (ACCCRN 2013)

City	Project	Sector	Issues faced by the city	Project brief	Climate impact of the project	Environmental and social co-benefits
Gorakhpur	Implementing and Promoting Ward-level Micro Resilience Planning	Water and sanitation	Flooding and water logging	Establish micro-level planning mechanisms in the Mahewa Ward in Gorakhpur city that address multiple sectors	Building local climate change resilience at the household level	Improved health, efficient drainage and solid waste management, water and sanitation
	Implementing and Promoting Adaptive Peri-urban Agriculture	Land Use	Strained natural resources of the city Variation in rainfall patterns and more intense storms degrading agricultural yield	Demonstrate the importance of ecosystem services of peri-urban agricultural land such as flood buffering for the city	Building resilience against climate risks such as rainfall variability and more intense/frequent storms	Increase in agricultural productivity and income of the farmers Enhancement of livelihood security and food security of the urban poor
Indore	Testing and Promoting Decentralized Systems for Different Water Sources and Uses	Water	Unsustainable water supply due to rapid urbanization Increase in disease incidence due to increase in storms and temperature particularly in the rainy season	Demonstrate alternative viable and sustainable models for cost-effective, reliable urban water management through community involvement	Developing resilience against rainfall variability, temperature increase and intense and frequent storms.	Provision of cheaper, high quality, and diverse local water supply options providing water security to economically weaker sections of the society
	Strengthening Vector-borne Disease Surveillance and Response Systems	Water & Sanitation	Perpetually water-logged areas, especially slums which cause vector-borne/water-borne diseases	Reduce the incidence of vector and water-borne disease outbreaks in Indore	Better coping mechanisms against temperature increase and intense storms	Establishment of an effective disease surveillance and response system

(continued)

Table 1 (continued)

City	Project	Sector	Issues faced by the city	Project brief	Climate impact of the project	Environmental and social co-benefits
			The existing system of monitoring vector/water-borne diseases overstretched and subject to errors			Reduction in treatment costs and reduced burden on health facilities
Indore	Peri-urban Lake Restoration to Create Emergency Water Management Options	Water	Water scarcity in Indore and intense and/or more frequent storms as well as an increase in dry periods Neglect of lakes in the peri-urban areas of the city	Conservation of lakes through better management of catchments which are undergoing rapid urbanization Increase the resilience of people to water scarcity, particularly during emergencies	Restoration of peri-urban lakes will help in increasing groundwater recharge and create buffer supplies during dry periods	Improved water quality in the restored lakes and the methodology used will inform policies on urban water body conservation
Indore and Surat	Promoting Cool Roof and Passive Ventilation for Indoor Temperature Comfort	Energy & Buildings	High temperatures and consequent increase in energy demand especially in summer Existing cooling methods use large amounts of water and energy	Promote indoor thermal comfort through no or low energy options	Reduction in indoor temperature by 2–3 °C reducing energy requirement for cooling	Decrease in ambient air temperature will reduce heat-related health problems
Surat	End-to-end Early Warning System	Water	Flooding of the city due to release of water from Ukai Dam. This was done with very little warning to the city	Develop a modelling system to improve reservoir operations for flood management at Ukai Dam. Setting up an end-to-end early warning system to help	Better ability of city administration to deal with floods due rainfall variability	Enhanced met-hydro reservoir models provide four days warning to city before flood water (from Ukai Dam) enters the city

(continued)

Table 1 (continued)

City	Project	Sector	Issues faced by the city	Project brief	Climate impact of the project	Environmental and social co-benefits
				city administration take timely action in case of extreme precipitation events		Construction of safer homes. City better able to protect its economy
Surat	Urban Health and Climate Resilience Centre	Health	City prone to vector-borne diseases, rainfall variability, flooding and potentially prolonged water logging	Establishing an Urban Health and Climate Resilience Centre with the primary goal of improving urban health management through evidence-based research	Establishment of the centre will help in recording health problems and minimizing health risks due to climate change	Improved prediction of epidemics which also prevents incidents leading to epidemics Provide support to Surat Municipal Corporation and its various departments to minimize the impacts of climate change on health

6 Conclusion

The world is urbanizing and India too is following this trend. With an urban population of over 377 million (in 2011), and 53 urban centres having a population of over a million, the focus of the Government of India is slowly but surely shifting to urban areas. Under the climate change scenario, this large urban population also makes urban areas more vulnerable to the consequences of climate variability. Urban areas use large amounts of resources to cater to the population living in them. The GHG emissions from urban areas are also large, generated mainly from motor vehicles, industries, energy use in buildings, and from urban waste. Therefore, the policies and programmes of the government should undoubtedly evaluate contributing sectors that will not only bring down the GHG emission levels. In addition, while sector strategies would be important, a co-benefits approach will make integrated interventions possible. The focus should be on both mitigation and adaptation measures. This will also help align environment and local issues. Figure 2 conceptually indicates how this alignment would work.

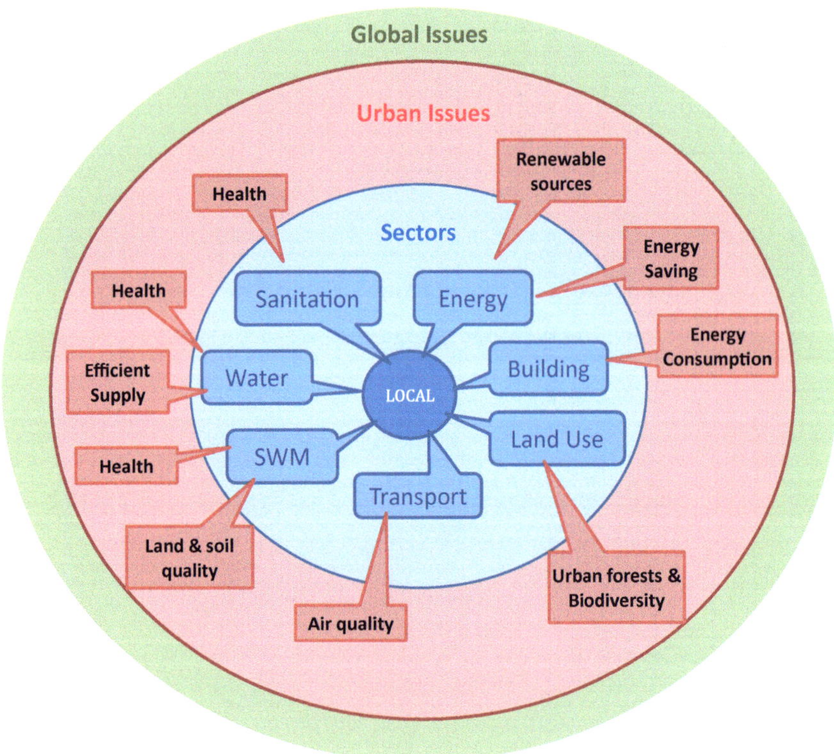

Fig. 2 Alignment of global and local environment issues

India needs massive investments in infrastructure for its economic growth and also to cater to its urban population. Under the Smart Cities Mission, India is also making its 100 cities 'smart'. This provides an opportunity to use the co-benefits approach, in fostering inter-linkages across different scales and sectors of development. These cities will need to provide infrastructure and services without affecting the environment negatively and, at the same time, have a positive impact on the economy, livelihoods and health of people. Integrating transport and land use, providing efficient public transportation system, promoting green buildings, renewable energy, improving management of solid and liquid wastes are some of the measures that can yield co-benefits which will make urban areas environmentally sustainable. India's GDP is ambitiously projected to grow at over 8% annually and so some of the co-benefits may be difficult to achieve. However, there is immense opportunity in following a low carbon growth path while still ensuring that the socio-economic well-being of the poor is not compromised.

References

ACCCRN. (2013). *ACCCRN city projects.* Bangkok, Thailand. ACCCRN-Asian Cities Climate Change Resilience Network. The Rockefeller Foundation.

Akbari, H. (2011). *Using cool roofs to reduce energy use, greenhouse gas emissions, and urban heat-island effects: Findings from an India experiment.* Lawrence Berkeley National Laboratory. http://www.google.co.in/url?sa=t&rct=j&q=&esrc=s&source=web&cd=1&cad=rja&uact=8&ved=0ahUKEwiL5-K0p77RAhUMMY8KHVCJAKgQFggZMAA&url=http%3A%2F%2Fwww.coolrooftoolkit.org%2Fwp-content%2Fuploads%2F2012%2F04%2FSatyam-project-in-India-IIIT.pdf&usg=AFQjCNHnBI3WN9AE1k_qIfZTYa3VHrNffQ&bvm=bv.144224172,d.c2I. Accessed on January 21, 2016.

Census of India. (2011). *Census of India 2011.* New Delhi: Government of India.

DMRC. (2014a). *Delhi metro becomes first metro project in the world to be registered with the prestigious gold standard foundation for its energy efficiency measures.* http://www.delhimetrorail.com/whatnew_details.aspx?id=uvsEhbp7AsisIlld. Accessed on January 15, 2016.

DMRC. (2014b). *UN Body Credits Delhi Metro-6.3 Lakh Carbon Credits for Modal Shift Project.* http://www.delhimetrorail.com/press_reldetails.aspx?id=746xECETA6Qlld. Accessed on January 15, 2016.

Doll, C. N. H., Dreyfus, M., Ahmad, S., & Balaban, O. (2013). Institutional framework for urban development with co-benefits: The Indian experience. *Journal of Cleaner Production, 58*(1), 121–129.

Goyal, P. (2003). Present scenario of air quality in Delhi: A case study of CNG implementation. *Atmospheric Environment, 37,* 5423–5431.

Hunsberger, C., & Evans, T. P. (2012). *GEO 5 Assessment Report.* Nairobi, Kenya: UNEP.

ICLEI. (2015). *ICLEI-ACCCRN process.* http://southasia.iclei.org/fileadmin/user_upload/documents/ICLEI_ACCCRN_Workbook_WORKBOOK.pdf. Accessed on January 11, 2016.

Kapshe, M., Kuriakose, P. N., Srivastava, G., & Surjan, A. (2013). Analysing the co-benefits: case of municipal sewage management at Surat, India. *Journal of Cleaner Production, 58*(1), 51–60. doi:10.1016/j.jclepro.2013.07.035.

Kathuria, V. (2004). Impact of CNG on vehicular pollution in Delhi: A note. *Transport Research Part D: Transport and Environment, 9,* 409–417.

Leiserowitz, A., & Thaker, J. (2012). *Climate change in the Indian mind*. Yale University. http://shaktifoundation.in/wp-content/uploads/2014/02/climate-change-indian-mind.pdf. Accessed on February 6, 2016.

MoEF. (2015). *India's intended nationally determined contribution: Working towards climate justice*. New Delhi: Ministry of Environment & Forests, Government of India.

Murty, M. N., Dhavala, K. K., Ghosh, M., & Singh, R. (2006). *Social cost-benefit analysis of Delhi Metro*. Delhi University Enclave, New Delhi: Behalf Plan. Comm. India Inst. Econ. Growth.

NAPCC. (2008). *National action plan for climate change 2008*. New Delhi: Ministry of Environment and Forests, Government of India.

NUSP–National Urban Sanitation Policy. (2008). *National urban sanitation policy*. New Delhi: Ministry of Urban Development.

NUTP–National Urban Transportation Policy. (2006). *National urban transportation policy*. New Delhi: Ministry of Urban Development.

Puppim de Oliveira, J. A., Doll, C. N. H., & Suwa, A. (2013a). *Urban development with climate co-benefits: Aligning climate, environmental and other development goals in cities*. Yokohama: United Nations University Institute of Advanced Studies (UNU-IAS), Japan.

Puppim de Oliveira, J. A., Doll, C. N. H., Balaban, O., Jiang, P., Dreyfus, M., Suwa, A., et al. (2013b). Green economy and governance in cities: assessing good governance in key urban economic processes. *Journal of Cleaner Production, 58*(1), 138–152.

Ravindra, K., Wauters, E., Tyagi, S. K., Mor, S., & Van Grieken, R. (2006). Assessment of air quality after the implementation of compressed natural gas (CNG) as fuel in public transport in Delhi, India. *Environmental Monitoring and Assessment, 115*, 405–417.

Sethi, M., & Mohapatra, S. (2013). Governance framework to mitigate climate change: Challenges in urbanising India. In H. Ha & T. Nath Dhakal (Eds.), *Governance approaches to mitigation of and adaptation to climate change in Asia* (pp. 200–230). Hampshire, UK: Palgrave Macmillan.

UN Habitat. (2011). *Global report on human settlements 2011: Cities and climate change*. London/Washington DC: United Nations Human Settlements Programme & Earthscan.

UNFCCC. (2011a). Green Building at Dehradun, Energy Demand. Oil and Natural Gas Corporation Limited.

UNFCCC. (2011b). Green Building at Kolkata, Energy Demand. Oil and Natural Gas Corporation Limited.

UNFCCC. (2012a). 5 MW Solar PV power project, Patan District, Gujarat, India, Energy industries (renewable/non-renewable sources). Surana Telecom and Power Limited.

UNFCCC. (2012b). Municipal Solid Waste management project at Aligarh, India, MSW. A2Z Waste Management (Aligarh) Limited.

UNFCCC. (2013). Rehabilitation of degraded wastelands at Dera Mandi in Southern District of National Capital Territory of Delhi through reforestation, afforestation and reforestation. New Delhi: Department Of Environment, Forests and Wildlife, National Capital Territory of New Delhi.

UNFCCC. (2014). Zero Energy Water Purifier CDM Project in India by Hindustan Unilever Limited, Energy Demand. Hindustan Unilever Limited.

URDPFI. (2014). *Urban and regional development plans formulation & implementation guidelines, Volume 1—First Draft Report*. New Delhi: Ministry of Urban Development, Government of India.

Part III
Co-benefits in Energy, Transport, Buildings, Waste and Biodiversity

Chapter 5
Co-benefits from the Energy Sector

Mahendra Sethi

Abstract With the onset of climate change on the horizon of human civilization, from the 1990s onwards, research in the interphase of economy, energy and environment (read emissions), collectively known as the E3 nexus is increasingly gaining focus in academics and policy-making. Since 2009, when the world became more urban than rural, there seems to be a growing interest to study how this E3 nexus operates in urban areas. In the absence of any previous investigations on E3 linkages in India, this study proceeds on the research premise that E3 challenges are most crucially played out in urban areas, that possibly seek concerted technological and policy-oriented interventions to produce climate co-benefits. The chapter reviews global literature to theoretically comprehend E3 linkages in urban settlements. This is followed by a detailed study on present conditions, issues and challenges of India's economy, energy and environment (emission) sectors. The research underscores the role of urban India in these three sectors and generates for the first time, superimposed scenarios of the nexus. The study culminates with a discussion on appropriate technology and policy/governance instruments that could possibly mitigate impacts and produce co-benefits in urban areas.

Keywords Economy · Energy · Emissions · Urban · Super-imposed E3
Technology and policy alternatives

1 Introduction

Climate co-benefits associated with the energy sector are known to be influenced by multifarious, complex and intertwined challenges in the area of economy, energy and environment (referred to with the acronym E3 and collectively termed as the 'nexus'). After the global energy crisis in the 1970s, inter-disciplinary studies in energy and economy have been in vogue. But with the onset of climate change on

M. Sethi (✉)
National Institute of Urban Affairs, New Delhi, India
e-mail: mahendrasethi@hotmail.com

M. Sethi and J.A. Puppim de Oliveira (eds.), *Mainstreaming Climate Co-Benefits in Indian Cities*, Exploring Urban Change in South Asia,
https://doi.org/10.1007/978-981-10-5816-5_5

the horizon of human civilization, from the 1990s onwards, research in the interphase of energy-environment (read emissions) and economy-emissions is increasingly gaining focus in global development, environment and geo-political discourse, national policy-making and academics. Recently in 2009, since the world became more urban than rural (UN DESA 2011), there seems to be a growing interest in global forums, multilateral organizations, national governments, research and advocacy groups, and associations of local governments to study how this E3 nexus is played out in urban areas. A variety of E3 circumstances in different countries and world regions raise diverse issues, challenges and alternatives for engagement. Table 1 captures the variations in the E3 situation across developed, developing and least developed countries, in their different stages of urbanization. It is noteworthy that India compares the emission intensity of its GDP (gross domestic product) in PPP (purchasing power parity) terms at present with those countries at a similar level of development; it is seen that their emissions then were far more than India's at present.

It has been widely recognized that cities could play a crucial role in mitigation and adaption of climate change (OECD 2009; World Bank 2010a, b; UN Habitat 2011). Depending upon the national situation, they are mandated to perform diverse civic functions like providing basic amenities of water, sanitation and electricity, public transportation, land use planning, housing, managing industries, municipal waste, air quality, and so on, that collectively influence the local E3 dynamics and cumulatively their national circumstances. A crucial aspect perhaps common to all the cities is that while they are expected to fuel their engines and exude economic

Table 1 State of economy, energy, emissions (carbon) and urbanization across various regional scales

		Urban	Economy	Energy	Environment
Regions	(1) Total population (in 000's) 2013	(2) Urban population (% of total population) 2011	(3) GDP/capita at PPP 2005 constant international dollars 2012	(4) Energy use (kg of oil equivalent/capita) 2010	(5) Carbon emissions (tons/capita) 2009
World	7,162,119	52.1	12,139	1852	4.7
More developed regions	1,252,805	77.7	35,743	4705	10.7
Less developed regions	5,909,315	46.5	6660	1185	3.1
Least developed countries	898,433	28.5	1604	339	0.3
Asia	4,298,723	45.0	7702	1341	3.8
India	1,252,140	31.3	3876	575	1.7

Source UN DESA (2013)

prosperity, at the same time they are hard-pressed to reduce emissions by pursuing sustainable strategies. In the absence of any previous investigations on this nexus in India, this study operates on the research premise that E3 challenges are most crucially played out in urban areas, that possibly seek concerted technology and policy-oriented interventions to produce climate co-benefits. The chapter particularly focuses on the role of energy systems in this relationship, that is, energy production (electricity generation from thermal plants) and consumption (in energy intensive sectors like cement, iron and steel, other industries, etc.) and limits from studying end of the pipe consumption and behaviour of the users.

The chapter reviews global literature to theoretically comprehend the status of the economy, energy and environment (emissions) in urban settlements in Sect. 2. With India as the case in point, a detailed study on India's economy, energy and greenhouse gas (GHG) emission and their relevance with respect to urban areas is reported in Sects. 3, 4 and 5 respectively. This helps to generate, for the first time, super-imposed scenarios of the three agents in Sect. 6. This is followed by a discussion in Sect. 7 on appropriate low-carbon technology and policy/economic instruments that could possibly mitigate climate impacts or produce co-benefits in human settlements, eventually followed by research conclusions and recommendations.

2 Literature Review: Status of the Economy, Energy and Emissions in Urban Settlements

Cities matter because they are large economies in themselves (World Bank 2010a, b). The Fifth Assessment Report (AR5) provides a clear and up-to-date review of the current state of scientific knowledge in this area (IPCC 2014). It reports that the concentration and scale of people, activities, and resources in urban areas fosters economic growth (Henderson et al. 1995; Fujita and Thisse 1996; Duranton and Puga 2004; Puga 2010), and in return, agglomeration economies made possible by the concentration of individuals and firms make cities ideal settings for innovation, jobs, and wealth creation (Rosenthal and Strange 2004; Carlino et al. 2007; Knudsen et al. 2008; Puga 2010). A precise estimate of the contribution of all urban areas to global GDP is not available. However, a downscaling of global GDP during the Global Energy Assessment (Grubler et al. 2007; GEA 2012) revealed that urban areas contribute about 80% of global GDP. Some studies show that urban economies generate more than 90% of global gross value (Gutman 2007; United Nations 2011). In a compilation by UN Habitat, big cities are shown to have a disproportionately high share of national GDP compared to their population (UN Habitat 2012). The importance of big cities is further underscored in another study that shows that the top 600 cities in terms of population generated 60% of global GDP in 2007 (McKinsey Global Institute 2011). World Bank 2010 reports that the world's 50 largest cities by population collectively house more than 500 million people and have combined economies of 9564 million $ PPP, second only to the

United States (14,204 million $ PPP), and larger than all of China (7903 million $ PPP) or Japan (4354 million $ PPP).

The environmental Kuznets curve is a hypothesized relationship between environmental quality and economic development. It suggests that various indicators of environmental degradation tend to get worse as modern economic growth occurs until average income reaches a certain point over the course of development (Shafik 1994). Thereafter, developed societies tend to invest more resources in reducing environmental impacts. However, there is little evidence that the relationship holds true for long-term pollutants, for natural resource or energy services. For example, energy, land and resource use—sometimes called the 'ecological footprint'—do not fall with rising income (EIA 2011). While the ratio of energy per real GDP has fallen, total energy use is still rising in most developed countries. Another example is the emission of many greenhouse gases, which continues to remain high in industrialized countries, in spite of several industrial activities being exported to the developing countries.

It could be argued with greater evidence and confidence that higher urban incomes are correlated with higher consumption of energy and GHG emissions (Kahn 2009; Satterthwaite 2009; Kennedy et al. 2009; Weisz and Steinberger 2010; Zheng et al. 2010; Hoornweg et al. 2011; Marcotullio et al. 2012). At the household level, studies in a variety of different countries (the Netherlands, India, Brazil, Denmark, Japan, and Australia) have also noted positive correlations between income and energy use (Vringer and Blok 1995; Cohen et al. 2005; Lenzen et al. 2006; Pachauri and Jiang 2008; Sahakian and Steinberger 2011). The Global Energy Assessment concluded that cities in non-Annex I countries generally have much higher levels of energy use compared to the national average, in contrast to cities in Annex I countries, which generally have lower energy use per capita than national averages. One reason for this inverse pattern is due to the significantly higher urban to rural income gradient in cities in non-Annex I countries compared to Annex I countries. That is, per capita incomes in non-Annex I cities tend to be multiple-fold higher than rural per capita incomes, thus leading to much higher energy use and resulting emissions.

The World Energy Outlook 2008 estimated urban energy related CO_2 emissions at 19.8 Gt, or 71% of the global total for the year 2006 (IEA 2008). The Global Energy Assessment provided a range of final urban energy use between 180 and 250 EJ with a central estimate of 240 EJ for the year 2005. This is equivalent to an urban share between 56 and 78% (central estimate, 76%) of global final energy use. Converting the GEA estimates on urban final energy (Grubler et al. 2012) into CO_2 emissions results in global urban energy related CO_2 emissions of 8.8–14.3 Gt (central estimate, 12.5Gt) which is between 53 and 87% (central estimate, 76%) of CO_2 emissions from global final energy use. Aside from these global assessments, there are some other attempts to estimate the total GHG (CO_2, CH_4, N_2O and SF_6) contribution of urban areas globally. While Satterthwaite (2008) performs a back of the envelope estimate to assess urban contributions to global warming as 30.5–40.8% of the total in 2004, from a production perspective; Marcotullio et al. (2013)

estimate the urban share of energy-related CO_2 emissions in 2000 is slightly lower than the GEA and IEA estimate, at 72%.

Economists differ in how they convert inputs such as energy (and thus emissions associated with energy consumption) into economic value. This efficiency is commonly measured as the ratio of emission to unit economic output (CO_2/GDP). Technology, a manifestation of energy or carbon intensity, is a prime driver of GHG emissions. Income and scale exert important influences on the mitigation potential for technologies. While lock-in may limit the rate of mitigation in mature cities, opportunity exists in rapidly growing cities to leapfrog to new technologies. For mature cities, technology is important due to agglomeration externalities, research and development (R&D) and knowledge concentration, and access to capital that facilitate the development and early deployment of low-carbon technologies (Grubler et al. 2012). On the other hand, for rapidly growing cities, the importance of technology as a driver may be low for systems with high capital requirements but high for less capital-intensive (e.g., some demand-side efficiency or distributed supply) systems. Hence, while certain high-cost, low-carbon technologies could be supported by or substituted with low-cost economic/policy instruments in urban agglomerations, a mix of technology and policy alternatives may hold potential to create multiple benefits in the longer run.

3 Economy

The Indian economy is the tenth largest in the world by nominal GDP and the third largest by purchasing power parity (IMF 2014). The country is one of the G-20 major economies, a member of BRICS and a developing economy that is among the top 20 global traders according to the WTO (2013). It has been argued that the post independence-era Indian economy (from 1947 to 1991) was a state-controlled mixed economy. It was an inward-looking, centrally planned (with five-year plans and annual plans), and import-substituting economic model that nationalized many sectors and failed to take advantage of the post-war trade expansion (Das 2006). After a fiscal crisis in 1991, India liberalized its economy to international trade and increasingly adopted free-market principles (OECD 2007; Goldman Sachs 2007). By 2008, India established itself as one of the world's emerging economies. In contrast to higher economic growth rates in the 2000s, in the range of 7–8%, it slowed to 4.7% for the 2013–14 fiscal year (LiveMint 2014). Projections by Indian Finance Ministry and International Monetary Fund for GDP growth for fiscal year 2014–2015 hover around 5.5–5.6% (GoI 2014; The Economic Times 2014). The agriculture sector is the largest employer in India's economy but contributes a declining share of its GDP (13.7% in 2012–2013) (The Economic Times 2013) refer (Fig. 1). The manufacturing industry has held a constant share in national economic contribution, while the fastest-growing, especially after 1991, has been the services sector—which includes construction, travel, trade, banking, telecom, software, information technologies, tourism, education, health care, and so on (GoI 2014).

India's gross national income per capita experienced high growth rates since 2002. It tripled from Rs. 19,040 in 2002–03 to Rs. 53,331 in 2010–11, averaging 13.7% growth over these eight years peaking 15.6% in 2010–11 (The Times of India 2012). However, growth in the inflation adjusted per-capita income of the nation slowed to 5.6% in 2010–11, down from 6.4% in the previous year. According to this World Bank's revised methodology, while the world had 872.3 million people below the new poverty line in 2013, 179.6 million of them lived in India (World Bank 2015).

With 1.2 billion people and the world's third largest economy by volume, India's recent growth and development has been one of the most significant achievements in recent times. India will soon have the largest and youngest workforce the world has ever seen. At the same time, the country is in the midst of a massive wave of urbanization as some 10 million people move to towns and cities each year in search of jobs. It is the largest rural–urban migration of this century. Massive investments will be needed to create jobs, housing, and infrastructure to meet soaring aspirations and make towns and cities more livable and green (World Bank 2014). It is noteworthy to mention that Urban India's contribution to national GDP has been rising steadily from 37.7% (1970–71) to 52% (1999–2000) to 63% in 2011, though the total expenditure on urban infrastructure is merely 1.59% of GDP (HPEC 2011). It is further forecasted that the urban share of GDP will be 75% by 2031. McKinsey Global Institute forecasts that capital investments to the tune of US$1.2 trillion is necessary to meet projected demands in Indian cities. Meanwhile, Government of India sponsored research estimates the investment for urban infrastructure over the 20-year period at Rs. 39.2 lakh crore (39.2 trillion) at

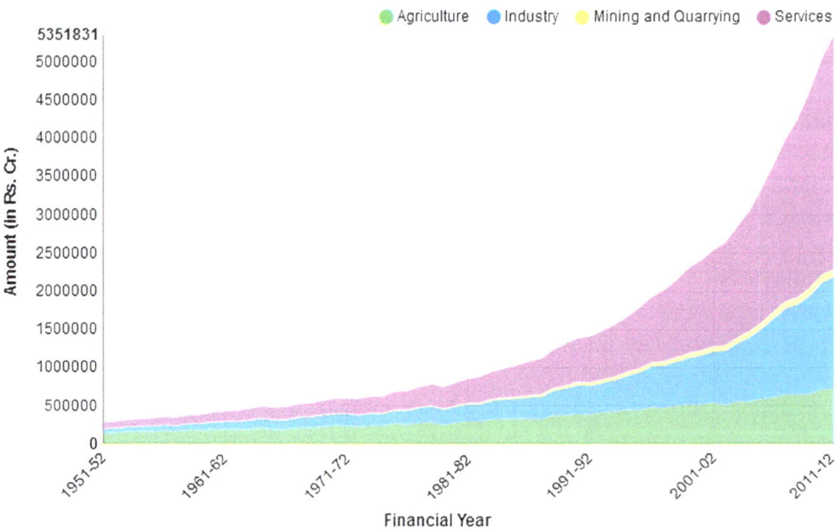

Fig. 1 Year-wise and Sector-wise contribution of GDP. *Source* Data.gov.in

2009–10 prices (HPEC 2011). By the sheer scale of Urban India's contribution to national economy and jobs, its role in E3 nexus could not be undermined.

4 Energy

In 2009, India had the third largest energy demand in the world after China and the United States and just ahead of Russia. As World Energy Outlook (WEO) 2011 shows, India's energy demand more than doubled from 319 million tons of oil equivalent (Mtoe) in 1990 to 669 Mtoe in 2009. The total primary energy consumption from crude oil (29.45%), natural gas (7.7%), coal (54.5%), nuclear energy (1.26%), hydro electricity (5.0%), wind power, biomass electricity and solar power was 595 Mtoe in 2013. In the same year, India's net imports were nearly 144.3 million tons of crude oil, 16 Mtoe of LNG and 95 Mtoe of coal totalling 255.3 Mtoe of primary energy which is equal to 42.9% of total primary energy consumption.

About 70% of India's electricity generation capacity is from fossil fuels, with coal accounting for 40% of India's total energy consumption followed by crude oil and natural gas at 28% and 6% respectively. Notably, India's per capita energy consumption (0.58 toe/capita) is still much lower than the world average (1.8), OECD (4.28), China (1.7) and Africa (0.67) in 2009 (IEA 2011b). It indicates that India's energy demand still has a long way to reach saturation (IEA 2012). The development of the power sector is closely tied with India's energy policy objectives of universal energy access and energy security. At 176 GW, India had already built the world's fifth largest installed capacity for power generation in 2009 (behind US, China, Japan, Russia) and almost tripled electricity generation from 289 terawatt hour (TWh) to 899 TWh from 1990 to 2009 (IEA 2011a). Yet its per capita electricity consumption remains at merely one-fifth of the world average. In 2012, total installed capacity in India was 199 GW and the generated electricity (excluding renewable) was 876 TWh in FY 2011/12 (CEA 2012). Coal represents 112 GW or 56% of total installed capacity and 71% of total generated electricity. Hydro is the second largest source, accounting for 20% of installed capacity and generating 15% of electricity. This is followed by natural gas (9 and 10%), nuclear (2 and 4%), diesel (1 and 0.2%) and renewable (12%). Renewable based electricity is not included in the Ministry of Power data for generated electricity (IEA 2012).

India made plans to enhance thermal power plant capacity significantly to produce 75,785 MW in the base-case scenario by 2017 (Planning Commission 2011). It is crucial for policy-makers to begin to assess the implications of continued use of coal for power generation in the context of the GHG mitigation challenge. Gas-based power is an attractive power generation option as the capital cost is low and the CO_2 emissions are only 0.4 kg per kWh. However, the cost of gas is usually much more than the cost of coal to generate one unit of electricity. Also there is uncertainty about availability of gas for power given the limited reserves and also its alternate use in fertilizer production and other sectors. It is therefore unlikely that gas can contribute a large share of electricity generation (IEA 2012). Meanwhile, India envisages to increase the contribution of nuclear power to overall electricity generation capacity from 4.2 to 9% within 25 years (Business Standard 2009). The country has five nuclear reactors under construction (third highest in the world) and plans to construct 18 additional nuclear reactors (second highest in the world) by 2025 (The Economist 2011). The pace of nuclear energy development has decreased because of safety concerns that surround the establishment of nuclear power plants and related public opposition. Hydropower, another important resource available to India, has been severely affected by the governments' inability to find appropriate rehabilitation and resettlement solutions and the associated ecological impacts (Srivastava and Mahajan 2011). India has the world's fifth largest wind power market (WEEVR 2011) and plans to add about 100 GW of solar power capacity by 2022 (Wall Street Journal 2010). According to a TERI analysis, one significant market opportunity that remains untapped in renewable energy is through installing roof-top solar PV in the urban pockets that depend on diesel generators for captive and back-up energy use. At an average of 2–5 KWp per roof, about 7000 MWp of solar rooftop potential in grid-connected mode can be developed in residential areas alone (Srivastava and Mahajan 2011).

As per the Government of India's future projections for 2020 based on supply options and fuel mix, for 8% and 9% economic growth scenarios, the supply-side CO_2 savings (from change in generation efficiency and mix) are therefore estimated to be 85 million tons in determined effort and 145 million tons in aggressive efforts, for 8% growth scenario, which is more close to the actual Indian economic growth (IEA 2012). The amount of energy India needs to sustain high economic growth of 8–9% per year over the next 25 years (until 2032) was addressed by Integrated Energy Policy (IEP) by looking beyond the traditional five-year cycle to determine how India could best meet its enormous energy demand. The IEP also addresses multifaceted energy problems the country must resolve to ensure efficient and sustainable use of energy, stating that 'India must pursue technologies that maximise energy efficiency, demand side management and conservation' (Planning Commission 2006). The IEP presents long-term goals for all energy sectors, not in an insulated and disconnected way, but rather in an 'integrated' and comprehensive manner. It also offers various scenarios based on different energy mixes and implementation of demand-side management.

But one of the key directions set for the long-term energy strategy is the validation of coal as a primary energy source for the long term and the necessity of ensuring coal supply with consistent quality. Currently, more than 70% of India's electricity generation is based on coal, and a significant fraction of new electricity supply in India is likely to be based on domestic coal derived from India's significant reserves (Chikkatur and Sagar 2009). Hence, it is crucial for policy-makers to begin to assess the implications of continued use of coal for power generation in the context of the GHG mitigation challenge. Findings from the Interim Report of the Expert Group on Low Carbon Strategies for Inclusive Growth (LCS), further reveal that specific CO_2 emission of all existing coal and lignite power plants is 1.1 kg per net KWh for 2007–08. Some of the old and less efficient coal power plants emit as high as 2 kg per KWh. Already, the Government of India requires large plants to be based on super critical technologies and in fact manufacturing capacity in India is being set up by a couple of private companies for such boilers. Given the growing uncertainty about coal (India imported 20% of its coal consumption in 2011) and increasing price of coal in the international market, super critical plants with 4–6% points higher efficiency of coal use, are economically attractive (Parikh and Parikh 2011). The LCS report emphasizes that coal is presently the lowest cost option and will continue to be the main power generation source in 2020 as well. To ensure energy security the present coal-based capacity needs to be expanded to 230 GW by 2020. Akin to IEP, the policy lays more emphasis on energy efficiency measures on the demand side. It offers limited scope for thermal energy alternatives, citing several technical, economic and regulatory challenges in carbon capture, sequestration and the option of gas-based power generation to be fully considered as a viable solution to mitigate climate change.

According to the International Energy Agency, more people lack access to electricity in India than in any other country (IEA 2007), a majority of them in all likelihood residing in rural areas that are unconnected with the grid. Energy access in India varies substantially depending upon your economic and location circumstances. In terms of lifestyle differences across household expenditure classes, the urban top 10% accounts for emissions of 4099 kg per capita per year, while the rural bottom 10% accounts for only 150 kg per capita per year (Parikh et al. 2009). It is important to underpin the role of urban areas in providing energy access and availability. From the perspective of energy production from thermal power plants, on urban–rural hierarchy, while 74 plants of 84,648 MW (85%) capacity are associated with urban areas (includes all megacities, million-plus cities, Class I to Class V towns), only 13 plants of 15,060 MW (15%) capacity are associated with rural areas (Sethi 2015).

Without any urban or rural differentiation, the Government of India has announced its goal of *Electricity for All*. This calls for some formidable efforts in energy generation and supply-side management. Between 2002 and 2015, the share of renewable grid capacity has increased over 6 times, from 2% (3.9 GW) to around 13% (36 GW). This momentum of a tenfold increase in the previous decade has recently led India to significantly scale up with the aim to achieve 175 GW renewable energy capacity in the next few years. This includes 60 GW for wind power, 100 GW for solar power, and 10 GW for biomass energy by 2022. The National Smart Grid Mission has been launched to bring efficiency in the power supply network and facilitate reduction in losses and outages. The Green Energy Corridor projects worth INR (Indian National Rupee) 380 billion (US$6 billion) are also being rolled out to ensure execution of renewable energy. A preoccupation towards macro-level policies demonstrate that there is plenty of scope to push off-grid solutions in enabling energy access that simultaneously delivers large-scale social benefits for the poor. At the global level, India has also decided to anchor the International Solar Alliance of about 120 countries located between the Tropic of Cancer and the Tropic of Capricorn to steer joint efforts through innovative policies, projects, programmes, capacity building measures and financial instruments to mobilize more than US$1000 billion of investments that are needed by 2030 for the massive deployment of affordable solar energy.

5 Emissions

In 2013, India was the fourth largest CO_2 emitting country/region (from fossil fuels and cement manufacturing) with 2070 $MTCO_2e$ (6% of the world share), after China: 10,330 (29%), the United States: 5300 (15%), and the European Union: 3740 (11%). It was well ahead of many industrialized nations like Russia: 1800 (5%) and Japan: 1360 (4%), all emission figures in $MTCO_2e$ (PBL 2014). India's emissions between 1990 and 2009 grew by a compound annual growth rate (CAGR) of 5.2% against 1.7% for the world. This is largely due to increased coal consumption, which represented 67% of the emissions increase from 1990 to 2009. India was a signatory to the United Nations Framework Convention on Climate Change (UNFCCC), but not obliged to contain its carbon emissions as non-Annex I country within the Kyoto Protocol. India always found it unacceptable to establish an internationally binding regime to curb carbon emissions, stating that most emissions were produced by developed countries and that India needs economic development and industrialization. India's per capita emissions are only one-third of the world average and 14% of per capita emissions of OECD member countries and more developing regions. India took a leading role in the G77 during the COP 15 held in Copenhagen in 2009, denouncing any attempt by industrialized countries to impose carbon reduction targets on developing countries, although it reconfirmed its goal to reduce carbon emissions per unit of GDP by 20–25% below 2005 levels by 2020.

At home, India follows the National Action Plan on Climate Change (NAPCC) prepared under the guidance and direction of the Prime Minister's Council on Climate Change. It was released in 2008 to achieve 'a sustainable development path that simultaneously advances economic and environmental objectives' (Press Information Bureau 2008). Similarly the 12th Five Year Plan recognizes the need to adopt low-carbon strategies, with carbon mitigation being an important co-benefit (Planning Commission 2013). This is followed by the final report on Low-Carbon Strategies for Inclusive Growth that adopts the message of GEA (2012) in identifying multiple benefits and development linkages for energy transformations. According to India's recent national communications of GHGs to UNFCCC, in 2007, India's GHG emissions by sources and removal by sinks was 1727.71 $MTCO_2e$. This includes emissions from the energy sector, industries, agriculture and waste and removals by the Land Use, Land Use Change and Forestry (LULUCF) sector. Without LULUCF, the GHG emissions were 1904.73 $MTCO_2e$. The GHGs covered include CO_2, CH_4 and N_2O. The distribution of emissions across sectors, the contributions by towns and cities in India, and their anticipated growth till 2020 in aggressive (9%) and determined (8%) efforts is shown in Table 2.

Table 2 The sectoral, spatial (urban) origin of India's GHG emissions along with projected growth rates

S. No.	Sector	(1) GHG emissions $MtCO_2e$ in 2007	(2) CAGR (%)	(3) Urban India baseline emissions $MtCO_2e$ in 2007	(4) Sectoral component of total urban India emissions (%)	(5) Projected emissions in 2020 with 8% GDP growth	
						Determined effort	Aggressive effort
1	Electricity	719.3	5.6	719.3	59.26	1428	1263
2	Residential	137.8	4.4	45.88	3.78		
3	Transport	142	4.5	70.98	5.85	435	413
4	Other energy	100.9	1.9	33.85	2.79		
5	Cement	129.9	6	129.92	10.70	336	294
6	Iron and steel	117.3	2	117.32	9.66	406	360
7	Other industry	165.3	2.2	38.92	3.21	300	240
8	Waste	57.73	7.3	57.73	4.76	163	146
		1570	2.9	1213.9	100.00	3537	3071

Source MoEF 2010 for (1), (2); Sethi and Mohapatra 2013 for (3), (4); Planning Commission 2011 for (5)

The largest portion of national emissions (38%) emanate from over a hundred thermal power plants generating 138,806.18 MW (CEA 2012) and emitting 719.31 $MTCO_2e$ (MoEF 2010). Estimates for India, based on location of where emissions are being generated indicate that urban areas contribute to almost two-thirds of the national GHG burdens (Sethi and Mohapatra 2013). The method of downscaling national GHG emissions is a component-wise derivative approach, essentially a top-down analysis. It assumes that as per a very broad definition of urban areas prevalent in India, thermal plants should qualify to be located in census towns, cities and urban agglomerations and significantly contribute to GHG emissions emanating from urban areas. A detailed study consisting of spatial analysis of 454 thermal units from over 100 plants shows that 87% emissions could be associated with urban and rapidly urbanizing areas, while only 13% emanate from rural areas (Sethi 2015). Thus it could be concluded with undoubted evidence that co-benefits of energy are strongly interlinked with cities.

6 Super-Imposed E3 Scenarios

There is a strong association between India's economic growth, energy consumption and production and GHG emissions while it is urbanizing. A comparison of urbanization with E3 indicators in Fig. 2 suggests that a unitary increase in urban

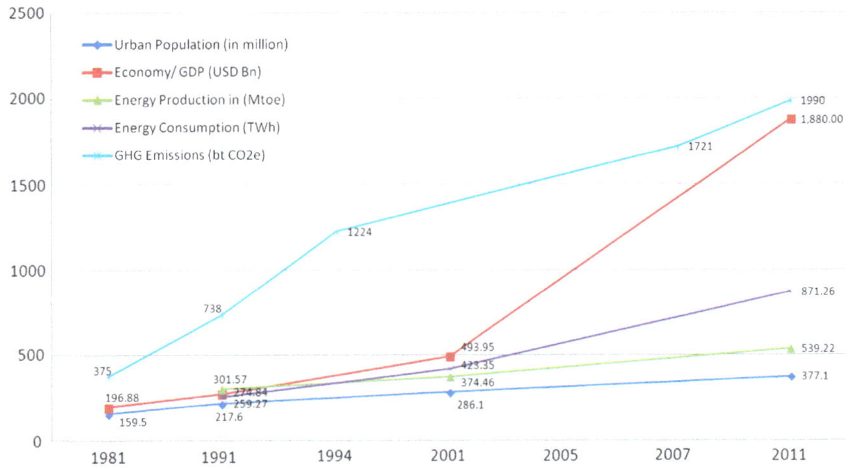

Fig. 2 Linkages between economy, energy (production and consumption) and emissions against urbanization in India (1981–2011). *Source* The time series analysis is generated by the author based on the following point data sources: *1*. Urban Population data from TCPO (2012), 'Data Highlights (Urban) based on Census of India', Town & Country Planning Organisation, Government of India, New Delhi. *2*. Energy Consumption and Production figures for 1991, 2001 and 2011 from IEA database available at http://www.iea.org/statistics/statisticssearch/report/?country=INDIA&product=indicators&year=1991. *3*. GHG emissions data (1994 and 2007) from MoEF (2010): 'India: Greenhouse Gas Emissions 2007', Indian Network for Climate Change Assessment, MoEF, India. Figures for 1981, 1991 are from World Bank indicators, see point 4 below. *4*. Economic baseline in current USD from World Bank indicators available on http://data.worldbank.org/indicator

population is approximately associated to growth in energy consumption and GHG emissions by a factor of 2.0 and to economic development by a factor of 3.0. Another important observation is that while overall GHG emissions have grown by 4.01%, from 1994–2007, growth of electricity production during the same period has been at 2.95%, in spite of the fact that urban India has been urbanizing at a consistent and relatively slower rate of 1.16% during 1971–2011. Urbanization is projected to be over double this pace (growth rate of 2.83%) from 2011 to 39, when most estimates consider India to be 50% urbanized, in terms of total population. This phenomenon of urban growth is consistently simultaneous with economic development, affluence, increasing energy demand for residential, commercial and industrial activities, energy production and rising GHG emissions.

It has been estimated that by 2030 and 2050 India's share of real GDP at 2005 PPPs in 2005 would be 11 and 16% of the World (OECD 2012). The New Policies Scenario (NPS), the central scenario of WEO 2011 projects that India's demand will continue to grow, reaching 1464 Mtoe in 2035, increasing by a compound annual growth rate (CAGR) of 3.1% from 2009 to 2035, which is more than double the growth in world's energy demand at a CAGR of 1.3% for the same period. Cumulatively, India's share in world energy demand increases from 5.5% in 2009 to 8.6% in 2035. The growth would come from all fuels. The above projections of economic development and energy consumption would critically affect future emissions.

Under NPS, India's carbon emissions will increase to 3535 $MtCO_2$ in 2035 at a CAGR of 3.2%, responsible for 8% of global emission of 43,320 $MtCO_2$. Emissions from coal combustion would be 2227 $MtCO_2$ or 63% of India's total emissions. It assumes that by then, India is projected to be the world's most populous nation with 1.511 billion people, India's per capita carbon emissions could reach 2.34 tCO_2/capita, which is higher than 1.7 tCO_2/capita at present but will be still substantially lower than the world average of 4.25, with China's per capita CO_2 emissions at 7.39 and the United States' at 12.03 in 2035. Such projections of a massive increase of carbon emissions in India raise concerns about their impact on global climate change (IEA 2012). Another study supported by the Indian government (MoEF 2009) formulates multiple models to forecast a future emission pathway. An average of results from four models presented suggests that GHG emissions in the next two decades are bound to become threefold, from 1727 (2007) to 5220 $MTCO_2e$ (2030–31), while per capita emissions will also become two-and-a-half times, from 1.5 (2007) to 5.6 tons of CO_2e (2030–31). Meanwhile the LCS Report pegs India's emissions at 3537 to 3071 $MTCO_2e$ in determined and aggressive low-carbon growth scenarios respectively.

In its submission to UNFCCC at COP 21 Paris, India has now declared that it aspires to achieve the ambitious target of reducing the emissions intensity of its GDP by 33–35% by 2030 from 2005 level (MoEF 2015). The sheer mass of urban population, its economic propensity, high energy consumption and generation of thermal energy, in comparison to their rural counterparts, cities bear a strong role and responsibility as agents to mitigate and adapt to climate change. It needs to be acknowledged here that GHG emissions from urban areas are not just governed by the historic E3 trends, but hold a strong tendency to be influenced by adoption of new technology and policy alternatives in future.

7 Technology and Policy Alternatives for Urban Areas

7.1 Technology Alternatives

Technological interventions are aimed at maximizing energy efficiency gains in various stages of energy generation and consumption (listed in Table 3). In the case of thermal energy plants, super critical and ultra super critical power plants improve combustion efficiency. Efforts should be made for their widespread adoption. Integrated Coal Gasification Combined Cycle is another promising technology, which can attain higher efficiencies and lower CO_2 emissions and also produce synthetic chemical fuels such as diesel and hydrogen. However, initial estimates under Indian conditions of high ash coal show very high auxiliary power

Table 3 A List of technological and policy alternatives to attain energy co-benefits in urban areas

Low-carbon emission technology	Policy, governance and economic instruments
Energy production, transmission and public distribution	
• Use of renewable energy like wind, solar and geo-thermal from local sources with decentralized micro-grids • Super critical and ultra super critical power plants improve combustion efficiency • Integrated Coal Gasification Combined Cycle can attain higher efficiencies and lower CO_2 emissions and also produce synthetic chemical fuels such as diesel and hydrogen • Introduction of Smart-grid devices and technologies in generation and transmission • Promotion of yeast/enzyme based conversion to high quality hydrocarbon fuels, advanced biomass gasification technologies, low temperature PEMFC for stationary power generation and energy storage technologies for bulk storage	• Installation of model WTE and renewable energy projects by ULBs • Promotion of energy-efficient appliances and equipment in transmission of distribution • Setting up of cap and trade mechanism between states, where cities can also trade • Transfer of electricity functions to ULBs or make distribution companies accountable • Incentives for projects to be installed for a ULB in city limits or near to captive plant on renewable energy, such as solar, wind, WTE • Campaigns for energy conservation • Energy audits and planning for all municipal services; energy saved is energy produced
Consumption in energy intensive sectors like industry	
• Shift towards DC arc technology for the electric furnace, LD convertor, cold rolling and slab casting in steel production • Use of newer smelt reduction technologies like COREX and FINEX. Coke Dry Quenching technique is a process that quenches carbonized coke using an inert gas • Waste heat recovery technologies and automated monitoring systems • Top pressure turbine (TPT), a power generation system for converting the physical energy of the high pressure blast furnace top gas • Use of Pulverized Coal Injection (PCI) over conventional coke • Installation of variable frequency drives, use of high-efficiency motors, pumps and blowers, improved insulation of furnaces and replacing electric heaters with fuel-fired heaters	• Provision of designated power supply to reduce load on captive (diesel generator) generation • Ease in grants/loans to industries for technology overhaul • Transfer of regulatory functions of SPCB to ULBs and industrial estates in urban limits • Cap and trade market with reduction targets for GHG emissions • Strengthening regulatory and market instruments for renewable energy certificates and renewable purchase obligation • Polluter pays, penalties on high emissions from captive plants/diesel generator sets • Phased overhaul of municipal infrastructure in industrial estates • Enable list of energy auditors and firms • Incentives of higher FAR, TDR for low emissions • Rebates on property tax for consistent compliance

consumption and hence the overall efficiency is comparable with sub-critical units at almost double the cost. In industrial sector (that roughly accounts for 25% of final energy consumption), obsolete technologies, scarce technological and financial alternatives contribute to low efficiencies, particularly in micro, small and medium enterprises (MSME). It has been argued that there is an urgent need for specific cluster programmes in technological up-gradation of 150 manufacturing clusters that are most energy intensive (Srivastava and Mahajan 2011). This shall entail need assessments, research, development, demonstration, advisory support and dissemination with the involvement of several stakeholders. In the residential sector (that accounts for 25% of final energy consumption), the market offers energy-efficient options in several household appliances like air-conditioners, refrigerators and lamps. The Bureau of Energy Efficiency (BEE) has introduced energy labelling and a certification programme to make consumers more aware, although there are no mechanisms to penalize manufacturers who produce sub-standard technology.

Meanwhile, energy-intensive industries like iron and steel require a series of technologies and production practices including a shift towards DC arc technology in electric furnaces, LD convertor, cold rolling and slab casting in steel production, use of newer smelt reduction technologies like COREX and FINEX. The Coke Dry Quenching technique is a process that quenches carbonized coke using an inert gas; the heated gas then used to produce electricity, creates energy benefits of 1 GJ/tcs over conventional wet quenching. Similarly, adoption of Top Pressure Turbine (TPT), a power generation system for converting the physical energy of the high pressure blast furnace top gas, also promises an energy saving of 0.6 GJ/tcs (NEDO 2008). In addition, several waste heat recovery technologies and automated monitoring systems offer additional co-benefits. Use of Pulverized Coal Injection (PCI) over conventional coke usage results in energy and emission savings. Next generation coke-making technologies (SCOPE 21) offer more flexibility in terms of coal resource quality and reduce energy and emissions intensity in the process. Further general energy saving practices like installation of variable frequency drives, use of high-efficiency motors, pumps and blowers, improved insulation of furnaces and replacing electric heaters with fuel-fired heaters could incrementally reduce energy usage in plants (World Bank 2010a, b).

There is scope for improvisations in the cement manufacturing sector through retrofitting of the existing plants. An increase in the blending percentage would directly offer options for emission reduction. Usage of waste materials (less carbon intensive than coal) for fuel substitution offers scope for substantial efficiency improvements in this sector (Cement Technology Roadmap 2009). Waste heat recovery and cogeneration can reduce emissions. Japanese plants have reported a potential emission saving of up to 0.06 MT CO_2/MT cement through co-generation (NEDO 2008). However, Indian plants utilize part of the process waste heat in drying of the feedstock materials and coal (which have a higher moisture content than the global average), thus resulting in a reduction in the co-generation potential. In spite of this characteristic, the absolute amount of waste heat generated in Indian plants is enough to provide for meaningful co-generation, more so with the increasing captive power costs. Other technology interventions like vertical roller mill technology, fluidized bed cement fired kiln system, and use of mineralizers could further reduce the carbon intensity of Indian cement production (CSI/ECRA 2009). The actual reduction achievable by a plant would depend on the feedstock quality and the 'mix' of technology and other interventions adopted.

Technological alternatives in renewable energy include several initiatives in wind energy (like development of smaller and efficient turbines, wind turbines for low wind regime, designs of offshore wind power plants), solar PV technologies (like those based on p-type silicon wafers and n-type silicon wafers, hetero junction with thin interfacial (HIT) module, back contact back junction (BCBJ) modules, crystalline silicon photovoltaic cells of >24% cell efficiency, high-efficiency concentrating PV (CPV), non-silicon based solar PV technologies), composite cylinders for on-board hydrogen storage, advanced biomass gasification technologies, low-temperature polymer electrolyte membrane fuel cell (PEMFC) for stationary power generation and for vehicular applications. There are some other useful energy storage technologies for bulk storage and renewable energy integration, frequency regulation, utility transmission and distribution applications and for community scale projects, yeast/enzyme based conversion to high quality hydrocarbon fuels, conversion of pre-treated biomass to fuels and chemicals; and gasification technologies like fluidized bed, plasma induced and so on, for power generation.

7.2 Policy Interventions

There are various modes to structure climate, clean energy and low-carbon policies, economic instruments and governance modes in cities. For instance, OECD (2009) and UN Habitat (2011) promote four types of governing modes: Self-governing, Regulation, Provision and Enabling. IEA (2009) groups policies as: Targets (that signal the goals intended by the authorities); Carrots (policies that offer subsidies and incentives); Sticks (restrictions, bans, taxes and any other forms of disincentives); Guidance (education and awareness campaigns, etc.) and Voluntary actions (cooperation and agreements between different stakeholders). With respect to mitigation inclusive urban governance for the energy sector in India, one can identify three major roles cities could perform: as efficient regulators, facilitators and leaders. There are obvious commonalities in all the above classifications, which this study recognizes, considers and identifies as possible policy alternatives for co-benefits in each sector (listed in Table 3).

A regulatory regime of Renewable Purchase Obligation (RPO) for energy generating units is a valuable policy to promote actions that address climate concern. Meanwhile for the most carbon emitting sectors in cities—electricity generation from thermal plants, cement, and iron and steel industry—the Perform-Achieve-Trade (PAT) mechanism envisaged under the Energy Conservation Act 2001 covers 478 units/facilities that account for more than 50% of the fossil fuel used in India and help reduce CO_2 emissions by 25 MT per year (MoEF 2010). Under the mechanism, select energy intensive industrial units and thermal power plants are mandated to reduce their energy consumption by a specified percentage. This has led to a decline of 4–5% in their specific energy consumption in 2015 as compared to that in 2012. The facilities that will achieve savings in excess of their mandated reduction would be issued Energy Savings Certificate (ESCerts), also known as Renewable Energy Certificates (REC), which would be tradable among specific entities. The scheme is to be widened to additional sectors like transport, railways and electricity distribution utilities and refineries.

In addition, policies would have to eventually rationalize pricing of energy products and services to affect their consumption. The objective should be to promote clean and efficient energy, reduce wasteful practices and reduce subsidies to people who are above the poverty line. It is equally relevant for petroleum products used for freight of essential commodities, public and private transport and cooking fuel, as well as electricity, used in household appliances, lighting, space conditioning, street illumination, municipal services, agricultural pumps and MSME. While petroleum products are being continuously decontrolled by the government, it still remains an issue of political discontentment. Subsidies in diesel, kerosene and liquefied petroleum gas need to be rationalized to target the most under-privileged. Effective demand management is the key that local governments should target. Municipal authorities need to think of effective regulatory mechanisms to discourage use of private transport, diesel vehicles as private modes and measures to ensure congestion charges and differential parking fees. Variable pricing could also be introduced in public transport to ensure effective demand management. Similarly, in the case of electricity consumption, excessive demand needs to be checked by introducing peak tariffs which are exorbitantly higher than off-peak ones.

8 Conclusions

In the absence of any previous investigations on E3 linkages in India, this study was conducted on the research premise that E3 challenges are most crucially played out in urban areas. It concludes with strong evidence that economy, energy and emissions individually and collectively as a nexus are increasingly asserting a major role in cities. While Indian cities host only 32% of the national population, they contribute to over 63% of the national economy, associated with 85% of energy generation from thermal plants and responsible for 66.5–70.3% of the national GHG emissions. This proportional contribution from Indian cities is quite significant in comparison to the global scenario, where urban areas are responsible for similar E3 impacts but with a larger 52% of world's urban population. Hence, Indian cities prove to be a breeding ground for highly intensive E3 linkages. This research thus affirms the high sub-national inequity in energy access and consumption and generation of emissions, between urban and rural areas, as highlighted from recent literature in the sub-continent by, for example, Parikh et al. (2009), IEA (2011a, b) and Sethi (2015). While there has been a slight increase in on-grid provision of energy, expansion of off-grid technologies can significantly improve energy access in backward areas and bring immense social benefits.

A super-imposed scenario of economy, energy generation and consumption, and emissions increasingly exhibits rapid growth with mutual inter-dependency. This further emphasizes the immediacy to take measures in low-carbon technology and economic or policy instruments to produce climate co-benefits in different energy generating and consuming sectors. Low-carbon technology alternatives are particularly inclined towards promoting renewables, enhancing energy efficiency, minimizing losses, avoidance or reduction in consumption, fuel efficiency, and so on. Certain measures like introducing energy efficient appliances and demand-side management are low-hanging co-benefits. Meanwhile, policy alternatives focus on a mix of alternatives like renewable energy obligations, energy pricing, carbon-tax, cap-and-trade, subsidies and intensive regulation. A synchronized national, state and local action could capture unlocked energy co-benefits in the E3 nexus. A review of India's major economic, energy and climate policy like the 12th Five-year Plan, Low Carbon Strategies for Inclusive Growth and NAPCC, reveal that pursuing a co-benefits approach serves as a guiding principle in meeting the E3 challenge. While major national policies do spell out technical and policy measures to deal with the situation, comprehensive planning, modelling of energy, economy and emission parameters along with responsive action by the state governments in India would go a long way in realizing urban-level climate co-benefits locked in the interphase of E3 nexus.

References

Business Standard. (2009, January 21). *Slowdown not to affect India's nuclear plans*. http://www.business-standard.com/article/economy-policy/slowdown-not-to-affect-india-s-nuclear-plans-109012100091_1.html. Accessed on February 13, 2016.

Carlino, G. A., Chatterjee, S., & Hunt, R. M. (2007). Urban density and the rate of invention. *Journal of Urban Economics, 61*, 389–419. doi:10.1016/j.jue.2006.08.003, (ISSN:00941190).

CEA. (2012). *Performance review of thermal power stations. Central electricity authority*. http://www.cea.nic.in/reports/yearly/thermal_perfm_review_rep/1112/complete_1112.pdf. Accessed on February 10, 2016.

Cement Technology Roadmap. (2009). *Carbon technology roadmap 2009: Carbon emissions reductions up to 2050*. Paris: IEA—WBSCD.

Chikkatur, A. P., & Sagar, A. D. (2009). *Coal initiative reports—White paper series: Positioning the Indian coal-power sector for carbon mitigation: Key policy options*. Cambridge, MA: School of Government, Harvard University.

Cohen, C., Lenzen, M., & Schaeffer, R. (2005). Energy requirements of households in Brazil. *Energy Policy, 33*, 555–562 (2005). doi:10.1016/j.enpol.2003.08.021. ISSN: 0301-4215.

CSI/ECRA. (2009). *CSI-ECRA—Technology papers: Development of state of the art techniques in cement manufacturing: Trying to look ahead*. World Business Council for Sustainable Development. http://www.wbcsdcement.org/pdf/technology/Technology%20papers.pdf. Accessed on March 12, 2016.

Das, G. (2006). The india model. *Foreign Affairs, 85*(4), p4.

Duranton, G., & Puga, D. (2004). Micro-foundations of urban agglomeration economies. In J. V. Henderson & J.-F. Thisse (Eds.), *Handbook of Regional and Urban Economics* (pp. 2063–2117). Elsevier.

EIA. (2011). Google public data US energy. *Energy Information Administration*. Accessed on December 17, 2015.

Fujita, M., & Thisse, J.-F. (1996). Economics of agglomeration. *Journal of the Japanese and International Economies, 10*, 339–378 (1996). doi:10.1006/jjie.1996.0021. ISSN: 08891583.

GEA. (2012). *Global energy assessment—Toward a sustainable future*. Cambridge, UK and New York, NY, USA: Cambridge University Press and Laxenburg, Austria: International Institute for Applied Systems Analysis.

GoI. (2014). India's fiscal budget 2012–13 (Chapter 10). *Government of India*. http://indiabudget.nic.in/es2012-13/echap-10.pdf. Accessed on December 15, 2014.

Goldman Sachs (2007). India's rising growth potential. *Goldman Sachs*. http://www.usindiafriendship.net/viewpoints1/Indias_Rising_Growth_Potential.pdf. Accessed on November 21, 2015.

Grubler, A., Bai, X., Buettner, T., Dhakal, S., Fisk, D., Ichinose, T., Keirstead, J., et al. (2012). Urban energy systems. In *Global energy assessment—Toward a sustainable future* (pp. 1307–1400). Cambridge, UK and New York, NY, USA: Cambridge University Press and Laxenburg, Austria: International Institute for Applied Systems Analysis.

Grubler, A., O'Neill, B., Riahi, K., Chirkov, V., Goujon, A., Kolp, P., et al. (2007). Regional, national, and spatially explicit scenarios of demographic and economic change based on SRES. *Technological Forecasting and Social Change, 74*, 980–1029 (2007). doi:10.1016/j.techfore.2006.05.023. ISSN: 0040-1625.

Gutman, P. (2007). Ecosystem services: Foundations for a new rural—urban compact. *Ecological Economics, 62*, 383–387 (2007). doi:10.1016/j.ecolecon.2007.02.027. ISSN: 0921-8009.

Henderson, J. V., Kuncoro, A. & Turner, M. (1995). Industrial development in cities. *Journal of Political Economy, 103*, 1067 - 1090 (1995). doi:10.1086/ 262013, ISSN:0022-3808.

Hoornweg, D., Sugar, L., & Gomez, C.L.T. (2011). Cities and greenhouse gas emissions: Moving forward. *Environment and Urbanization, 23*, 207–227 (2011). doi:10.1177/0956247810392270.

HPEC (2011). *Report on Indian urban infrastructure and services*. New Delhi: The High Powered Expert Committee on Urban Infrastructure, Government of India.

IEA. (2007). *World energy outlook: China and India Insights*. Paris: OECD/IEA Publications.

IEA. (2008). World Energy Outlook 2008 Edition. *International Energy Agency*. http://www.worldenergyoutlook.org/media/weowebsite/2008-1994/weo2008.pdf. Accessed on October 11, 2015.

IEA. (2009). *Cities, towns & renewable energy. Yes in my front yard*. Paris: OECD/IEA Publications.

IEA. (2011a). *World energy outlook*. Paris: OECD/IEA Publications.

IEA. (2011b). *Key world energy statistics*. Paris: OECD/IEA Publications.

IEA. (2012). *Understanding energy challenges in India—Policies, players and issues*. Paris: OECD/IEA Publications.

IMF. (2014). India. *International Monetary Fund*. http://www.imf.org/external/pubs/ft/weo/2014/01/weodata/weorept.aspx?pr.x=79&pr.y=7&sy=2012&ey=2015&scsm=1&ssd=1&sort=country&ds=.&br=1&c=534&s=NGDPD%2CNGDPDPC%2CPPPGDP%2CPPPPC&grp=0&a=. Accessed on April 8, 2015.

IPCC. (2014). Climate change 2014: Mitigation of climate change. In O. Edenhofer, R. Pichs-Madruga, Y. Sokona, E. Farahani, S. Kadner, & K. Seyboth, et al. (Eds.), *Contribution of Working Group III to the Fifth Assessment Report of the Intergovernmental Panel on Climate Change*. Cambridge: Cambridge University Press, NY, USA: United Kingdom and New York.

Kahn, M.E. (2009). Urban Growth and Climate Change. *Annual Review of Resource Economics, 1*, 333–350 (2009). doi:10.1146/annurev.resource.050708.144249.

Kennedy, C., Steinberger, J., Gasson, B., Hansen, Y., Hillman, T., & Havránek, M., et al. (2009). Greenhouse gas emissions from global cities. *Environmental Science & Technology, 43*, 7297–7302 (2009). doi:10.1021/es900213p. ISSN: 0013-936X.

Knudsen, B., Florida, R., Stolarick, K., & Gates, G. (2008). Density and Creativity in U.S. Regions. *Annals of the Association of American Geographers, 98*, 461–478. doi:10.1080/00045600701851150, (ISSN:0004-5608, 1467-8306).

Lenzen, M., Wier, M., Cohen, C., Hayami, H., Pachauri, S., & Schaeffer, R. (2006). A comparative multivariate analysis of household energy requirements in Australia, Brazil, Denmark, India and Japan. *Energy, 31,* 181–207 (2006). doi:10.1016/ j.energy.2005.01.009, ISSN: 0360-5442.

LiveMint. (2014, June 1). RBI governor Raghuram Rajan meets Narendra Modi. *LiveMint.* http://www.livemint.com/Politics/Jh1N8RUWG7J5BKZfTSyYZO/RBI-governor-Raghuram-Rajan-meets-Narendra-Modi.html. Accessed on March 18, 2015.

Marcotullio, P.J., Sarzynski, A., Albrecht, J., & Schulz, N. (2012). The geography of urban greenhouse gas emissions in Asia: A regional analysis. *Global Environmental Change, 22*, 944–958 (2012). doi:10.1016/j.gloenvcha.2012.07.002.

Marcotullio, P.J., Sarzynski, A., Albrecht, J., Schulz, N., & Garcia, J. (2013). The geography of global urban greenhouse gas emissions: An exploratory analysis. *Climatic Change*, *121*, 621–634 (2013). doi:10.1007/s10584-013-0977-z.

McKinsey Global Institute. (2011). Urban world: Mapping the economic power of cities. *McKinsey Global Institute*. http://www.mckinsey.com/global-themes/urbanization/urban-world-mapping-the-economic-power-of-cities#0. Accessed on September 27, 2015.

MoEF. (2009). *India's GHG emissions profile*. New Delhi: Climate Modeling Forum, Ministry of Environment and Forests, Government of India.

MoEF. (2010). *India: Greenhouse gas emissions 2007* (pp. 1–54). New Delhi: Indian Network for Climate Change Assessment, Ministry of Environment & Forests, Government of India.

MoEF. (2015). *India's intended nationally determined contribution: Working towards climate justice*. New Delhi: Ministry of Environment & Forests, Government of India.

NEDO. (2008). *Global warming countermeasures 2008 revised edition: Japanese technologies for energy savings/GHG emissions reduction*. New Energy and Industrial Technology Development Organization. http://www.google.co.in/url?sa=t&rct=j&q=&esrc=s&source=web&cd=2&cad=rja&uact=8&ved=0ahUKEwj5_azHlrzRAhXJvY8KHeGRAUwQFggtMAE&url=http%3A%2F%2Fietd.iipnetwork.org%2Fsites%2Fietp%2Ffiles%2FJapanese%2520Technologies%2520for%2520Energy%2520Saving.pdf&usg=AFQjCNESRhMO7pglnzXZB_y_1SLslo6_2Q&bvm=bv.144210762,d.c2I. Accessed on.

OECD. (2007). *Economic survey of India 2007: Policy brief*. Paris: OECD/IEA Publications.

OECD. (2009). *Cities, climate change and multilevel governance* (pp. 30–44). OECD Environmental Working Papers No 14. Paris: OECD Publishing.

OECD. (2012). *OECD economic outlook, 2012 (1)*. Paris: OECD Publishing.

Pachauri, S., & Jiang, L. (2008). The household energy transition in India and China. *Energy Policy, 36*, 4022–4035 (2008). doi:10.1016/j.enpol.2008.06.016.

Parikh, J., & Parikh, K. (2011). India's energy needs and low carbon options. *Energy, 36*, 3650–3658. doi:10.1016/j.energy.2011.01.046.

Parikh, J., Panda, M., Ganesh-Kumar, A., & Singh V. (2009). CO_2 emissions structure of Indian economy. *Energy, 34*(8), 1024–1031 (2009). doi:10.1016/j.energy.2009.02.014.

PBL. (2014). *Trends in global* CO_2 *emissions*; *2012 report*. The Hague: PBL Netherlands Environmental Assessment Agency; Ispra: Joint Research Centre.

Press Information Bureau. (2008, June 30). *PM releases national action plan on climate change*. http://pib.nic.in/newsite/erelease.aspx?relid=39898, Accessed on December 11, 2015.

Planning Commission. (2006). *Integrated energy policy: Report of the expert committee*. New Delhi: Planning Commission, Government of India.

Planning Commission. (2011). *Interim report of the expert group on low carbon strategies for inclusive growth*. New Delhi: Planning Commission, Government of India.

Planning Commission. (2013). *Final report of the expert group on low carbon strategies for inclusive growth*. New Delhi: Planning Commission, Government of India.

Puga, D. (2010). The magnitude and causes of agglomeration economies. *Journal of Regional Science, 50*, 203–219 (2010). doi:10.1111/j.1467-9787.2009.00657.x.

Rosenthal, S. S., & Strange, W. C. (2004). Evidence on the nature and sources of agglomeration economies. In J. V. Henderson & J. -F. Thisse (Eds.), *Handbook of Regional and Urban Economics*. New York: North-Holland; New York, NY, USA: Elsevier, Amsterdam. pp.2119–2171 (ISBN:9780444509673).

Sahakian, M.D., & Steinberger, J.K. (2011). Energy reduction through a deeper understanding of household consumption. Journal *of Industrial Ecology, 15*, 31–48 (2011). doi:10.1111/j.1530-9290.2010.00305.x. ISSN: 1530-9290.

Satterthwaite, D. (2008). Cities' contribution to global warming: Notes on the allocation of greenhouse gas emissions. *Environment & Urbanization, 20*(2), 539–549.

Satterthwaite, D. (2009). The implications of population growth and urbanization for climate change. *Environment and Urbanization, 21*(2), 545–567.

Sethi, M., & Mohapatra, S. (2013). Governance framework to mitigate climate change: Challenges in urbanising India. In H. Ha & T. N. Dhakal (Eds.), *Governance approaches to mitigation of and adaptation to climate change in Asia* (pp. 200–230). Hampshire, UK: Palgrave Macmillan.

Sethi, M. (2015). Location of greenhouse gases (GHG) emissions from thermal power plants in India along the urban-rural continuum. *Journal of Cleaner Production, 103,* 586–600. doi:10.1016/j.jclepro.2014.10.067.

Shafik, N. (1994). Economic development and environmental quality: An econometric analysis. *Oxford Economic Papers, 46*(October), 757–773.

Srivastava, L., & Mahajan, D. (2011). India's energy challenges: Defining a framework of action. *Management in Government, 4* (2011).

The Economist. (2011, March 24). Going nuclear. *The Economist*. http://www.economist.com/blogs/dailychart/2011/03/global_nuclear_power. Accessed on September 6, 2015.

The Times of India. (2012, May 2). Homes become more affordable in last 10 years. *The Times of India*. http://timesofindia.indiatimes.com/business/india-business/Homes-become-more-affordable-in-last-10-years/articleshow/12958960.cms. Accessed on January 14, 2016.

The Economic Times. (2013, August 30). Agriculture's share in GDP declines to 13.7% in 2012–13. *The Economic Times*. Accessed on October 28, 2014.

The Economic Times. (2014, October 8). IMF projects India's GDP to pick up in 2015 to 6.4%. *The Economic Times*. Accessed on October 23, 2014.

United Nations. (2011). National accounts main aggregates database. *United Nations Statistics Division*. https://unstats.un.org/unsd/snaama/Introduction.asp. Accessed on December 11, 2015.

UN Habitat. (2011). *Global report on human settlements 2011: Cities and climate change*. London/Washington DC: United Nations Human Settlements Programme & Earthscan.

Habitat, U. N. (2012). *State of the world cities report*. New York: Routledge.

UN DESA. (2011). Urban population development and the environment wallchart. United Nations, Department of Economic and Social Affairs, Population Division. http://www.un.org/en/development/desa/population/publications/urbanization/urbanPopDevEnv_2011.shtml. Accessed on November 11, 2015.

UN DESA. (2013). Urban population development and the environment wallchart. United Nations, Department of Economic and Social Affairs, Population Division. http://www.un.org/en/development/desa/population/publications/pdf/development/pde_wallchart_2013.pdf. Accessed on March 6, 2016.

Vringer, K., & Blok, K. (1995). The direct and indirect energy requirements of households in the Netherlands. *Energy Policy, 23*, 893–910 (1995). doi:10.1016/0301-4215(95)00072-Q.

(The) Wall Street Journal. (2010, December 29). *India raises renewable energy target fourfold.* http://www.wsj.com/articles/SB10001424052970203513204576048870791325278. Accessed on March 19, 2016.

WEEVR—Wind Energy and Electric Vehicle Review (2011, April 12). *Wind energy in India: Chinese company sold 125 wind turbines of 2MW.* http://www.evwind.es/2011/04/12/wind-energy-in-india-chinese-company-sold-125-wind-turbines-of-2-mw/11182. Accessed on April 3, 2016.

Weisz, H., & Steinberger, J. K. (2010). Reducing energy and material flows in cities. *Current Opinion in Environmental Sustainability, 2*, 185–192 (2010). doi:10.1016/j.cosust.2010.05.010.

World Bank. (2010a). *Cities and climate change: An urgent agenda*. New York: The World Bank.

World Bank. (2010b). *Energy intensive sectors of the Indian economy: Path to low carbon development*. New York: The World Bank.

World Bank. (2014). India country overview 2013. *World Bank*. http://www.worldbank.org/en/country/india/overview. Accessed on July 28, 2015.

The World Bank. (2015). A measured approach to ending poverty and boosting shared. *World Bank*. https://sustainabledevelopment.un.org/content/documents/1792A%20Measured%20Approach%202015.pdf. Accessed on February 23, 2016.

WTO. (2013). *World trade report 2013*. Geneva: World Trade Organization.

Zheng, S., Wang, R., Glaeser, E.L., & Kahn, M.E. (2010). The greenness of China: household carbon dioxide emissions and urban development. *Journal of Economic Geography 11*, 761–792 (2010). doi:10.1093/jeg/lbq031.

Chapter 6
Co-benefits of Urban Transport

Sudhir Gota and Alvin Mejia

Abstract Climate co-benefits are often not the primary drivers for choosing policies and investments in the transportation sector. However, the growing awareness of the costs of climate change, the importance of the sector in the global greenhouse gas emissions profile and the growing opportunities provided by the emerging climate instruments have opened up a more holistic paradigm in assessing transportation investments. With growing negative externalities from the transport sector, limited availability of funds and long-term impacts of transport investments, there is an urgent need to maximize benefits through the integration of multiple objectives including climate concerns in the assessment of policies and projects. It is often assumed that applying the co-benefits approach to the transport sector is difficult, needs a lot of resources and often is not straightforward. This chapter tries to break this myth. It establishes a case for quantifying co-benefits and presents a specific case study on the Chennai Metro Project. It shows how quantifying multiple co-benefits from transport projects could be simple yet effective in understanding the economic viability, and long-term impacts of the project. The adoption of the co-benefits approach in the assessment of transport investments aims to reveal a wider range of costs and benefits of the alternatives which are currently not captured in existing assessment approaches. In doing so, a clearer picture of the impacts of the different alternatives is generated and thus better investment decisions can be made.

Keywords Quantifying co-benefits · Transport projects
TEEMP models · Chennai metro · Economic analysis · Scaling up

S. Gota (✉)
Independent Consultants, Bangalore, India
e-mail: sudhirgota@gmail.com

A. Mejia
Metro Manila, Philippines

M. Sethi and J.A. Puppim de Oliveira (eds.), *Mainstreaming Climate Co-Benefits in Indian Cities*, Exploring Urban Change in South Asia,
https://doi.org/10.1007/978-981-10-5816-5_6

1 Introduction

After electricity and heat production, transport contributes to the largest share (14% of the total) greenhouse gas emissions from fossil fuels (IPCC 2014). In the case of India, of the gross GHG emission of 1904.73 $mtCO_2e$, 7.5% (142.04) is attributable to the transport sector. Within the transport sector, the predominant contribution to GHG emissions comes from the road sector. The four wheelers and two wheelers are the biggest contributors to GHG emissions. The aviation sector contributes around 10% followed by railways at about 7%. Urban areas have a strong linkage with the road sector and its resultant GHGs (MoEF 2010). Policies and actions in the transport sector have long-term consequences, both desired and undesired. It is often believed that although infrastructure investments are expensive, it is even more expensive to skimp on infrastructure (U.S. Department of Treasury and Council of Economic Advisers 2010). This is partly true, although a better perspective is that although infrastructure investments are expensive, it is even more expensive to take the wrong investment decision. Transport infrastructure can be very expensive to build; however, once built, these investments will have a long-lasting impact. Considering limited resources and the multidimensional pull from different sectors for funds, it is necessary to invest in the right investments. While making this decision, analysts often compare the main project benefits with a baseline that is an appropriate counterfactual to the project investment, and this is at the core of existing frameworks for assessing the economic impacts of transport projects.

Transport projects are implemented to provide better transport services and not mainly for addressing climate change or air pollution issues, although these can be important factors in steering transport systems towards a more sustainable direction. It is important not to neglect these consequences, that is, the potential influence of climate change on future transport policies and projects. Apart from satisfying the primary objective of transport service provision, any other positive impacts that occur in addition to the intended primary policy goal could be considered as co- benefits.

Generally, transport benefits can be categorized into the following, as shown in Fig. 1 (Leather and CAI-Asia 2009).

Primary benefits: These are the primary objectives of transport policies and projects, which in many cases pertain to the reduction in traffic congestion (measured as increased vehicle speed, level of service) and the reduction in vehicle operation costs, depending on the specific policy or project.

Direct co-benefits: There are other benefits—aside from the primary benefits—that directly result from transport policies or projects. For example, travel time savings, reduced GHG emissions, reduced air pollutant emissions, reduced noise levels, improved equity, improved safety, and reduced fuel subsidies paid by governments. These direct co-benefits are sometimes accidental but in the case of a well-developed strategic capacity of policy-makers, aiming for such impacts can be intentional.

Indirect co-benefits: These are benefits that indirectly result from transport policies or projects, that is, as a consequence brought about by the primary

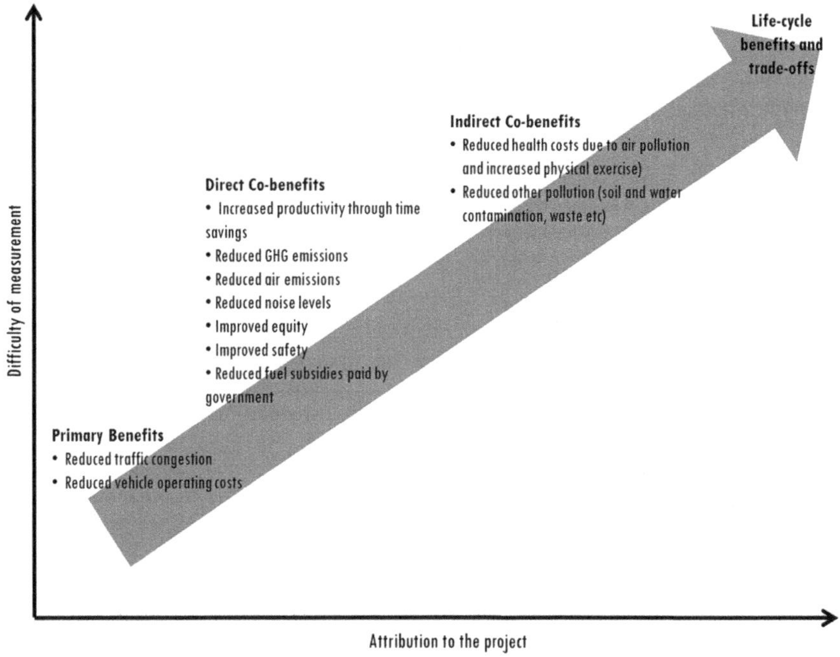

Fig. 1 Co-benefits of transport projects and difficulty in measurement. *Source* CAI-Asia Center (2009)

co-benefits. For example, improved public health and reduced health costs due to reduced air pollution or increased physical exercise, and reduced other pollution (to soil, water and as waste).

The concept of co-benefits itself has been defined in several ways by different organizations and individual authors who have studied the matter. The Intergovernmental Panel on Climate Change (IPCC), in its third Assessment Report (AR 3), refers to co-benefits as the intended positive benefits of a policy and distinguishes it from unintended positive side effects or ancillary benefits. The United States Environmental Protection Agency (USEPA) provides a basic definition of co-benefits as the sum of positive outcomes associated with multiple, simultaneous emissions reductions (USEPA 2005). The Institute for Global Environmental Strategies (IGES) defines a co-benefits approach as a win-win strategy aimed at capturing both development and climate benefits in a policy or measure (IGES 2010). It is clear that the concept of 'co-benefits' differs based on the primary focus of organizations that have used it in the past. This chapter would like to refer to the co-benefits of transportation projects as the benefits that are either intended or unintended but are outside the main objectives (transportation-related) of the project.

In many developing countries, co-benefits is a widely misunderstood terminology. The term is often viewed with a 'cart that is placed in front of the horse' syndrome—that is, prioritizing one benefit over other in decision-making. It is important to note that taking a co-benefits approach is not about prioritizing a single benefit over the others, but about enabling a more holistic approach in evaluating transport investments by encouraging a more thorough analysis of the impacts of the investments alternatives. For example, Maizlish et al. (2013) established that while low-carbon driving reduced carbon emissions by 33.5% and the cardiorespiratory disease burden by less than 1%, non-motorized transport improvement (a walking and cycling increase from 4 to 22 min) reduced the burden of cardiovascular disease and diabetes by 14%, increased the traffic injury burden by 39%, and decreased carbon emissions by 14%. Clearly, co-benefits do matter in decision-making. Further, Reynolds (2010) established that for certain strategies, air pollution benefits may be offset by an increase in greenhouse gas emissions due to implementation of certain policies. It is thus important to understand the comprehensive impacts of implementing such strategies.

Transport investments under the sustainable transport paradigm requires less amounts of investments to cater to the same transport demand while mitigating emissions and generating higher economic outputs. With growing transport-related externalities costing 5–10% of the gross domestic product (GDP), the economic and environmental consequences of bad transport investments can be irreversible. It is clear that there is a need to look 'beyond the fence' and consider all significant implications of the investment. This chapter investigates various challenges related to co-benefits quantification, focusing on climate, air pollution mitigation and safety co-benefits and its potential to identify appropriate measures for implementation (co-benefits considered are emissions—carbon dioxide (CO_2), particulate matter (PM), nitrogen oxide (NO_x), black carbon—and accident fatality reductions).

Sustainable transport has an important cross-cutting role in sustainable development (SLoCaT 2015). For example, the Open Working Group (OWG) on Sustainable Development Goals (SDGs) acknowledges the contribution that sustainable transport can make in realizing commonly agreed upon goals on food security, health, energy, infrastructure development, urban development, sustainable consumption and production as well as climate change. Indeed, the fact that transport-related targets are included in seven out of the seventeen proposed SDGs illustrates the important cross-cutting role that transport has in sustainable development. The transport sector would not be able to facilitate this role if co-benefits were not considered in the decision-making.

2 Challenges to the Application of a Co-benefits Approach in Transport Projects

The measurement of co-benefits in the transport sector can be difficult, costly and time-consuming, compared to the measurement of the direct benefits of transport policies and projects (Leather and CAI-Asia 2009). Co-benefits are often considered as a 'secondary', albeit often explicit, benefit of good transport systems. Co-benefits are often not quantified and considered in transport investments. The main challenges are the following:

1. **Lack of methodology, data and awareness**

In the past, institutions involved in the assessment and selection of transport investments (e.g. development banks, local and national governments) have not used a holistic co-benefits approach (e.g. non-inclusion of co-benefits such as emissions changes brought about by projects) in quantifying costs and benefits, citing the lack of an accurate methodology and data for computing emissions (ADB 2010; Williams et al. 2012). The existing systems for collecting and analysing data required to fully understand the impacts of transport sector investments in say, air pollution and GHG emissions, are not adequate nor standardized. For some projects with comparable elements where co-benefits were quantified, it was found that differences in GHG mitigation calculations was in order of magnitude (Rogers 2010).

A detailed review of transport emission methodologies developed with the primary goal to measure CO_2 emissions (SLoCaT 2015) establishes that co-benefits are often neglected. Roughly one-quarter to one-third of methodologies can be used to assess co-benefits of proposed transport mitigation measures. SLoCaT's analysis indicates that nearly 40% of GHG methodologies quantify air pollutants such as particulate matter (PM) and Nitrous oxides (NO_x). Only about 15% of methodologies quantify short-lived climate pollutants (SLCPs) like methane and black carbon, which can be critical factors in shifting the balance of a transport project from infeasible to feasible. Additional co-benefits captured in the methodologies include fuel savings (27%), road safety (13%) and travel time (8%).

Furthermore, the limited awareness, knowledge and capacity with regard to the co-benefits approach among policy-makers, transport planners and engineers could contribute to low applications of the co-benefit approach in the transport sector. The following quote from India's Economic Analysis Evaluation Manual for Highway Projects shows the general recognition of this gap (Indian Roads Congress 1993). "This manual quantifies highway development benefits into road user benefits (i.e. vehicle operating cost savings, value of travel time savings, accident cost savings and savings in maintenance costs) and social benefits (improvement in environmental standards, land value, health, education, agriculture, trade etc.)." Further, it mentions that at the present state of knowledge in the country, it is possible to monetarily quantify only the direct road user benefits. Only when sufficient research is carried out on other aspects can the full quantification of benefits be possible.

2. **Lack of policy, institutional and stakeholder support**

In cases where co-benefits are considered, the concept is mainly used in post-project assessment and not as part of the planning process (e.g. analysis on Japanese ODA loans). This may be attributed to the fact that the concept of co-benefits has only been introduced fairly recently in the transport sector and perhaps also due to the poor documentation of existing case studies. Although there are some methodologies that exist, the co-benefits approach has not been mainstreamed into the transport sector (Leather and Clean Air Asia 2009). Further, there is a severe lack of participation by diverse stakeholders and institutions in project decision making and implementation, which hinders the identification of the full range of co-benefits associated with such projects.

For example, the introduction of bus rapid transit (BRT) projects in Delhi, Pune and Indore have struggled because of opposition from car owners and a lack of institutional coordination (Yale Environment 360, 2014). In many of these projects, co-benefits were not adequately highlighted. For example, between 2005 and 2009, a four-year survey of news articles on the Ahmadabad BRT (Gota 2009) indicated that out of 221 articles, there was not even a single article on the environmental benefits of the said BRTS and only 12% of articles included discussions on social, safety and security impacts of the BRTS system. Clearly, if co-benefits are not quantified and acknowledged, support from diverse stakeholders would be limited.

3. **It is often perceived that co-benefits are difficult to quantify**

Applying the co-benefit concept to the transport sector is not straightforward and has been viewed as a difficult, costly and time-consuming activity, compared to the measurement of the direct benefits of transport policies and projects (reduced traffic congestion and vehicle operation cost). The situation is worse for secondary co-benefits, such as reduced health costs resulting from reduced air pollution, and the co-benefits across the life-cycle of a transport policy or project if such a life-cycle approach were to be considered. Further, limited empirical research restricts the development of sketch level tools which could enable easier quantification of co-benefit impacts, particularly for second-order co-benefits. Further, it has been argued that simplification will necessarily present trade-offs—ease of use will come at the expense of analytical precision (IGES 2011).

In cases where several different co-benefits can be quantified (e.g. climate co-benefits, air pollution mitigation co-benefits, other development co-benefits), it is still not certain that the quantified co-benefits will be able to aid in the investment decisions and the selection of the right project to be implemented. The quantified co-benefits can either be taken into the assessment process either by transforming these co-benefits into a standardized unit through monetization or by assigning weights to the different co-benefits and employ a multi-criteria analysis of the project alternatives. Both approaches are generally perceived as challenging and vulnerable to bias and lack of information.

4. **Co-benefits currently do not matter in cost-benefit analysis/economic analysis**

A cost-benefit study of the World Bank of transport projects indicates that transport projects in general have a high median rate of return (30% in 2008) (IEG World Bank 2010). Transport, among all the sectors, was found to have one of the highest rates of return. Moreover, a recent evaluation report from the Asian Development Bank confirms that the transport sector is one of the best performing sectors over the past 20 years in terms of their success rates (ADB 2013).

It can be argued that traditional benefits alone would make the majority of transport projects economically viable. It has been reported that the monetized impact of CO_2 emissions reduction on economic analysis is marginal and may not influence the decision-making for projects having high IRR. But in comparing investment alternatives, as well as in assessing projects whose IRR are around 11–13% (can either be feasible or unfeasible), the value of the co-benefits could be critical (ADB 2010). A good example of this could be the Delhi Metro. Research on the Delhi Metro has suggested that accounting for benefits from the reduction of urban air pollution due to the Delhi Metro has increased the economic rate of return by only 1.4% while the metro has an economic rate of return of 21% (Murty et al. 2006). Similarly, a BRT study in Metro Manila, Philippines states that the value of the quantified CO_2, PM and NO_x reduction due to the project constituted only 0.22% of total benefits (IGES 2011).

Considering the above arguments, it is clear that there is no incentive for project officers to quantify co-benefits during project appraisals or during evaluations. However, improved knowledge on the value function attached to co-benefits analysis should give co-benefits inclusion a boost. Appropriate monetary values for reducing a ton of CO_2, PM or NO_x should capture the wider benefits that such reductions might have (e.g. capturing avoided morbidity costs as well as reflecting other values dictated by markets such as for CO_2). Moreover, certain impacts might not be possible to monetize, and thus are often left out of the decision-making process. However, these somehow have to be discussed and ways of including them must be explored.[1]

Also, up to now, other co-benefits such as the avoidance of the infrastructure requirements for the counterfactual baseline scenario are still not included in existing co-benefits analysis methods. The inclusion of the additional costs of providing road space to accommodate the expected increases in motorized vehicle travel demand in the absence of a sustainable project will have immense impacts on the way we quantify the costs of the baseline scenario.

3 TEEMP Models: Tools for Quantifying Co-benefits from Transport Projects

In order to simplify the co-benefits quantification process and to reduce the required resources for such a process, Clean Air Asia, together with partners such as the Institute for Transport Development Policy (ITDP), ADB, Cambridge

[1]Such as the use of a multi-criteria analysis utilizing weights for different impacts as provided by affected stakeholders.

Systematics, and the United Nations Environment Programme (UNEP) Global Environment Facility Scientific and Technical Advisory Panel (GEF-STAP), developed Excel-based, free-of-charge spreadsheet models, collectively called the 'Transport Emissions Evaluation Models for Projects' (TEEMP). The TEEMP tools were initially developed for evaluating the emissions impacts of ADB's transport project and have been modified and extended for GEF projects. In the current form, TEEMP can be easily applied for evaluating co-benefits from transport projects without requiring lot of data or investments (ADB 2010; GEF-STAP 2010).

The TEEMP tools are 'sketch' models which enable the estimation of emissions in both 'project' and 'no-project' scenarios and can be used for evaluating short- to long-term impacts of transport projects. TEEMP primarily evaluates the impacts of transport projects on CO_2 emissions and to some extent air pollutant emissions (PM and NO_x) using data gathered during project feasibility and actual operations. Co-benefits such as travel time savings, fuel savings, CO_2/PM/NO_x emissions and accident savings are quantified in these TEEMP tools. The main objective of TEEMP tools was to support sustainable transport policies to improve air quality and mitigate climate change, which have traditionally been separately conceived and implemented due to lack of a uniform tool.

The following projects can be assessed using the tools inside the TEEMP suite: roads including urban and rural roads, and expressways, mass rapid transit, light rail transit, and/or metro systems, bus rapid transits, bikeways and bike-sharing, pedestrian improvement projects, railway projects and combination of projects in a city (TEEMP City). Table 1 summarizes types of co-benefits which could be quantified in TEEMP suite of tools.

The TEEMP tools are based on data that are essentially collected or collated during project economic evaluation. Figure 2 shows the basic philosophy of TEEMP. Various interventions can be evaluated and compared with the 'no improvement' case. In each model, the 'business as usual' scenario refers to a scenario where no project is implemented, or no modifications to the existing situation are done, while 'intervention' refers to a scenario where the project is implemented (such as a BRT system). Emission savings are quantified under the assumption that no major improvements would have happened without the project. By using the same activity data across different models, the impacts of various projects can be gauged (Fig. 2).

The TEEMP models are anchored on the ASIF philosophy which connects Activity (passenger and freight travel) with Structure (shares by mode and vehicle type) with Intensity (fuel efficiency) and with Fuel type and appropriate emission factors (Schipper and Mariel-Lilliu 1999). The models borrow heavily from detailed methodologies such as COPERT and allow users to put in default emission factors. The impact of age (age deterioration factors), grade and temperature are not considered in the developed methodology/model to reduce the assumptions.

Table 1 Co-benefits assessed in the various TEEMP models

TEEMP Tool	Fuel and travel time savings	CO_2 emissions	PM and NO_x emissions	Noise	Fatalities and injuries	Increased productivity	Land use impacts	Improved equity	Heath impact
BRT	✓	✓	✓	NA	✓	NA	✓	NA	NA
Metro	✓	✓	✓	NA	✓	NA	✓	NA	NA
Roadways	✓	✓	✓	NA	NA	NA	NA	NA	NA
Railways	✓	✓	✓	NA	NA	NA	NA	NA	NA
Bikeways	✓	✓	✓	NA	NA	NA	NA	NA	NA
Bike share	✓	✓	✓	NA	NA	NA	NA	NA	NA
Walkability	✓	✓	✓	NA	NA	NA	NA	NA	NA
City	✓	✓	✓	NA	✓	NA	NA	✓	NA

Note ✓ indicates quantification possible and NA indicates quantification not possible in TEEMP model

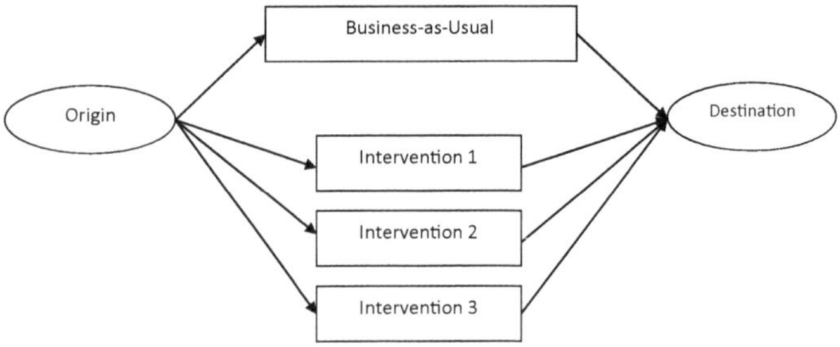

Fig. 2 Basic structure of transport emissions evaluation model for projects

3.1 Case Study: Chennai Metro

Globally, metro rail systems exist in nearly 190 cities with over 11,200 km of rail (Global Mass Transit 2014). Based on the estimates of total infrastructure built and ridership carried, a metro carries 3.6 million passengers/km/year, on average. A metro system can provide benefits via:

1. Improved fuel-use efficiency through efficient public transport
2. Mode switching due to the availability of a more efficient and attractive public transport system
3. Occupancy increase
4. Speed increase
5. Potentially a fuel switch to low carbon fuels.

The increased demand for metro systems in cities around the world is mainly due to high benefits generated by the system. The benefits of these systems are often due to mode shift and efficient travel that are asssociated with high occupancies and speeds. By shifting the projected vehicular trips to more efficient modes and comparing it to a 'no-project' scenario, savings on parameters such as travel time, fuel consumption and emissions are established. In a counterfactual scenario where the metro is not constructed, the users would have been forced to use the existing modes and thus the estimated savings result from comparing both scenarios with and without the metro.

Chennai city is home to nearly 5 million residents. The city, as with other cities in the region, is undergoing rapid motorization and a rapid decline in the public transport trip mode share. Figure 3 indicates the drastic reduction in sustainable transport trip mode shares, that is, walking, cycling and public transport. Increasing motorization levels have increased the transport-related externalities and thus in order to improve public transport facilities in Chennai city, the Government of Tamil Nadu has decided to implement the Chennai Metro Rail Project. The proposed Chennai Metro consists of two corridors:

Corridor I: Corridor 1 starts from Washermanpet and ends at Chennai Airport. The dead end to dead end length of the corridor is 23.085 km, out of which 14.3 km is

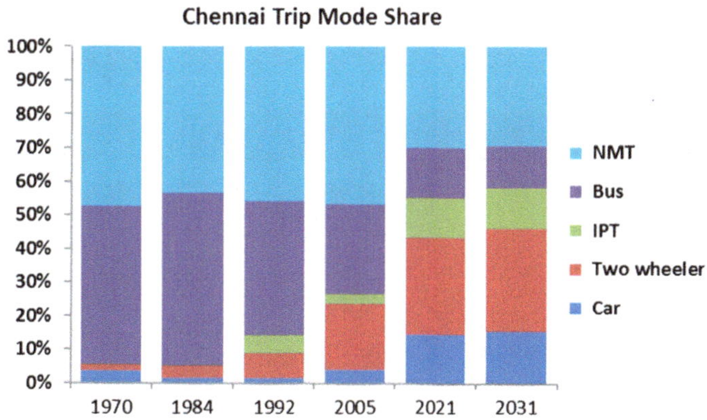

Fig. 3 Historical and projected mode shares (%) in Chennai

underground, 8.785 km is elevated. A total of 18 stations have been planned along this corridor.

Corridor II: Corridor 2 starts from Chennai Central and ends at St Thomas Mount. The dead end to dead end corridor length is 21.961 km. The corridor is partly underground and partly elevated. A total of 18 stations have been planned along this corridor.

The per kilometre cost of the project (including interest during construction and commitment charges to the Japan Bank for International Cooperation (JBIC) and excluding state taxes and cost of vacant state government land) is approximately US $76 million/km.

The analysis was carried out with a refined version of the TEEMP MRT model. This model captures the impact of Mass Rapid Transit systems on CO_2, PM and NO_x emissions by quantifying the construction, operation and traffic impacts of projected MRT riders (Fig. 4). The main input for this analysis is the projected ridership over 20 years of project implementation. This data, combined with 'shift' scenarios (considering the modes that would have been taken by the passengers if the MRT was not constructed) and other data such as those related to construction, fuel efficiencies and emission factors were used in calculating the potential impacts of the project when compared with the no project scenario.

The construction emissions are computed based on the project data on the usage of construction materials (e.g. cement, steel and bitumen) and emission factors (CO_2/amount of material used) from reviewed sustainability reports of manufacturing companies that produces the aforementioned construction materials. The operation emissions can be estimated either using top-down (i.e. measuring the electricity consumption) or by using bottom-up (i.e. using ridership forecasts and emission factors). In this project, both traction and non-traction annual electricity consumption values are available and hence utilized for computations. The detailed project report suggests that the annual energy consumption is assessed around

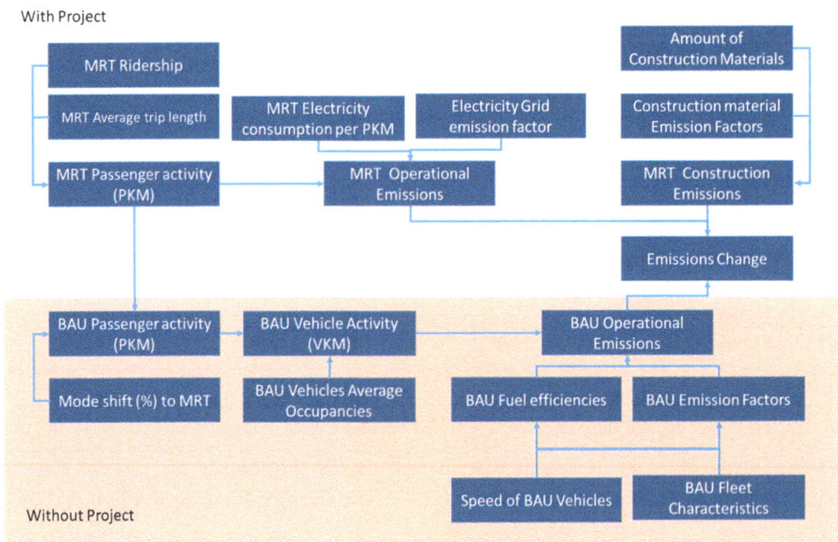

Fig. 4 Emissions calculation framework for the TEEMP MRT model. *Note* The analyses for the other co-benefits utilize the differences in the other parameters inside the model, for example, differences in VKT translate into differences in fatalities and injuries, differences in speeds translate into travel time savings, etc.

Table 2 Trip mode shares of MRT riders in the absence of the metro

Trip mode share of MRT riders	2014 (%)	2023 (%)	2033 (%)
Walking/cycling/no trip	31.0	30.0	29.4
3-wheelers	8.00	12.00	12.00
2-wheelers	20.00	29.00	31.00
Car	10.00	14.00	15.00
Bus	31.00	15.00	12.48

111.8 million units for Corridor I and 52.0 million units for Corridor II in initial years (2011), which will be about 158.2 and 92.3 million units respectively in the year 2026. Without the metro, the supposed passengers are assumed to travel with existing modes as shown in Table 2 (based on literature on trip patterns from the MOUD study and metro detailed project report).

The main inputs for calculating the impact on carbon emissions are summarized in Tables 3 and 4.

Considering only traction electricity consumption, the carbon footprint of the project (i.e. construction and operation emissions) is around 1.1 million tons (CO_2 emissions) for 20 years of analysis. The construction emissions considering only the material consumption is significant, that is, equivalent to more than 20 years of operation emissions.

Table 3 Traffic characteristics

	2014	2023	2033
Ridership/day ('000)	676.764	960.5272	1351.081
Average trip length (km)	8.3	8.3	8.3
Route length (km)	43.8	43.8	43.8
Average occupancy			
3-wheelers	1.8	1.8	1.8
2-wheelers	1	1	1
Car	1.8	1.8	1.8
Bus	45	40	35
Average trip length (km)			
3-wheelers	8.3	8.3	8.3
2-wheelers	8.3	8.3	8.3
Car	8.3	8.3	8.3
Bus	8.3	8.3	8.3

Table 4 Data for economic analysis[a]

Project construction cost (US$ m)	1850.5
Cost of CO_2 (US$)/ton	100
Cost of PM (US$)/ton	30,000
Cost of NO_x (US$)/ton	7,000
Cost of time (US$/h)	2
Cost of fatality (US$/person)	30,000
Cost of injury (US$/person)	5,000
Annual inflation (%)	0.08
Discount rate (%)	0.12
Annual maintenance (% of construction)	0.005

[a]Most are default values from the standard TEEMP MRT model provided at http://cleanairinitiative.org/portal/TEEMPTool

Considering traction and non-traction (station, commercial shops) electricity consumption, the carbon footprint increases to 1.5 million tons (CO_2 emissions). The construction emissions considering only the material consumption is equivalent to nearly 15 years of operation (traction and non-traction) emissions.

Considering only traction energy consumption and 20 years of duration, the analysis indicates that the project will result in the following savings (Figs. 5, 6 and 7):

1. CO_2 savings of 3300 tons/year/km
2. PM emissions savings of 1.2 tons/km/year
3. NO_x emissions savings of 18 tons/km/year
4. Avoided total fatal accidents of 2,900 in 20 years
5. 25,000 vehicle kilometre travel savings in 20 years.

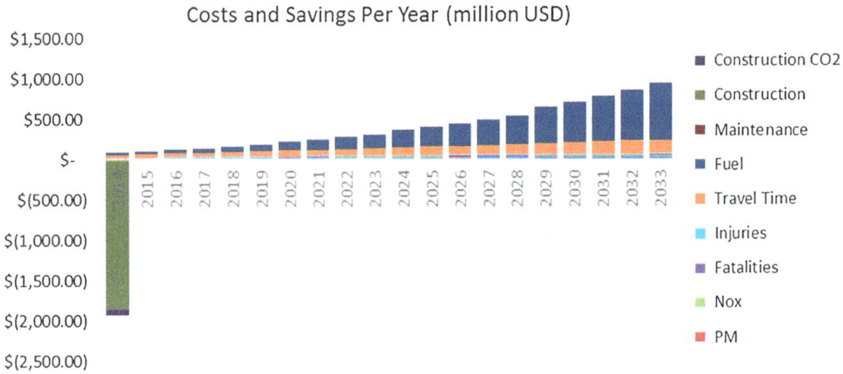

Fig. 5 Annual costs and savings for the Chennai Metro using TEEMP MRT

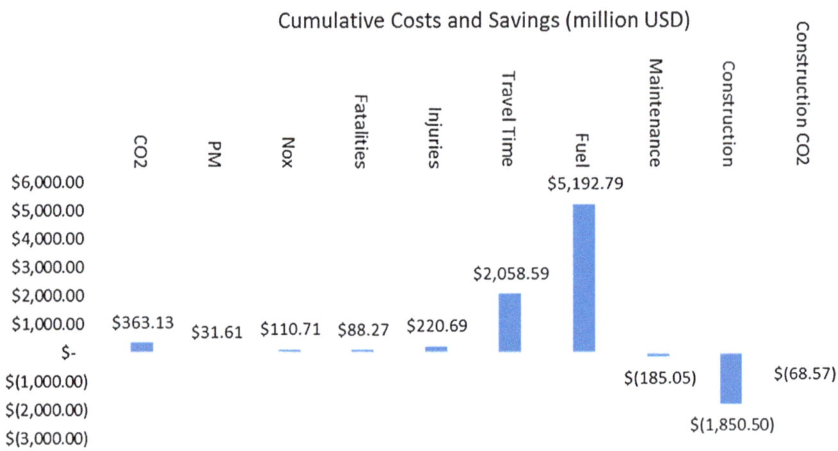

Fig. 6 Cumulative costs and savings for the Chennai Metro using TEEMP MRT

Fig. 7 Percentage share of monetized co-benefits from the Chennai Metro

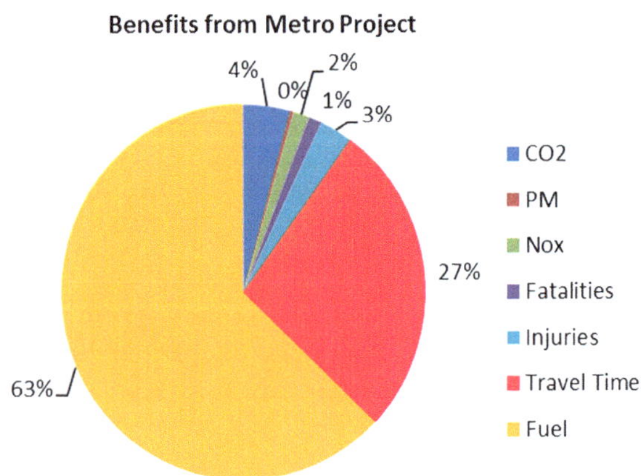

The vehicle-kilometre savings has a significant value associated with it. It is not only a key factor on emissions but also in terms of infrastructure. This aspect is highlighted later.

Traditional benefits of fuel and travel time constitute nearly 90% of savings in monetary terms. While considering cost for PM, NO_x and CO_2, the monetized co-benefits constitute only 10% of the savings. Clearly, the impact of co-benefits appears to be marginal. However, there are some important factors that need to be highlighted.

1. The quantum of savings derived from the project is huge. In 20 years of its life-cycle, the project saves nearly 3 million tons of CO_2 emissions. This is equivalent to 35% of total road transport CO_2 emissions in Chennai in 2011 (Guttikunda and Mohan 2014).
2. In terms of particulate matter emissions, the project saves nearly 1000 tons. This is equivalent to the removal of about 3000 Pre-Euro heavy duty trucks (25 tons) travelling for 10 years of the life-cycle for 60,000 km annually.
3. In terms of NO_x emissions, the project saves nearly 15,800 tons of NO_x emissions. This is equivalent to the removal of about 2000 Pre-Euro heavy duty trucks (25 tons) travelling for 10 years of the life-cycle.
4. The project economic analysis indicates an IRR of 12%. Removing co-benefits from analysis reduces IRR to 7%, reducing the feasibility of the project.
5. The project over its life-cycle reduces nearly 2,900 accident fatalities which is more than two years of Chennai's total accident fatalities.
6. Annual health benefits amount to US$2.8 million due to improved air quality using the aircount calculator (Abt Associates 2010). This online tool is provided by Abt Associates to help developing cities to calculate the health benefits of air quality improvements and the economic value of those benefits.
7. An estimated 13 avoided deaths annually because of improved air quality.
8. Improving accessibility to stations—Clean Air Asia's walkability surveys have shown that public transport facilities generally have poor accessibility, and many potential walkers would shift modes rather than walk in case of no improvement (Clean Air Asia 2011). In an interview survey with a sample size of 1,900 people, 62% indicated that they will shift to other modes (20% to cars and 22% to two-wheelers) if no improvement is carried out in future. In order to simulate this impact, it was considered that poor accessibility to stations would result in a conservative 5% reduction in proposed ridership (World Bank 2007).[2] This assumption is only to show the sensitivity of accessibility on ridership projections and can be conservative, as experience from the Delhi and Bangkok metros has shown that the majority of ingress and egress trips are by

[2]Revealed passenger behaviour indicates that people are generally willing to walk up to about 5 min to access a station, which indicates that they may typically walk up to a little over 400 metres at an average walking pace of 5 km/h. Reducing passengers' perception of the walk-in time by 5% could increase walk-in demand by 10% because it expands the area from which people perceive they are within their 5 min limit.

walking. This conservative assumption has a significant impact on the emissions savings (reduction of 12%). The project IRR becomes only 6% if co-benefits are neglected.

9. In many cases, metro actual ridership is only a fraction of the projected ridership. For example, for the Delhi metro, actual ridership was only 25% of projected ridership (Tiwari 2011). If the ridership decreases by 75%, the projects does not provide any reductions of carbon emissions and only marginal reductions in air pollutants. Clearly, not all expensive public transport projects are economically and financially feasible and provide co-benefits.
10. Consider a scenario where the authorities decide to include bikeways improvement along with this massive investment of US$2 billion metro project to increase co-benefits. Assuming that bikeways are constructed for 50 km, analysis by the TEEMP bikeways model indicates that for similar traffic characteristics, nearly 0.6 million tons of CO_2, 600 kg of PM and 5500 kg of NO_x could be additionally saved over project lifetime.
11. The cost of quantifying co-benefits is negligible. This case study was completed with the budget of 0.01% of cost of building 1 km of metro line. A more comprehensive health impact assessment would cost only 0.02% of total cost or 2–4% of project planning studies cost (Gota and Mejia 2013).

3.2 *Infrastructure Savings*

A key co-benefit missing in the discussions up to now is the savings accrued from the avoidance of new infrastructure required in the counterfactual baseline scenario to accommodate the projected increase in motorized transport demand in the absence of the metro project. Recognized methodologies for public transport CO_2 emission quantification tools such as those used in Clean Development Mechanism (CDM), Global Environment Facility (GEF), Japan International Cooperation Agency (JICA), Clean Technology Fund (CTF), and so on, emphasize mode shift impact calculations without clarifying how people would travel in the 'without project' scenario, with limited road space for baseline computations over the next 20 years. Currently, these aforementioned methodologies do not consider this aspect. Investments in a metro project, for example, prevents massive road expansion activities required to cater to the growing private vehicles required in case of the 'no metro' scenario. It has been calculated that the metro project would cumulatively save 25,000 million vehicle kilometres. In the absence of a metro system, that is, a city without an effective and prioritized public transport system, this increase in vehicular traffic would lead to slowing of traffic speeds, congestion and this would subsequently force cities to widen roads to cater for increased traffic

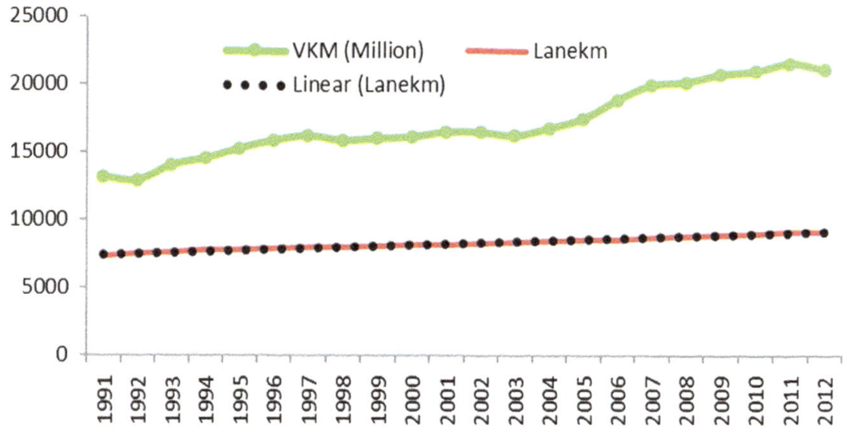

Fig. 8 Singapore infrastructure expansion with growing traffic movement

growth. This aspect is neglected currently in the baseline calculations as it is assumed that people will keep travelling the same way in the future irrespective of the infrastructure.

Since past road widening experience is not available from these cities, the data from the experience in Singapore is considered in order to make the most conservative calculation of this infrastructure requirement in the absence of a BRT/Metro system. Singapore does not follow the traditional approach of increasing road space with increasing congestion. From 1991 to 2012, vehicles have increased by 2.7% annually while vehicle kilometre travel has increased by 2.2% and road supply (lane kilometres) has only increased by 1% as per the statistics of Land Transport Authority-Singapore.[3] Roads currently take up 12% of total land area in Singapore. There is limited land supply, so road widening is only carried out when all other options fail. In Singapore, each lane-kilometre accommodates around 5,600 VKT/day/lane (Fig. 8). This ratio is applied to the project above and infrastructure requirements are estimated considering the lifetime vehicle travel reduction.

The Chennai Metro project helps prevent construction of a 21 lane roadway which may have been required in absence of the metro (the avoided road space by the metro is approximately 3 square metres per rider). Metro construction is often considered to be very expensive. However, comparing the construction costs of a metro with the alternate roadway expansion costs reveals that there is not much of a difference in terms of the costs between metro construction, and roadway expansion. However, considering the co-benefits involved, metro projects provide very high co-benefits. In the without metro scenario (road expansion), additional motor traffic is induced as public transport deteriorates and new roads with extra capacity

[3]http://www.lta.gov.sg/content/ltaweb/en.html.

attract increased use of vehicles. This leads to a vicious cycle of motorization. Avoiding this vicious cycle of induced motorized travel by high penetration of the public transport network would have exponential impacts on the quantification of co-benefits and these should not be missed out. If such impacts are not factored into the analysis, the Chennai Metro on an average provides a saving of 3,300 tons of CO_2/km/year. This should be a very conservative estimate as road expansion impacts on the baseline is not considered. The expansion of roads would result in increased induced traffic, thereby increasing the carbon footprint over the baseline. ADB (2010) has estimated that, in general, expressway projects increase CO_2 emissions by one-fifth to one-half or more over their 20-year lifetime compared with business-as-usual because of effects on induced travel that overwhelm the short-term benefits of curbing low-efficiency congested traffic. Clearly, the co-benefits of the metro projects would multiply, considering induced motorized travel increase in the 'without metro project' scenario.

4 Scaling up Benefits at National Level

India is urbanizing very fast and it has been estimated that India's urban population would exceed 550 million by 2030. It is also expected that the share of economic activity in urban areas would increase from 56% of GDP in 1990 to about 75% in 2020. Such a high impact of rising income in urban areas and urbanization on motorization is pronounced with experts suggesting an increase in vehicle ownership levels in the coming decades, with the growth nearly twice as fast as per capita income in India (Dargay et al. 2007). With more than 550 million urban people in 2030, India's urban sprawl would become the second largest urban system in the world (Assocham India 2011).

Recently, India submitted its Intended Nationally Determined Contribution (INDC) which communicated to the UNFCCC secretariat its country-level commitments and strategies to reduce carbon emissions and increase resilience for the post-2020 period. India committed to reduce the emissions intensity of its GDP by 33–35% by 2030 from the 2005 level (MOEF 2015). This target is founded on its earlier established voluntary goal of reducing the emissions intensity of its GDP by 20–25%, over 2005 levels, by 2020. In order to achieve this target, the transport sector was considered as a priority mitigation sector. Within the transport sector, urban transport mitigation measures are considered critical.

In order to provide a new direction and vision for the policy-makers and to increase the attention towards sustainable urban transport, the Government of India implemented the National Urban Transport Policy and Jawaharlal Nehru National Urban Renewal Mission (JNNURM) scheme. The JNNURM offers carrot and stick policy incentives where the national government offers transport (infrastructure and rolling stock) and asks city governments to reform the urban transport sector and take public transport improvement measures.

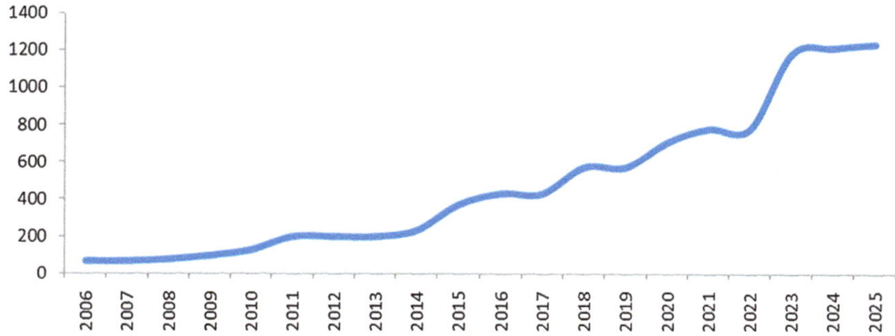

Fig. 9 Cumulative kilometres (Metro) at national scale

The majority of public transport investments could go towards BRTS and metro systems. Based on the evaluation of current city mobility plans, nearly 20 cities have planned to build a cumulative length of 1230 km at an estimated conservative cost of US$32 billion (MOUD 2014) (Fig. 9). By making appropriate assumptions for the implementation schedule based on discussions with experts and the comprehensive mobility plan of cities, the savings calculated for the Chennai Metro are extrapolated and scaled up to determine the national level impact of metro construction. This sketch analysis's basic purpose is to showcase the magnitude of co-benefits generated from such projects.

Between 2006 and 2025, the following cumulative co-benefits are estimated:

1. Reduction of 31 million tons of CO_2, that is, annual saving of 1.5 million tons of CO_2
2. Reduction of 11,000 tons of PM, that is, annual saving of 550 tons of PM
3. Reduction of 170,000 tons of NO_x, that is, annual savings of 8,500 tons of NO_x
4. Annual savings of 1,500 accident fatalities
5. Annual savings of 350 tons of black carbon
6. 127 avoided deaths annually because of improved air quality
7. Health savings of US$26 million/year.[4]

These estimates closely match with the values given by a more comprehensive study by the Akbar et al. (2014) which estimated the impact on co-benefits on scaling up the public transport system in the form of BRT which can be implemented at a fraction of metro costs. The World Bank estimated[5] the impacts of 1000 km of new BRT corridor deployed in 20 or more Indian cities:

1. 1100–1350 reduced traffic fatalities per year
2. 1.9–2.3 million tons/year of CO_2 emissions reduction
3. Annual savings of 300 tons of black carbon

---

[4]Using aircount calculator for Delhi air pollution—http://aircounts.com/aircounts.html.

[5]The author was part of the modelling team for the World Bank study on integration of short-lived climate pollutants in World Bank activities (World Bank 2013).

4. US$6.4–8.1 billion in macroeconomic benefits (over 20 years)
5. 50,000–90,000 short-term jobs rising to 128,000 permanent new jobs
6. More than 175 avoided deaths annually in India because of improved air quality
7. More than US$500,000 in annual avoided crop losses because of air pollution
8. 500 million h/year of time savings because of shortened trips.

Clearly, public transport projects generate significant magnitude of co-benefits that should not be neglected but prioritized in the decision-making process.

5 Challenges During Co-benefits Quantification

Sketch level analysis tools such as the Transport Emissions Evaluation Model for Projects give us the ability to get a glimpse of the potential impacts of sustainable transport projects on different aspects aside from financial costs and savings brought about by transport projects. The biggest challenge faced the conduct of the quantification was limited availability of data. Many data parameters were assumed in the absence of project specific data. Furthermore, there is lot of ambiguity and misconception with regard to the transport baseline. In case a particular type of project is not implemented, a suitable alternative, strategy or funding would be found by the authorities. However, this alternative strategy would depend on local historical road-widening experience and city characteristics. In the absence of this crucial element, experience from Singapore is considered in the avoided infrastructure quantification. Singapore, has already implemented several sustainable transport policies, that is, major emphasis is on curbing travel demand by road pricing, controlling ownership, better and effective public transport facilities. Thus, it could only be considered as a conservative example of road expansion within a city. In the actual scenario, the impact would be much higher.

6 Conclusion

The application of a holistic co-benefits analysis is clearly needed in building a business case for sustainable transport projects. Current methods and approaches in assessing investment alternatives are not adequate for capturing the wider range of costs and benefits of the different alternatives. The concept of co-benefits allows us to have a more comprehensive paradigm in analysing transport investments. Transport co-benefits such as emissions savings (climate pollutants and traditional air pollutants) may not necessarily form a big portion of the monetized savings, but can provide that needed increment to make projects more feasible. Moreover, a more holistic analysis of the costs of the 'baseline' scenario (e.g. accounting for the infrastructure requirements to accommodate the growth in private motorized travel in cities) will provide additional arguments for choosing alternatives that promote public transport and non-motorized transport. Broader benefits—which are more difficult to quantify and, moreover, harder to monetize—should also be taken into consideration, such as improved quality of life, equity and social cohesion. The call for the adoption of a co-benefits approach in the transport sector is merely a call for the adoption of a more holistic approach to assess and evaluate investments in the sector. It is a call for a smarter approach to move towards sustainability in the sector.

References

Abt Associates. (2010). *Aircounts*. http://aircounts.com/aircounts.html. Accessed on June 25, 2014.

ADB-Asian Development Bank. (2010). *Reducing carbon emission from transport projects*. http://www.adb.org/sites/default/files/EKB-REG-2010-16_0.pdf. Accessed on May 15, 2015.

ADB-Asian Development Bank. (2013). *2013 annual evaluation review*. http://www.adb.org/sites/default/files/2013-aer.pdf. Accessed on June 7, 2015.

Akbar, S., Kleiman, G., Menon, S., & Segafredo, L. (2014). Climate-smart development: Adding up the benefits of actions that help build prosperity, end poverty and combat climate change: Main report. The World Bank.

Assocham India. (2011). *Study on urbanizing India and mega metro network vision for the emerging cities of India–2030*. http://www.assocham.org/arb/general/Metro-Study_Update.pdf . Accessed on June 12, 2015.

CAI-Asia- Clean Air Initiative for Asian Cities. (2011). *Walkability in Indian cities*. http://cleanairinitiative.org/portal/sites/default/files/Walkability-India_SEP.pdf. Accessed on June 9, 2015.

Dargay, J., Gately, D., & Sommer, M. (2007). *Vehicle ownership and income growth, worldwide: 1960–2030*. http://www.econ.nyu.edu/dept/courses/gately/DGS_Vehicle%20Ownership_2007.pdf. Accessed on June 17, 2015.

Environment 360. (2014). *In New Delhi, A rough road for bus rapid transit systems*. http://e360.yale.edu/feature/in_new_delhi_a_rough_road_for_bus_rapid_transit_systems/2763/. Accessed on June 6, 2015.

Gota, S., & Mejia, A. (2013). *Assessing co-benefits from BRT Projects. A presentation delivered at the transportation and low emissions development strategies workshop in Manila, September 30, 2013*. http://esci-ksp.org/wp/wp-content/uploads/2014/04/SESSION-2-Sudhir-Gota-Assessing-Co-benefits-from-BRT-ProjectsR.pdf. Accessed on June 22, 2015.

Global Mass Transit. (2014). *Global mass transit*. http://www.globalmasstransit.net/index.php. Accessed on June 25, 2015.

Guttikunda, S., & Mohan, D. (2014). Re-fueling road transport for better air quality in India. *Energy Policy, 68*, 556–561.

IEG World Bank—Independent Evaluation Group World Bank. (2010). *Cost-benefit analysis of world bank projects*. http://ieg.worldbankgroup.org/Data/reports/chapters/cba_full_report1.pdf. Accessed on August 7, 2015.

IGES. (2011). *Mainstreaming transport co-benefits approach: A guide to evaluating transport projects*. http://pub.iges.or.jp/modules/envirolib/upload/3209/attach/transport%20co-benefits%20guideline.pdf. Accessed on July 2, 2015.

IGES. (2010). *Fact sheet no. 1 what are co-benefits?* http://pub.iges.or.jp/modules/envirolib/upload/3378/attach/acp_factsheet_1_what_co-benefits.pdf. Accessed on 11 July 2015.

Indian Roads Congress. (1993). *Manual on economic analysis of highway projects*. Special Publication 30 (Revised 2007). New Delhi: Indian Roads Congress.

IPCC. (2014). Intergovernmental panel on climate change. In Core Writing Team, R. K. Pachauri, & L. A. Meyer (Eds.), *Climate change 2014: Synthesis report. Contribution of working groups I, II and III to the fifth assessment report of the intergovernmental panel on climate change*. Geneva, Switzerland: IPCC.

Leather, J., & CAI-Asia. (2009). *Rethinking transport and climate change*. ADB sustainable development working paper series. No. 10. http://www.adb.org/sites/default/files/pub/2009/ADB-WP10-Rethinking-Transport-Climate-Change.pdf. Accessed on July 25, 2015.

Maizlish, N., Woodcock, J., Co, S., Ostro, B., Fanai, A., & Fairley, D. (2013). Health cobenefits and transportation-related reductions in greenhouse gas emissions in the San Francisco Bay Area. *American Journal of Public Health, 103*(4), 703–709. doi:10.2105/AJPH.2012.300939.

MoUD-Ministry of Urban Development. (2014). *Metro rail projects & RRTS—status*. http://moud.gov.in/sites/upload_files/moud/files/Status_of_Metro_Rail_Projects_&_RRTS.pdf, Metro Rail Projects & RRTS—Status, Ministry of Urban Development, Government of India. Accessed on July 17, 2015.

MoEF. (2010). *India: Greenhouse gas emissions 2007*. New Delhi: Indian Network for Climate Change Assessment, Ministry of Environment & Forests, Government of India, pp. 1–54.

MoEF. (2015). *India's intended nationally determined contribution: Working towards climate justice*. New Delhi: Ministry of Environment & Forests, Government of India.

Murty, M., Dhavala, K., Ghosh, M., & Singh, R. (2006). *Social cost-benefit analysis of Delhi Metro*. http://mpra.ub.uni-muenchen.de/1658/1/MPRA_paper_1658.pdf. Accessed on 3 August 2015.

SLoCaT-Partnership of Low Carbon Transport. (2014). *Development of a post-2015 results framework on sustainable transport*. http://slocat.net/sites/default/files/u10/draft_final_report_for_st_results_framework_july_25th_july.pdf. Accessed on August 11, 2015.

SLoCaT-Partnership of Low Carbon Transport. (2015). *SLoCaT analysis demonstrates growing body of greenhouse gas emission measurement methodologies for transport*. http://slocat.net/news/1452. Accessed on 9 August 2015.

Reynolds, C. (2010). Strategies to reduce transportation emissions in India: Identifying air quality andclimate co-benefits for the developing world. (Doctoral thesis, The University of British Columbia).Retrieved from https://open.library.ubc.ca/cIRcle/collections/ubctheses/24/items/1.0071479

Rogers, J. (2010). GEF STAP presentation: Assessment of GHG mitigation from land transport projects. https://www.thegef.org/gef/sites/thegef.org/files/documents/C.39.Inf_.16%20STAP%20-%20Manual%20for%20Calculating%20Greenhouse%20Gas%20Benefits_0.pdf. Accessed on 15 August 2015.

Schipper, L., & Marie-Lilliu, C. (1999).*Transportation and CO_2 emissions: Flexing the link—A Path for the World Bank*. http://www-wds.worldbank.org/servlet/WDSContentServer/WDSP/IB/2000/02/09/000094946_00012505400755/Rendered/PDF/multi_page.pdf. Accessed on August 21, 2015.

GEF-STAP-Scientific and Technical Advisory Panel of the Global Environment Facility. (2010). *Manual for calculating greenhouse gas benefits for global environmental facility transportation projects*. https://www.thegef.org/sites/default/files/publications/GEF_CalculatingGHGbenefits_webCD_1.pdf. Accessed on August 12, 2015.

Gota, S. (2009). *Media study of Ahmadabad bus rapid transit system*. http://cai-asia.blogspot.in/2009/07/media-study-of-ahmedabad-brts.html. Accessed on July 21, 2015.

Tiwari, G. (2011). *Metro systems in India: Case study DMRC*. In A presentation delivered at the Promoting Low Carbon Transport in India National Workshop, October 19–20, 2011. Delhi. Available at http://www.unep.org/transport/lowcarbon/publication/pdf/presentations/2011/Oct%2020/Session%20V/Metro_Systems_in_India_Case_Study_DMRC_Tiwari.pdf. Accessed on September 2, 2015.

U.S. Department of Treasury and the Council of Economic Advisers. (2010). *An economic analysis of infrastructure investment.* http://www.whitehouse.gov/sites/default/files/infrastructure_investment_report.pdf. Accessed on September 7, 2015.

United States Environmental Protection Agency. (2005). Integrated environmental strategies (IES) program. In *International conference on atmosphere protection. CGE Training Workshop on mitigation assessments*, Seoul, Korea. http://pub.iges.or.jp/modules/envirolib/upload/3378/attach/acp_factsheet_1_what_co-benefits.pdf. Accessed on June 25, 2015.

Williams, C., Hasanbeigi, A., Price, L., & Wu, G. (2012). *International experience with quantifying the co-benefits of energy efficiency and greenhouse gas mitigation programs and policies.* https://eaei.lbl.gov/sites/all/files/LBL_5924E_Co-benefits.Sep_.2012.pdf.

World Bank. (2007). *Strategic Urban Transport Policy Directions for Bangkok.* Available at http://siteresources.worldbank.org/INTTHAILAND/Resources/333200-1177475763598/2007june_bkk-urban-transport-directions.pdf.

World Bank. (2013). *Integration of Short-Lived Climate Pollutants in World Bank Activities.* Available at http://www-wds.worldbank.org/external/default/WDSContentServer/WDSP/IB/2013/08/19/000333037_20130819113818/Rendered/PDF/804810WP0G80Re00Box0379805B00OUO090.pdf.

Chapter 7
Co-benefits from Buildings and Construction

A. Narender

Abstract This chapter assesses the role of the building and construction sector on climate and the environment and the impact of climate friendly policies in achieving co-benefits in Indian cities. The urban climate co-benefits approach may be defined as implementation of climate friendly policies in cities by tackling both global and local environmental problems together while simultaneously contributing to solutions for local development needs. The building and construction sector plays a significant role in creating employment and income and at the same time contributes significantly to greenhouse gas emissions. Hence it offers opportunities for climate co-benefits to reduce GHG emissions, mitigate climate change impact and promote economic growth and employment. The slums and informal housing sector require special attention as it is more vulnerable to climate change impact. The chapter has reviewed Indian policies in the building sector which include the Energy Conservation Act 2001, Energy Conservation Building Code 2007, National Building Code, National Mission for Sustainable Habitat, rating systems for green buildings and so on. The chapter provides a brief assessment of the impact of these policies and makes suggestions for strengthening the same.

Keywords Buildings and construction · GHG emissions
Climate Co-benefits · Informal settlements · Energy conservation
GRIHA · LEED

A. Narender (✉)
Consultant to Various Organizations, Hyderabad, India
e-mail: anarender10@gmail.com

M. Sethi and J.A. Puppim de Oliveira (eds.), *Mainstreaming Climate Co-Benefits in Indian Cities*, Exploring Urban Change in South Asia,
https://doi.org/10.1007/978-981-10-5816-5_7

1 Introduction

Urban areas are the engines of economic growth and productivity. While proactive urban development policies and projects have the potential to promote economic growth, they could be accompanied by climate change impacts and environmental decline due to energy and resource-intensive strategies. This in turn could negate the developmental benefits. Hence it is increasingly being realized that the development policies and projects in urban areas could be formulated to align with developmental, climate change and environmental objectives to ensure better environmental outcomes, mitigation and adaptation of climate change impacts along with growth and development. This calls for a shift in policy formulation and implementation from energy and resource-intensive consumption approach to conservation approach. The development trajectory adopted earlier that led to generation of greenhouse gases could be modified to pursue a low carbon pathway having both developmental and climate change benefits. This new approach is termed as a 'co-benefits approach' with simultaneous and multiple benefits to climate change, environment and development. There is a need to evolve urban development and sectoral policies such as water supply, transport and waste management to promote climate co-benefits along with developmental objectives.

Urbanization in India is increasing over the years resulting in increased building and construction activity to meet the growing demand. This has led to intensive use of energy and building materials which cause high carbon emissions. It is possible for Indian cities to adopt climate friendly policies simultaneously meeting the demand in the buildings and construction sector. This research assesses the role of the building and construction sector on climate and environment and the potential of climate friendly policies in achieving co-benefits in Indian cities. Accordingly, this chapter first presents the theoretical or conceptual relationship between urban development and climate change co-benefits (distinguishing the co-benefits approach from the conventional one), assesses the relationship between building sector and greenhouse gas emission, reviews the building sector policies having the potential for climate co-benefits, assesses the impact of these policies and makes suggestions for strengthening the building sector policy framework with climate co-benefits in Indian cities.

2 Urban Development and Climate Co-benefits Framework

The urban climate co-benefits approach may be defined as the implementation of climate friendly policies in cities by tackling both global and local environmental problems together, while simultaneously contributing to solutions for local development needs (Puppim de Oliveira et al. 2013a: 7). The co-benefits approach enables the implementation of local policies that can achieve multiple benefits at the local level while facilitating the reduction of greenhouse gas emissions at the global level. The co-benefits policies are considered to be more effective in cities. The co-benefits in cities may be realized through pro-climate policies such as composting of municipal solid waste, an efficient public transport system and efficient design of buildings. Designing and implementing policies in these sectors in Indian cities with a climate co-benefits approach can achieve significant climate change, environmental and developmental outcomes at the global and local level. A co-benefits approach to city development enables achievement of development goals along with improved environment and quality of life. Based on Puppim de Oliveira et al. (2013b), a framework of climate co-benefits approach in cities is presented in Fig. 1.

The climate co-benefits approach is not conflicting with the city development approach. On the contrary, it contributes to improved efficiency and performance of various sectors. The co-benefits approach can be adopted in designing and implementing sectoral policies related to buildings, energy, transportation, water supply and waste sectors. Government policy and financing at the national, state and local levels could effectively promote such an approach. Incentives could be provided to projects and initiatives with co-benefits. Budgets from different tiers of government could be combined to promote co-benefit initiatives. Private sector financing and citizen participation could be incentivized to support such initiatives (Puppim de Oliveira et al. 2013a: 28). Bilateral and multilateral agencies could encourage co-benefit sectoral policies by making financial support conditional for implementation of such policies. International city networks could promote sharing of information and city to city cooperation. Dedicated institutional mechanisms and capacity could be introduced at the city level. A key requirement for promoting a co-benefits approach is to measure, assess and quantify such benefits, which is not

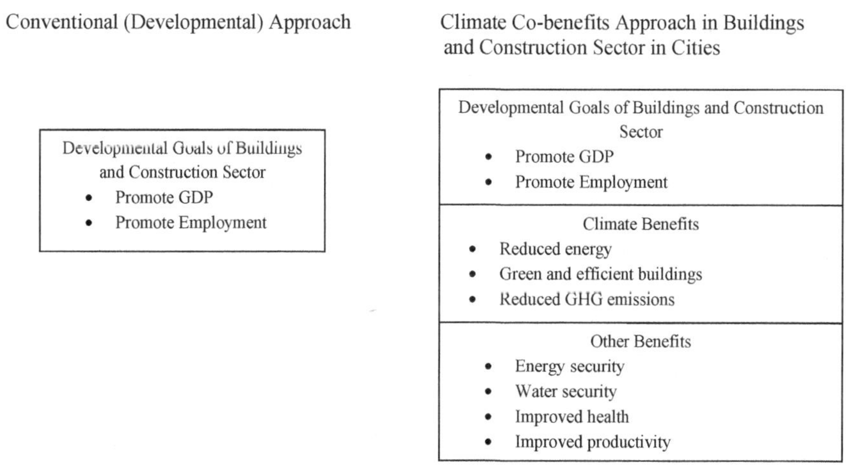

Fig. 1 Conceptual relationship between conventional (developmental) approach and climate co-benefits approach in buildings and construction sector in cities

an easy task (Puppim de Oliveira et al. 2013a: 33). Cities in developed countries have implemented several sectoral policies with a co-benefits approach and benefitted from the same. The Tokyo Metropolitan Government is stated to be the first city government in the world to seek to cap emissions from buildings while Shanghai has brought out energy efficiency improvements in the building sector (Puppim de Oliveira et al. 2013a: 10 and 17).

The climate co-benefits approach in cities, including in the buildings and construction sector, can contribute to the achievement of sustainable development goals adopted by the world leaders at the United Nations Sustainable Development Summit on 25 September 2015 as part of the 2030 Agenda for Sustainable Development (United Nations 2015a). The 2030 agenda for sustainable development includes a set of 17 Sustainable Development Goals (SDGs) to eradicate poverty, inequality and injustice and address climate change issues by 2030. Goal 13 focuses on taking urgent action to combat climate change and its impacts. Another key goal among SDGs is Goal 11 which aims to make cities and human settlements inclusive, safe, resilient and sustainable. The climate co-benefits approach in cities can directly contribute to Goal 11 and more particularly to the following targets:

> By 2020, substantially increase the number of cities and human settlements adopting and implementing integrated policies and plans towards inclusion, resource efficiency, mitigation and adaptation to climate change, resilience to disasters, and develop and implement in line with Sendai Framework for Disaster Risk Reduction 2015–30, holistic disaster risk management at all levels.
>
> Support least developed countries, including through financial and technical assistance, in building sustainable and resilient buildings utilizing local material (United Nations 2015a).

The United Nations Framework Convention on Climate Change (UNFCCC) Conference of the Parties Agreement (COP), also known as Paris Agreement 2015, has welcomed the adoption of the 2030 agenda on sustainable development by the United Nations General Assembly, particularly Goal 13 on taking urgent action to combat climate change and its impacts. The COP 2015 required the member countries to communicate ‘intended nationally determined contributions’ (INDCs) towards achieving the objective of the Convention which is to strengthen global response to climate change in the context of sustainable development and eradication of poverty (United Nations 2015b).

The Government of India has communicated its INDCs to UNFCCC COP 2015 which include, among others, the following (MoEF 2015):

- To reduce the emission intensity of GDP by 33–35% by 2030 from 2005 level
- To put forward and further propagate a healthy and sustainable way of living based on traditions and values of conservation and moderation
- To adopt a climate friendly and cleaner path than the one followed hitherto by others at a corresponding level of economic development.

The climate co-benefit framework for the building and construction sector in cities in India can significantly contribute to the realization of the above INDCs.

3 Contribution of Building Sector to GHG Emissions

The Inter Governmental Panel for Climate Change (IPCC) reports in 2007 and 2014 have calculated the global GHG emissions by sectors for the years 2004 and 2010 respectively and found that the energy sector is the largest contributor, followed by industry. The energy sector cuts across all other sectors and the cumulative contribution of energy from all the sectors is estimated to be around 50%. Table 1 presents the sectoral contributions to GHG.

The IPCC 2014 Working Group 3 report has made the following observations with regard to building sector contribution to greenhouse gas emissions (IPCC 2014, Working Group 3: 18, 23–24).

1. GHG emissions are projected to grow in all sectors except for net CO_2 emissions in AFOLU sector
2. Infrastructure developments and long-lived products that lock societies into GHG intensive emissions pathways may be difficult or very costly challenge, reinforcing the importance of early action for ambitious mitigation
3. Energy supply sector emissions are expected to continue to be a major source of GHG emissions, ultimately accounting for significant increases in indirect emissions from electricity use in buildings and industry sectors
4. The building sector accounted for 32% of final direct and indirect energy use in 2010 and the energy demand is projected to double, while CO_2 emissions are projected to increase by 50–150% by 2050
5. Recent advances in technologies, know-how and policies provide opportunities to stabilize or reduce energy demand in the building sector by 2050
6. Adoption of very low energy building codes for new buildings and retrofitting of existing buildings constitute the key strategy for the building sector

Table 1 Global GHG emissions by sector, 2004 and 2010

Sector	Contribution to GHG 2004 (in %)	Contribution to GHG 2010 (in %)	Distribution of GHG from electricity by sectors (2010)
Transport	13.1	14.0	0.3
Energy supply	25.9	34.6	1.4
Industry	19.4	21.0	11.0
Residential and commercial buildings	7.9	6.4	12.0
Agriculture, forestry and other land uses	30.9	24.0	0.87
Waste and waste water	2.8	–	–
Total	100.0	100.0	25.0

Source Data feeds from IPCC (2007: 5) and IPCC (2014, Working Group Report 3: 8)

7. Changes in life style, culture and behaviour could significantly reduce energy demand in buildings by up to 20% in the short term and by up to 50% in the medium term
8. Integrating traditional life styles and design into building practices and architecture could reduce energy demand
9. Co-benefits from a GHG emission reduction strategy in the building sector include energy cost savings, improved energy security, health, environmental outcomes, work place productivity, fuel poverty reductions and net employment gains
10. Strong barriers such as split incentives as in the case of owners and tenants, fragmented markets and inadequate access to information and financing hinder market-based uptake of cost effective opportunities and need to be overcome by policy interventions addressing all stages of the building and appliance life cycle
11. Well-designed and implemented building codes and appliance standards would be among the most environmentally and cost-effective instruments for emission reductions in the building sector.

In India, the extent of GHG emissions by sector are calculated by the Indian Network for Climate Change Assessment (INCCA) report prepared for the Ministry of Environment and Forests. According to this report, GHG emissions from the energy, industry, agriculture, and waste sectors constituted 58, 22, 17 and 3% of the net carbon emissions respectively. Following the global trends, the electricity sector contributes nearly 38% of GHG emissions followed by industry at 22%, agriculture at 18%, transport and residential sectors at 7% each and waste at 3% (MoEF 2010: ii). Thus, it can be seen that around 82% of GHG emissions are from non-agricultural sectors which have a link with urban areas. Table 2 presents the sectoral contributions to GHG emissions in India.

Table 2 A comparison of GHG emissions in million tonnes of carbon emissions by sector in India between 1994 and 2007

Sector	1994	2007
Electricity	28.4	37.8
Transport	6.4	7.5
Residential	6.3	7.2
Other energy	6.3	5.3
Cement	4.9	6.8
Iron and steel	7.2	6.2
Other industry	10.0	8.7
Agriculture	27.6	17.6
Waste	1.9	3.0
Total	100	100

Source Ministry of Environment and Forests (2010: ii)

Table 3 GHG emissions from energy sector (2007) in India

Sub-sector of energy	Share of GHG (in %)
Electricity	65.4
Transport	12.9
Residential/commercial	12.6
Petroleum refining/solid fuels	3.1
Agriculture/fisheries	3.1
Fugitive emissions	2.9
Total emissions from energy	100 (1100.06 tonnes)

Source Ministry of Environment and Forests (2010: iv)

Table 3 throws light on sub-sectoral contributions to GHG emissions within the electricity sector and it can be seen from the table that around 65% of GHG emissions is contributed by the electricity sector itself, followed by transport sector and residential/commercials sectors which contribute around 13% each.

4 Opportunities for Climate Co-benefits in the Buildings and Construction Sector

The buildings and construction sector plays a key role in promoting national income and employment generation. It is estimated that the buildings and construction sector contributes around 5–10% of employment and 5–15% of the gross domestic product (GDP) of a country (UNEP 2007: 1). In India, the construction sector has accounted for 8.2% of national GDP employing 41 million people in 2011–2012 and is expected to become the third largest construction sector in the world by 2018 (GIZ undated: 7). Unlike agriculture, the share of the buildings and construction sector to employment and GDP is expected to increase in the coming years. However, the key problems associated with the growth of this sector are increasing energy consumption and greenhouse gas emissions due to high energy and demand. India is experiencing huge energy shortages which are likely to go up in future. Coal is the major source of supply and bears a high potential for generation of greenhouse gas emissions. The Integrated Energy Policy 2006 of India indicated that the power generation capacity must increase to 800,000 MW by 2031–32 from the current level of 160,000 MW. To meet additional capacity, the policy has suggested to pursue available fuel options and forms of energy including conventional and non-conventional sources, an expanding energy resource base, seeking new and emerging energy sources, pursuing technologies that maximize energy efficiency, demand-side management and conservation. The policy has suggested energy efficiency in all sectors as one of the key strategies to reduce greenhouse gas emissions in the economy (Planning Commission 2006: xiii and xxix). The building sector can play a key role in promoting energy efficient technologies and energy conservation.

The share of the building sector to total commercial energy consumption in India has increased from 14% in the 1970s to approximately 33% by 2005 with an increase of 8% per year (UNEP 2010: 6). Hence a key challenge is to take advantage of the potential of the buildings and construction sector for increased employment and GDP at the same time as mitigating the impact of greenhouse gas emissions by reducing energy consumption. The sector has a large presence in urban areas and policies with climate co-benefits could contribute to this objective by primarily targeting energy consumption at all stages. The energy consumption in the building sector takes place over the life-cycle of a building covering the following phases (UNEP 2007: 3).

1. Manufacturing of building products and components
2. Transportation of building products and components to construction site
3. Construction of building
4. Operational phase of building or time for which building is put in use for social and economic purposes
5. Demolition and recycling of the building at the end of its operational phase or when it becomes unfit for use (dilapidation condition).

The key phase in the life-cycle of a building is the operational phase, which consumes nearly 80% of energy for purposes such as lighting, cooling, heating and so on. The electrical appliances used in the operational phase also consume significant energy. Hence energy conservation during the operational phase is assigned highest priority. Both energy conservation and efficiency policies as well as design of green buildings focus on the operational phase. The building materials also have the potential for energy conservation. The conventional building products and components are mainly cement, steel, bricks and lime and they all consume significant energy. Use of traditional and green building products such as those recycled from waste could reduce consumption of energy. The COP 15 has presented the following six key messages and four priorities for the buildings and construction sector for reduction of GHG emissions (UNEP 2009: 4):

1. The building sector has the most potential for delivering significant and cost effective GHG emission reductions
2. Countries will not meet emission reduction targets without supporting energy efficiency gains in the building sector
3. Proven policies, technologies and knowledge exist to deliver deep cuts in building related GHG emissions
4. The building industry is committed to action and in many countries is already playing a key role
5. Significant co-benefits including employment will be created by policies that encourage energy efficient and low emission building activity
6. Failure to encourage energy efficient and emission reduction measures in new buildings and retrofitting old buildings will lock them into poor performance for decades.

Four priorities of COP 15 are:

1. Prioritize the building sector as key to meeting national GHG emission reduction targets
2. Energy efficiency and GHG emission reduction programmes in the building sector must be recognized as National Appropriate Mitigation Action (NAMA)
3. A Clean Development Mechanism (CDM) must be reformed to support investment in energy efficient building programmes in developing countries
4. Develop base lines for building-related GHG emissions using a consistent international approach to performance monitoring and reporting.

The buildings and construction sector also offers opportunities for climate co-benefits in the areas of water conservation and waste management in addition to energy conservation. India has pursued policies in promoting energy conservation in buildings as well as green buildings. Green buildings include many other emission reduction initiatives such as rain water harvesting, water conservation, composting and waste water recycling in addition to energy conservation.

5 Informal Settlements, Slums and Climate Co-benefits

The above section has presented the opportunities of climate co-benefits in the building sector which relate to the formal housing sector. However, slums and informal housing settlements in cities, particularly in the developing world, experience a different set of issues and merit special attention. According to the UN Habitat Global Report on Human Settlements 2011 that focused on the theme of Cities and Climate Change, 32.7% of the global population in 2010 lived in slums, which are very vulnerable to climate change due to structures of substandard quality, lack of basic services, overcrowding, social exclusion, location on hazardous areas with steep slope in flood plains, lack of adequate drainage system and so on (UN Habitat 2011: 87).

According to a guidance document on climate change adaptation in cities by the World Bank, one billion people across the world live in informal settlements, lacking access to improved water and sanitation, sufficient living area, structural quality and durability of dwelling and security of tenure and are constantly exposed to the risks of climate-related disasters such as storms, floods, landslides, heat waves and droughts (World Bank 2011: 13). The majority of the urban poor in Indian cities live in slums and informal housing settlements that are characterized by poor access to basic services, low quality of housing structure, inadequate space and overcrowding within the house. The population living in slums in urban India has gone up from 52.37 million in 2001 to 65.49 million in 2011 (Narender 2013: 2). The formal building sector policies have limited relevance to the slums. Households living in slums are more vulnerable to climate change impacts as compared to those living in formal houses. The UN Habitat report has identified the following public health impact of climate change (UN Habitat 2011: 78–79).

1. Increased incidence of heat stress and heat related mortality
2. Significant increases in respiratory related diseases and hospitalization
3. Death, injury and property loss due to heavy storms and flooding
4. Disruption of health services during storms and flooding due to power outages and water shortages
5. Increased incidence of contamination diseases due to water contamination during disasters
6. Post-disaster illnesses such as post-traumatic disaster, anxiety, stress, grief and depression.

The UN noted that health impacts, both immediate and long term, tend to hit the poorest urban residents the hardest, in part because they often lack mobility, resources and insurance, occupy the highest-risk areas of cities and have low access to assets (UN Habitat 2011: 79). According to a report by the World Health Organization, a key issue in developing countries is the rapid growth of unregulated peri-urban slums and informal urban dwellings, where building codes are rarely enforced, leaving residents without safe shelter, adequate utility infrastructure, or access to public transport services (WHO 2011: 97). The report noted that housing and the built environment has a profound impact on human health where healthy housing can significantly decrease communicable and non-communicable disease risks (WHO 2011: 9). The report observed that high prevalence of pneumonia, diarrhoea, malaria, measles, TB and other major infectious diseases is often due to poor living conditions rather than income levels per se (WHO 2011: 38). Slums and informal settlements contribute to climate change due to low-density and poor use of urban space which exacerbate the urban heat island effect and make infrastructure delivery more energy-intensive and costly. The report suggested to implement low-carbon, low-cost and health co-benefit 'packages' such as promoting the use of solar and renewable energy in slums and informal settlements which can yield immediate benefits to slum areas (WHO 2011: 116). The WHO report has made the following recommendation to achieve climate co-benefits in slums and informal housing sector (WHO 2011: 85):

1. Redevelopment of housing for the poor by adopting low-carbon designs that are resilient to extreme heat, cold, rain, storms and drought and low-carbon household energy solutions that improve indoor and outdoor air quality
2. Providing equitable access to low-carbon public transport and walking and cycling networks that generate multiple health benefits
3. Providing equitable access to safe drinking water, improved sanitation and waste disposal through grid expansion or small community water harvesting systems
4. Early provision of services in the housing development phase by clustering housing around services to promote more 'compact' urban development

5. Adopting solar energy solutions including solar photovoltaic electricity, lights and appliances and solar water heating
6. Ensuring better house screening, natural ventilation and household water management.

6 Building Sector Policy with Climate Co-benefits in India

6.1 Energy Conservation Act 2001 and Bureau of Energy Efficiency

The Government of India has formulated the Energy Conservation Act 2001 (ECA 2001) to provide for efficient use of energy and its conservation. The key provisions include creation of the Bureau of Energy Efficiency (BEE), formulation of energy conservation building code, energy audit, labelling of equipment and formulation of energy consumption standards. The ECA 2001 has defined the powers of both central and state governments to facilitate and enforce efficient use of energy and its conservation. It has enable the state governments to create a State Energy Conservation Fund for the purposes of promoting the efficient use of energy and its conservation within the states. The ECA 2001 has envisaged promotion of energy efficiency by assigning a wide variety of functions to the BEE and central government.

The powers of the central government under the ECA 2001 include the following:

1. Specify the norms and standards for processes, equipment and appliances and for designated consumers
2. Display of labels on equipment as per regulations
3. Direct energy intensive industries or any designated energy consumers to get conducted energy audits by accredited agencies
4. Direct any designated consumer to furnish information, appoint energy managers and prepare a scheme for energy efficiency as may be necessary
5. Prescribe energy conservation building codes and amend as required
6. Promote awareness, disseminate information and undertake training
7. Encourage preferential treatment for use of energy efficient equipment or appliances.

The powers of state government under the ECA 2001 include the following:

1. Amend the energy conservation building codes by notification and in consultation with BEE
2. Direct every owner or occupier of a building to comply with the energy conservation building codes

3. Any designated energy consumers to get energy audits conducted as per regulations
4. Designate any agency to coordinate, regulate and enforce provisions of this ECA 2001 within the state
5. Direct any designated consumer to furnish information, appoint energy mangers and prepare a scheme for energy efficiency as may be necessary
6. Promote awareness, disseminate information and undertake training
7. Encourage preferential treatment for use of energy efficient equipment or appliances
8. Constitute a State Energy Conservation Fund for promoting energy efficiency within the state
9. The designated agency may appoint as many inspection officers after expiry of five years from the commencement of the ECA 2001 for the purpose of carrying out field inspections, checking equipment, verifying data, etc.

The powers of the BEE as envisaged by the ECA 2001 include, among others, the following:

1. Make recommendations to the central government with regard to norms for processes and energy consumption standards and particulars to be displayed on the labels of equipment and appliances
2. Take suitable steps to prescribe guidelines for energy conservation building codes
3. Take all measures related to awareness creation and dissemination of information on energy efficiency and conservation
4. Undertake training, research and development and promote consultancy services in the field of energy conservation
5. Develop and promote testing, certification procedures and facilities
6. Promote pilot projects, use of energy efficient processes equipment, devices and systems and innovative financing of energy efficiency projects.

The Government of India set up BEE in 2002 under the ECA 2001 with the mission to assist in developing policies and strategies with a thrust on self-regulation and marketing principles within the framework of the ECA 2001 to achieve the objective of reducing energy intensity in the Indian economy. Thus BEE is envisaged as an advisory agency to the central and state government as well as an implementing agency. The key sectors that are the focus of BEE include lighting, building, appliances, agriculture, municipality, industries and small and medium enterprises. The key policy areas include an energy conservation building code and a national mission for enhancing energy efficiency. To this end, the BEE has facilitated the drafting of an Energy Conservation Building Code (ECBC) in India and formulated a draft outline of a national mission for enhancing energy efficiency. The BEE has promoted the labelling of appliances across the sectors and initiated several awareness campaigns on energy conservation.

6.2 Energy Conservation Building Code (ECBC)

As discussed in the previous section, formulating energy conservation building codes is a key provision of the ECA 2001. These entail norms and standards of energy consumption expressed in terms of per square metre of the area of a building in accordance with its geographic location. The first ECBC was formulated in 2007 by the Ministry of Power and BEE and the code covered building aesthetics, the building envelope, mechanical systems and equipment including the heating, ventilating and air conditioning (HVAC) system, interior and exterior lighting system, service hot water, electrical power and motors including thermal comfort in non-centrally air conditioned/heated buildings. The purpose of the code is to provide minimum requirements for the energy-efficient design and construction of buildings. The code is made applicable to buildings or building complexes that have a connected load of 500 kW or greater or a contract demand of 600 kVA or greater. According to the ECBC, buildings or building complexes having a conditioned area of 1000 sq. m or more will fall under this category. The code would become mandatory as and when it is notified by the central government or state government in the Official Gazette under clause (p) of Section 14 or clause (a) of Section 15 of the ECA 2001. To begin with, the compliance with the provisions of ECBC is made voluntary. According to the National Mission on Sustainable Habitat, mandatory application of ECBC could help reduce energy consumption by about 1.7 billion units of electricity per year (MoUD 2008: 15).

6.3 National Mission for Enhanced Energy Efficiency

The Government of India formulated the National Action Plan on Climate Change (NAPCC) in 2008 with the objective of promoting sustainable development and reducing GHG emissions by devising efficient and cost effective strategies, deploying appropriate technologies and engineering new and innovative forms of market, regulatory and voluntary mechanisms (NAPCC 2008). The NAPCC has suggested setting up eight national missions, namely the National Solar Mission, National Mission for Enhanced Energy Efficiency (NMEEE), National Mission on Sustainable Habitat, National Water Mission, National Mission for Sustaining Himalayan Eco-system, National Mission for Green India, National Mission for Sustainable Agriculture and National Mission for Strategic Knowledge for Climate Change. The Ministry of Power and Bureau of Energy Efficiency (2008) were assigned the task of preparing and implementing NMEEE and a Steering Committee was constituted for the purpose. The Steering Committee prepared a draft outline for NMEE which recommended the introduction of market-based mechanisms for energy efficiency, promoting innovative measures to make energy-efficient products more affordable, evolving mechanisms for financing of demand side management programmes in all sectors and developing fiscal instruments for promoting energy efficiency. All these measures tend to have a significant effect on promoting energy efficiency in the construction and building sectors.

6.4 National Mission for Sustainable Habitat

The National Mission for Sustainable Habitat (NMSH) was formulated by the Ministry of Urban Development (MoUD) in 2008 under the National Action Plan for Climate Change. The NMSH has articulated various strategies for adaptation and mitigation of climate change and sustainable habitat covering the following aspects:

1. Extension of the energy conservation building code
2. Better urban planning
3. Urban transport management and modal shift to public transport
4. Recycling of material and urban waste management including solid waste and storm water
5. Better water supply management including water resource and drinking water
6. Promotion of energy efficiency in residential and commercial sectors
7. Energy efficient and green buildings.

The NMSH has emphasized a great deal on the strategies related to the building and construction sector. The report highlighted the role of the National Building Code 2005 (NBC 2005), preparing the ECBC and introducing a rating system for green buildings as part of the key energy efficiency initiatives in the building sector. The report highlighted the Centre for Environmental Sciences and Engineering Building in IIT, Kanpur; India's first platinum-rated green building in Hyderabad; the Eco-housing Programme of Municipal Corporation of Greater Mumbai; and Bio-diversity Conservation India Limited (BCIL) eco-housing projects in Bangalore, as emerging good practices in Indian cities.

6.5 National Urban Housing and Habitat Policy

The National Urban Housing and Habitat Policy (2007) was formulated with the goal of providing 'affordable housing for all', with special emphasis on vulnerable sections of society such as scheduled castes/scheduled tribes, backward classes, minorities and the urban poor. The emphasis was on provision of social housing for the EWS/LIG categories so that they are fully incorporated into the mainstream of ecologically well-balanced urban development. The ultimate goal of this policy was to ensure sustainable development of all urban settlements, providing basic civic amenities to enable better quality of life for all urban citizens. The policy suggested several measures including green buildings, energy conservation, encouraging the use of renewable energy resources, promoting research and development in energy conservation practices in housing sector, promoting low energy intensive and renewable energy-based construction techniques. It also recommended for suitable provisions in building by-laws for innovative and renewable energy practices, particularly solar water heating systems in residential and commercial buildings.

6.6 Policies for Improvement and Redevelopment of Slums and Informal Settlements

According to the National Housing and Habitat Policy 2007, it is of critical significance that the housing stock in slums is improved through urban renewal, in situ slum improvement and development of new housing stock accompanied with high quality provision of basic services. The Government of India, through the Ministry of Urban Development (MoUD) and the Ministry of Housing and Urban Poverty Alleviation (MoHUPA), has implemented various policies for slum improvement in the past few decades. These include Environmental Improvement of Urban Slums (EIUS), Urban Basic Services to the Poor (UBSP), National Slum Development Programme (NSDP), Basic Services to the Poor Programme (BSUP), Jawaharlal Nehru National Urban Renewal Mission Programme (JNNURM), Integrated Housing and Slum Development Programme (IHSDP) and the Rajiv Awas Yojana Programme (RAY) (Narender 2013: 9). An Agenda towards making Indian cities slum free is emerging and this has the potential to reduce climate change impact and promote climate co-benefits. The following strategies are articulated towards slum free cities (Narender 2013: 24–25):

1. Conceptualizing and implementing schemes for slum redevelopment and rehabilitation
2. Allocating land for housing the urban poor
3. Promoting rental housing for the urban poor
4. Providing institutional credit and interest subsidies to the poor for affordable housing
5. Encouraging beneficiary contribution and participation
6. Promoting low cost housing and building materials.

6.7 National Building Code (NBC)

The National Building Code (NBC) was first published by the Bureau of Indian Standards (BIS) in 1970. It was first revised in 1983 followed by a second revision in 2005. The NBC was prepared to unify the building regulations throughout the country for use by government departments, municipal bodies and other construction agencies. While the 1983 revision brought in provisions related to development control rules, green belts, landscaping, special provisions for low income housing, fire safety regulations for high rise buildings, earthquake resistant code, and so on, the NBC 2005 revision included latest knowledge and innovations in building construction and special attention was given to energy conservation and sustainable development. The NBC mandated for planning, design and installation of electrical installations, air-conditioning and heating in accordance with part VIII on building services of the document. The key provisions of the NBC for promoting energy efficiency and conservation include (MoUD 2008: 15):

1. Use of pozzolanas (such as fly-ash, rice husk ash, metakaoline, silica fume, ground granulated blast furnace slag etc.) in concrete production
2. Day light integration (indoor lighting levels to be met via day lighting)
3. Artificial lighting requirements (levels) for indoor spaces
4. Ventilation standards (natural and mechanical) for optimal human health and well-being
5. Electrical standards (minimum power factor, allowances for diversity, etc.)
6. Select HVAC design norms.

In addition to energy conservation, the 2005 revision of the NBC has emphasized an integrated approach and sustainable development covering optimum utilization of renewable energy, rain water harvesting, integrated waste management, earth quake resistance, and so on.

6.8 Green Rating for Integrated Habitat Assessment (GRIHA) System

The Ministry of New and Renewable Energy has developed the Green Rating for Integrated Habitat Assessment (GRIHA) as a national rating system for green buildings. The objective of GRIHA is to evaluate the environmental performance of a building holistically over its entire life-cycle based on accepted energy and environmental principles. The green rated buildings are expected to benefit the community at large by reducing GHG emissions, improving energy security and reducing the stress on natural resources. The system was initially conceived and developed by The Energy and Resources Institute (TERI) as TERI-GRIHA. The system takes into account the provisions of the NBC (2005) and ECBC (2007) and other by-laws. There are 34 indicators with a total of 100 points under the GRIHA system, covering the areas of health and well-being, materials and resources, water efficiency, energy efficiency and renewable energy, sustainable site planning and solid waste management. The buildings are awarded a one to five star rating based on the points scored. The Centre for Environmental Sciences and Engineering at IIT Kanpur is the first five-star rated building under GRIHA. An integrated approach to this building is stated to have resulted in 59% of energy savings (UNEP 2010: 24). Around 350 buildings with 1.02 million square metres of space are reported to have registered for rating under GRIHA (GIZ, undated: 8).

6.9 Leadership in Energy and Environmental Design (LEED) India Rating System

The Confederation of Indian Industry (CII)-Sohrabji Godrej Green Business Centre has established the Indian Green Building Council (IGBC) in 2001 as a consensus driven not-for-profit organization representing the building industry. It encourages builders, developers, owners, architects and consultants to design and construct green buildings to achieve enhanced economic and environmental performance. The Leadership in Energy and Environmental Design (LEED) rating system for green buildings was introduced by the IGBC in 2003. The IGBC introduced a revised version (version 3.0) of its new building rating system in 2014. The LEED rating system assesses green features of buildings in the areas of sustainable architecture and design, site selection and planning, water conservation, energy efficiency, building materials and resources, indoor environmental quality, and innovation and development. The rating provided to buildings cover categories such as certified (good practices), silver (best practices), gold (outstanding performance), platinum (national excellence) and super platinum (global leadership) representing lower to higher scale of rating (IGBC 2014: 7). The CII-Sohrabji Green Business Centre building in Hyderabad was the first platinum-rated green building in India. In 2003 around 20,000 square metres of building area was rated under the LEED system which had gone up to 207 million square meters covering 2771 buildings by August 2014.

6.10 Urban and Regional Development Plan Formulation and Implementation Guidelines 2014

The Institute of Town Planners India (ITPI) framed the Urban Development Plan Formulation and Implementation (UDPFI) guidelines in 1996 as the first national level planning guidelines to provide a framework for plan preparation and implementation process at the national, state and local scales. The Ministry of Urban Development felt the need to revise these guidelines and brought out draft Urban and Regional Development Plan Formulation and Implementation (URDPFI) Guidelines in 2014 in two volumes to address the emerging needs and requirements related to inclusive planning, sustainable habitat, preparation of comprehensive mobility plans, service level benchmarking and so on. Chapter 6 of Volume 1 of the URDPFI provided guidelines on sustainability and urban development including impact of climate change, environment policies of statutory obligations and planning for disaster management which promote climate co-benefits in the buildings sector. The sustainability guidelines suggest the following measures for promoting energy efficiency in buildings (URDPFI 2014: 225).

1. Meeting 15% of the total external lighting load in residential buildings and 5% in non-residential buildings through renewable energy
2. Development of city level energy efficiency and renewable energy policy options as done by Nagpur and Bhubaneswar cities
3. Adopting energy efficient development control regulations, building by-laws and ECBC in designing private and public buildings
4. Adopting Building Performance Certification and Rating Systems such as GRIHA.

The sustainability guidelines emphasized the need for green buildings to mitigate climate change impact and suggested the following measures (URDPFI 2014: 228–229):

1. Adopting the concept of green building which uses less water, optimizes energy efficiency, conserves natural resources, generates less waste and provides healthier spaces for occupants as compared to conventional building
2. Sustainable site development including efficient land use, habitat preservation and restoration, efficient transportation management and efficient use of locally available materials and resources
3. Use of water in a self-sustainable manner through reducing, recycling and reusing strategies including rain water harvesting
4. Reducing energy consumption of infrastructural equipment, on-site power generation using renewable energy technologies and clean fuels
5. Effective waste management strategies by segregating waste at source and promoting reuse of products and materials
6. Adopting norms for indoor environment quality as suggested by MoEF and various rating systems such as GRIHA and LEED.

The UDRPFI guidelines articulated for protecting the interests of the urban poor by reserving space, extending legal title and ownership and suggested master plans and development plans to take this component into account. It is suggested that planners should allocate space with serious assessment of the requirements of the urban poor after assessing the ground realities with regard to location of vendors, informal activities, slums and need for in situ redevelopment and upgradation (URDPFI 2014: 20). Pursuing sustainable guidelines of the UDRPFI will undoubtedly generate several human health, social and environmental co-benefits at the local level.

7 Impact of Building Sector Policies with Climate Co-benefits in India

As seen from the previous section a fragmented policy framework with climate co-benefits has evolved in India since 2001. The effort in formulating policies or an institutional framework such as ECA (2001) and BEE and ECBC (2007) are seen as significant steps towards both energy efficiency and climate change mitigation. According to a study by the World Bank, an ECBC compliant building is expected to use 39% less energy of 110 kWh/m^2/year as compared to the national benchmark of 180 kWh/m^2/year (Liu et al. 2010: 73). The study noted that several case studies on ECBC compliant buildings have demonstrated about 50% energy savings over the baseline design with initial cost escalation of 10–15% and payback times of 5–7 years (Liu et al. 2010: 130).

While ECBC compliance has high potential for energy savings, ensuring its adoption by states in India poses a key challenge. A study for the US Department of Energy observed that six years after enactment of the ECBC only two states and one union territory had formally adopted ECBC and six states were in the process of adopting it. The key barriers identified in this regard are lack of awareness among stakeholders, inadequate capacity and resources, lack of established effective legal mechanisms, non-inclusion of the ECBC in local building by-laws and lack of systematic approach to measure and verify compliance and energy savings to promote confidence in the ECBC. The study observed that the United States and China have well-developed compliance evaluation systems which could be adopted by various tiers of governments in India (Yu et al. 2014: 2). Both GRIHA and LEED rating systems have included compliance with ECBC as one of the criteria which would encourage their application. Various state governments are in the process of adopting energy efficiency measures in their departments.

The green buildings contribute a wide range of developmental and climate co-benefits. The Indian cities are required to achieve Service Level Benchmarks prescribed by the Ministry of Urban Development in the areas of water supply and sanitation, particularly to reduce non-revenue water and promote processing and treatment of waste. Green buildings contribute to these goals in addition to achieving reduction in energy consumption. Green buildings are 25–40% energy efficient, 20–30% water efficient, reuse 70–100% of treated waste water and reduce construction and demolition waste by 50%. Around 350 buildings with 1.02 million square metres of space are reported to have registered for rating under GRIHA while the building area rated under the LEED system had gone up to 207 million square meters across 2771 buildings in August 2014. This indicates a significant impact. Government buildings offer a huge potential for promoting green practices. The Delhi government has made moves in this direction by deciding to apply ECBC in all its buildings, initiating a conversion of 15 into green buildings (Kumar and Sahni 2013: 251). Similarly, the Public Works Department of Punjab Government issued an Request for Proposal for implementing energy efficiency in 67 government buildings. Some of the Indian states are in the process of formulating policies. For example, the Government of Kerala has brought out a draft green building policy, which aims to achieve improved efficiency in the use of energy, water and sanitation services. Thus, the energy efficiency and green building policies in India have a huge potential to achieve climate co-benefits in the buildings and construction sector in India.

Meanwhile, a comprehensive framework is emerging to make Indian cities slum-free with provision of high quality of housing stock and services. Cities are encouraged to undertake a comprehensive survey of slum locations to identify tenable and untenable slums. The tenable slums are those located on non-hazardous sites while untenable slums are located on hazardous sites including river banks and flood prone areas. The slum-free strategy provides for relocation of untenable slums to safer areas. This will significantly reduce flood risks. Both tenable and untenable slums seek better quality of housing stock and services by adopting higher densities and sustainable habitat principles. These measures are expected to mitigate climate change impacts on slums and informal settlements on the one hand and reduce GHG emissions on the other.

8 Suggestions for Strengthening the Climate Co-benefits Framework in Buildings and Construction Sector in Urban India

In addition to the United States and China, India is one of the first few countries to develop a comprehensive framework for promoting energy efficiency and green buildings to ensure climate co-benefits. However, this framework is not currently effective due to inadequate adoption and implementation of policies at the state and city level. For example, only a few states have adopted the ECBC code and many are yet to do so. Many urban local bodies are yet to formulate or amend the building by-laws. States and cities should make it mandatory to adopt the ECBC in all new public and private buildings and also in retrofitting the large existing buildings. The sustainable habitat principles articulated by the NBC (2005) and URDPFI Guidelines (2014) should be strictly adopted in plan formulation and implementation by the state and city governments.

National urban policies and programmes that finance various services have a huge potential in facilitating implementation of energy efficiency and green building policies in Indian cities. The Ministry of Urban Development is formulating a Smart City Scheme covering 100 cities to make them efficient innovative and smart to deliver world-class services by adopting innovative technologies. The MoUD has earmarked Rs. 70,000 million for the year 2014–15 for preparatory and planning activities. It is suggested that the Smart City Scheme guidelines make it conditional for cities to comply with the ECBC and green building rating for accessing funds under the scheme. The eligible cities are required to prepare a Smart City Development Plan (SCDP) and it is suggested that the SCDP should incorporate the climate co-benefits framework and strategies for implementation.

The MoUD is also expected to introduce an urban mission scheme along the lines of the Jawaharlal Nehru National Urban Renewal Mission for cities not covered by the Smart City Scheme. It is suggested that the guidelines of these schemes should also make it conditional for adoption of ECBC and green buildings by cities. Each city should develop an action plan for identifying and incorporating climate co-benefits into urban development strategies including the buildings and construction sector. There is a need to create awareness and build capacities among the officials of urban local bodies on these issues. Methods for assessing the climate co-benefits in the building sector and informal housing sector should be put in place. The health impact assessment (HIA) systems and environmental impact assessment (EIA) systems could be applied to measure the health and environmental co-benefits in the building sector. The HIA system seems to have been widely applied in Europe and United Kingdom in the housing sector (WHO 2011: 93). The house rating systems should be broad based and measure not only energy efficiencies but also health and safety aspects along the lines of the Housing Health and Safety Rating System developed in 2001 in the United Kingdom that assesses potential risks to health and safety from any deficiencies identified in a dwelling, including those related to design and construction, and the Health Housing Index developed in New Zealand (WHO 2011: 96). In addition, a compliance evaluation system for ECBC and green buildings should be put in place at the city and state level to assess the progress, measure the impact and capture the benefits.

References

Bureau of Indian Standards. (2005). *National building code of India 2005*. New Delhi: Government of India.

GIZ. (n.d.). *Promoting sustainable and inclusive growth in emerging economies: Green buildings*. Berlin: Economic Policy Forum.

Indian Green Building Council. (2014). *IGBC green new buildings rating system version 3.0—Abridged reference guide*. New Delhi: Confederation of Indian Industry (CII)-Sohrabji Godrej Green Business Centre.

IPCC. (2007). Intergovernmental panel on climate change. In *Climate change 2007: Synthesis report. Climate change—Summary for policy makers*. Cambridge: Cambridge University Press.

IPCC. (2014). Summary for policymakers. In O. Edenhofer, R. Pichs-Madruga, Y. Sokona, E. Farahani, S. Kadner, K. Seyboth, A. Adler, I. Baum, S. Brunner, P. Eickemeier, B. Kriemann, J. Savolainen, S. Schlömer, C. von Stechow, T. Zwickel, & J. C. Minx (Eds.), *Climate change 2014: Mitigation of climate change. Contribution of working group iii to the fifth assessment report of the intergovernmental panel on climate change*. Cambridge University Press, Cambridge, United Kingdom and New York, NY, USA.

Kumar, P., & Sahni, L. (2013). Green buildings in India: A path ahead for sustainable environment. *International Journal of Scientific Research, 2*(3).

Liu, F., Meyer, S. A., & Hogan, F. J. (2010). *Mainstreaming building energy efficiency codes in developing countries: Global experiences and lessons from early adopters. World Bank working paper no 204*. Washington, DC: World Bank.

Ministry of Housing and Urban Poverty Alleviation. (2007). *National housing and habitat policy 2007*. New Delhi: Ministry of Housing and Urban Poverty Alleviation, Government of India.
Ministry of Law, Justice and Company Affairs. (2001). *The energy conservation act, 2001*. New Delhi: Government of India.
Ministry of Power & Bureau of Energy Efficiency. (2007). *Energy conservation building code 2007*. New Delhi: Government of India.
Ministry of Power & Bureau of Energy Efficiency. (2008). *National mission for enhanced energy efficiency*. New Delhi: Draft Mission Document: Implementation Framework, Government of India.
Ministry of Urban Development. (2008). *National mission on sustainable habitat*. New Delhi: Ministry of Urban Development, Government of India.
MoEF. (2010). *India: Greenhouse gas emissions 2007* (pp. 1–54). New Delhi: Indian Network for Climate Change Assessment, Ministry of Environment & Forests, Government of India.
MoEF. (2015). *India's intended nationally determined contribution: Working towards climate justice*. New Delhi: Ministry of Environment & Forests, Government of India.
NAPCC. (2008). *National action plan for climate change 2008*. New Delhi: Ministry of Environment and Forests, Government of India.
Narender, A. (2013). Towards slum free cities in India: policies and issues. *Nagarlok, 45*(4), 12–40.
Planning Commission. (2006). *Integrated energy policy: Report of the expert committee*. New Delhi: Planning Commission, Government of India.
Puppim de Oliveira, J. A., Doll, C. N. H., & Suwa, A. (2013a). *Urban development with climate co-benefits: Aligning climate, environmental and other development goals in cities* (pp. 10–17). Yokohama: United Nations University Institute of Advanced Studies (UNU-IAS), Japan.
Puppim de Oliveira, J. A., Doll, C. N. H., Kurniawan, T. A., Geng, Y., Kapshe, M., & Huisingh, D. (2013b). Promoting win–win situations in climate change mitigation, local environmental quality and development in Asian cities through co-benefits, *Journal of Cleaner Production, 58*(1), 1–6.
United Nations Environment Programme. (2007). *Buildings and climate change: Status, challenges and opportunities*. Nairobi, Kenya: UNEP.
United Nations Environment Programme. (2009). *Buildings and climate change: Summary for decision-makers*. Nairobi, Kenya: UNEP-SBCI, Sustainable Buildings and Climate Initiative.
United Nations Environment Programme. (2010). *The 'state of play' of sustainable buildings in India*. Nairobi, Kenya: UNEP-SBCI, Sustainable Buildings and Climate Initiative.
UN Habitat. (2011). *Global report on human settlements 2011: Cities and climate change*. London, Washington, DC: United Nations Human Settlements Programme & Earthscan.
United Nations. (2015a). Transforming our world: The 2030 agenda for sustainable development, A/RES/70/1. https://sustainabledevelopment.un.org/post2015/transformingourworld/publication. Accessed on December 15, 2015.
United Nations. (2015b). Conference of parties, December 2015, FCCC/CP/2015/L.9. United Nations Framework Convention on Climate Change. http://unfccc.int/resource/docs/2015/cop21/eng/l09.pdf. Accessed on January 24, 2016.
URDPFI. (2014). *Urban and regional development plans formulation & implementation guidelines, volume 1-first draft report*. New Delhi: Ministry of Urban Development, Government of India.
World Bank. (2011). *Guide to climate change adaptation in cities*. Washington, DC: World Bank.
World Health Organization. (2011). *Health in the green economy: Health co-benefits of climate change mitigation—housing sector*. Geneva, Switzerland: World Health Organization.
Yu, S., Evans, M., & Delgado, A. (2014). *Building energy efficiency in India: Compliance evaluation of energy conservation building code*. Oak Ridge, USA: Pacific Northwest Laboratory & U.S. Department of Energy.

Chapter 8
Co-benefits of Waste to Energy

Amit Chatterjee, Manmohan Kapshe, Binayak Choudhury and Shomit Bade

Abstract Rapidly increasing urbanization and economic development in India has impacted the quantity of waste generation. Presently, only 68% of Municipal Solid Waste (MSW) generated in the country is collected of which 28% is treated by the municipal authorities. In the case of wastewater about 30% of the wastewater generated from major cities of India is treated. Untreated waste leads to an adverse impact on public health and also creates various environmental problems including pollution of air, water and land resources. There have been efforts by the respective governments for initiating action on converting waste to energy but these initiatives are fragmented and have not been integrated into national policy frameworks. Further, waste to energy involves higher costs and a relatively higher degree of expertise amongst the governments about integration of various technologies. The main aim of the paper is to document and analyse waste to energy initiatives both in solid waste and waste water so as to identify and disseminate innovative urban practices. Among the various initiatives taken for 'waste to energy' in India, three cases are selected, namely: biogas-fertilizer plants (BGFPs) for generation, purification/enrichment, bottling and piped distribution of biogas in Talwade, Nasik district of Maharashtra; organic MSW-based decentralized biomethanation plant at Pune city of Maharashtra and methane recovery and power generation from sewage treatment plants (STPs) in Surat city of Gujarat. This paper made an attempt to analyse and document the innovative urban practices in India on 'waste to energy' that can be replicated at other cities and thus offers an implementable solution to the problem faced by many Indian cities.

Keywords Waste management · Co-Benefits · Surat
Legal and policy framework · MSW generation & treatment · Waste to energy

---

A. Chatterjee (✉) · B. Choudhury · S. Bade
Department of Planning, School of Planning and Architecture (SPA), Bhopal, India
e-mail: amit.chatterjee@spabhopal.ac.in

M. Kapshe
Department of Architecture and Planning, Maulana Azad National Institute of Technology (MANIT), Bhopal, India

M. Sethi and J.A. Puppim de Oliveira (eds.), *Mainstreaming Climate Co-Benefits in Indian Cities*, Exploring Urban Change in South Asia,
https://doi.org/10.1007/978-981-10-5816-5_8

1 Introduction

Rapid urbanization, growing population and a change in lifestyle of the people in India has led to a steady growth in the quantum of waste generation. The provisional population of India as per the Census of India (2011) was 1.21 billion. For the first time since Independence, the absolute increase in population is more in urban areas than in rural areas. Between 2001 and 2011, the number of people living in urban areas increased from 286 to 377 million, a net addition of 91 million. In comparison, the rural population increased by 90 million, up from 743 million in 2001 to 833 million in 2011. The level of urbanization increased from 27.81% in the 2001 Census to 31.16% in the 2011 Census. Thus, the proportion of the rural population declined from 72.19 to 68.84%. As a consequence of fast urbanization, the large urban centres are becoming larger and many smaller towns and cities are experiencing spatial growth. In 2011, there were 53 urban agglomerations and cities with a population of one million or more, which were 35 in 2001 and 23 in 1991 (Census of India 2011).

Over the years, this rapid urbanization has put a lot of stress on the existing infrastructure of the cities and as a result waste (both sewage[1]and solid waste[2]) generated from urban areas has also increased. Especially in large cities, waste management and disposal is one of the main concerns. Urban local bodies (ULBs) in India render solid waste management (SWM) services. Though it is a mandatory civic service, it does not receive the priority it deserves and services are poor, which can be catastrophic in cases where it results in an epidemic. There have been incidences of such epidemics in the past. Before independence, such epidemics some times have even resulted in depopulation in cities. In independent India, most notable being the outbreak of pleague in Surat that had been commonly associated with unhygenic living conditions in the city. Such episodes of epidemics make it clear that population growth in urban areas has put tremendous pressure on public agencies in relation to managing urban infrastructure and services. ULBs in Indian cities have failed to cope with the growing need for waste management services as evident from the decrepit and overstretched existing infrastructures. Many major cities like Delhi, Mumbai (Bombay), Chennai (Madras), Lucknow, Bhopal, Patna, Surat, and Bengaluru (Bangalore) are facing environmental problems related to municipal solid waste management (MSWM) across all aspects—collection, transportation, treatment and disposal.

In the last few decades, waste management has emerged as a major environmental problem that needs to be addressed across countries. There has been a major thrust in the recent past to identify and develop alternative energy sources to

[1]In India 'sewage' and 'wastewater' is usually carried together. Therefore, throughout this paper 'sewage' and 'wastewater' are used interchangeably as the context demands. Further, the pipes that are used to carry sewage are termed 'sewer' and the system of sewers is called 'sewerage'.

[2]'Solid waste' can be defined as non-liquid material that no longer has any value to the person who is responsible for it. The words *rubbish, garbage, trash* or *refuse* are often used as synonyms when talking about 'solid waste' (Zhu et al. 2008).

consider 'waste as resource', and accordingly several efforts have been made the world over to address waste-related issues. In India, too, several initiatives have been taken up by the ULBs in waste sectors that include: (i) solid waste management (segregation of waste at source, improved collection and transport, methane avoidance and waste to energy projects); and (ii) liquid waste management through low energy sewage treatment and methane recovery (MoUD 2012). The Ministry of New and Renewable Energy (MNRE) of the Government of India has been aggressively promoting the concept of biomass/Municipal Solid Waste (MSW)-based power plants in the 5–15 MW range. During the year 2008–09, a new initiative was taken by MNRE for technology demonstration on biogas bottling projects in the entrepreneurial mode for installation of medium-size mixed-feed biogas plants for generation, purification and bottling of biogas. Installation of such plants aims at the production of compressed natural gas (CNG) quality of compressed biogas (CBG) to be used as vehicular fuel in addition to meeting stationary and motive power, electricity generation and thermal application, and so on, in a decentralized manner through the establishment of a sustainable business model in this sector. There is huge potential for the installation of such plants in various areas. Under the demonstration phase, the Ministry has sanctioned a central financial assistance for a limited number of such projects for implementation following an entrepreneurial mode in different states, namely Chhattisgarh, Gujarat, Haryana, Karnataka, Maharashtra, Punjab, Madhya Pradesh, Andhra Pradesh and Rajasthan (MNRE 2010).

There have been efforts by the government for initiating action on converting waste to energy but these initiatives are fragmented and have not been integrated into national policy frameworks. This has been due to limited resource availability to the ULBs to support experimentation with innovative urban practices and the inability to identify and disseminate good policy practices to convert waste as a resource. The main objective of this chapter is to strengthen the policy-making process for combating the issue of waste by identifying, documenting and analysing the initiatives of waste to energy. Among the various initiatives taken for 'waste to energy' in India, the three cases are selected; namely: the biogas-fertilizer plants (BGFP) for generation, purification/enrichment, bottling and piped distribution of biogas at Nasik; the organic MSW based decentralized biomethanation plant at Pune; and the Methane recovery and power generation from sewage treatment plant at Surat. These recent initiatives hold the potential to be replicated in other cities and thereby provide an implementable solution to the problem faced by many Indian cities.

The next section of the chapter introduces waste management, the Co-Benefits and enabling the legal and policy framework in the Indian perspective. The third section focuses on technology options and case studies of MSW in Indian context. The fourth section describes estimation of greenhouse gas (GHG) reduction and documentation of Co-Benefits for the case of wastewater management in Surat followed by conclusions in the last section.

2 Waste Management in India

There has been a significant increase in the generation of MSW in urban India over the last few decades. The MSW volume is expected to increase significantly in the near future as the country strives to attain an industrialized nation status by the year 2020 (Sharma and Shah 2005; Shekdar et al. 1992). The quantity of MSW generated depends on a number of factors such as food habits, the standard of living, the degree of commercial activities and seasons. With increasing urbanization and changing lifestyles, Indian cities now generate eight times more MSW than they did in 1947. Currently, of the estimated 62 million tonnes of MSW generated annually by 377 million people in urban areas (based on an average per capita generation of 450 g per person per day), more than 80% is disposed of indiscriminately at designated/non-designated dump yards in an unhygienic manner by the ULBs, leading to problems of health and environmental degradation. The untapped waste has a potential of generating 439 MW of power from 32,890 TPD (tonnes per day) of combustible wastes including refuse derived fuel (RDF), 1.3 million cubic metres of biogas per day or 72 MW of electricity from biogas and 5.4 million metric tonnes of compost annually to support agriculture. The existing policies, programmes and management structure do not adequately address the imminent challenge of managing this waste, which is projected to be 165 million tonnes by 2031 and 436 million tonnes by 2050. Further, if the current 62 million tonnes annual generation of MSW continues to be dumped without treatment; it will need 340,000 m^3 of landfill space every day (1240 ha per year). Considering the projected waste generation of 165 million tonnes by 2031, the requirement of land for setting up a landfill for 20 years (considering a 10 m high waste pile) could be as high as 66,000 ha of precious land, which our country cannot afford. As per the Ministry of Urban Development (MoUD), 70 million tonnes of waste is generated currently in urban centres. The Central Pollution Control Board (CPCB) has reported that 133,760 metric tonnes (MT) of waste are generated (Table 1) daily in urban areas in the country (CPCB 2013). There are thus conflicting data about the quantum of waste actually generated in urban areas in the country, principally because there is no system of periodically collecting and updating a country-wide database on quantity and composition of waste (Planning Commission 2014).

Considering that the volume of waste is expected to increase by 5% per year on account of increases in the population and change in lifestyle of the people, it is assumed that urban India will generate 276,342 TPD by 2021, 450,132 TPD by 2031 and 1,195,000 TPD by 2050. The Central Pollution Control Board (CPCB) report also reveals that only 68% of the MSW generated in the country is collected, of which 28% is treated by the ULBs. Thus, merely 19% of the total waste generated is currently treated (Planning Commission 2014). The recently launched 'Swachh Bharat Mission' (Clean India Mission) by the Government of India took an initiative of making the country clean and litter free with scientific solid waste

Table 1 Estimated[a] waste generation in the country state-wise

Sr. No.	Name of the state/UT	Municipal solid waste (MT/day), 1999–2000			Municipal solid waste (MT/day), 2012
		Class I towns	Class II towns	Total	
1	Andaman & Nicobar	–	–	–	70
2	Andhra Pradesh	3943	433	4376	11,500
3	Arunachal Pradesh	–	–	–	181
4	Assam	196	89	285	650
5	Bihar	1479	340	1819	1670
6	Chandigarh	200	–	200	340
7	Chhattisgarh	–	–	–	1896
8	Daman Diu & Dadra	–	–	–	85
9	Delhi	4000	–	4000	7500
10	Goa	–	–	–	183
11	Gujarat	–	–	–	8336
12	Haryana	3805	427	4232	3490
13	Himachal Pradesh	623	102	725	1370
14	Jammu & Kashmir	35	–	35	1792
15	Jharkhand	–	–	–	4450
16	Karnataka	3118	160	3278	6500
17	Kerala	1220	78	1298	1576
18	Lakshadweep	–	–	–	21
19	Maharashtra	8589	510	9099	17,000
20	Manipur	40	–	40	176
21	Meghalaya	35	–	35	268
22	Mizoram	46	–	46	552
23	Madhya Pradesh	2286	398	2684	5079
24	Nagaland	–	–	–	270
25	Orissa	646	9	655	2383
26	Pondicherry	60	9	69	495
27	Punjab	1001	265	1266	3853
28	Rajasthan	1768	198	1966	5037
29	Sikkim	–	–	–	40
30	Tamil Nadu	5021	382	5403	14,532

(continued)

Table 1 (continued)

Sr. No.	Name of the state/UT	Municipal solid waste (MT/day), 1999–2000			Municipal solid waste (MT/day), 2012
		Class I towns	Class II towns	Total	
31	Tripura	33	–	33	360
32	Uttar Pradesh	5515	445	5960	19,180
33	Uttaranchal	–	–	–	1251
34	West Bengal	4475	146	4621	8674
Total MSW		48,134	3991	52,125	133,760

Source Data feeds from CPCB, Annual Report (2013)

[a]Does not include waste picked up by *Kabadiwala* (informal sector scrap dealer who purchases marketable waste material from households) at the doorstep and the rag pickers

management in about 4041 towns covering a population of 306 million. Accordingly, the Government proposes to invest significantly in SWM projects over the years and has provided INR 25 billion (US$397 million) as grants in aid to states and ULBs through public-private partnerships (INDC 2015).

In the case of wastewater based on the projected population of 1093 million, the CPCB has estimated that in the year 2051 around 132,000 million litres wastewater would be generated per day in India. Further, the studies carried out by the CPCB to assess the status of wastewater generation and treatment in Class I towns (towns with a population of at least 100,000 persons) and Class II towns (towns with a population ranging between 49,999 and 99,999) during 1978–1979, 1989–1990, 1994–1995 and 2003–2004 (CPCB published three reports in three decades), indicate that about 26,254 million litres per day (MLD) of wastewater was generated in the 921 Class I and Class II towns in India (housing more than 70% of the urban population). According to the CPCB's 2008 statistics, in 498 Class I towns, the total water supplied was estimated to be 44,769.05 MLD and per capita per day water supply was estimated to be 179.02 l. In 410 Class II towns the total water supplied was estimated to be 3,324.83 MLD and per capita per day water supply was estimated to be 120.79 l. The estimated sewage generation from Class I and Class II towns together is 38,254 MLD, out of which only 11,787 MLD (30%) is being treated with a capacity gap of 26,467 MLD—see Table 2 (CPCB 2008).

Recent initiatives by the Government of India on wastewater management would cover an additional population of 41 million and enhance recycling and reuse of treated water. There are about 816 Sewage Treatment Plants (522 operational and rest at different stages of construction and planning) having a combined capacity of 23,277 million litres per day across 28 States and Union Territories. Standards related to effluent discharge have been modified for over 2000 industries focused on reducing the quantity of wastewater generation, conservation of water, promotion of Zero Liquid Discharge (ZLD) and use of treated effluent for irrigation (INDC 2015).

Table 2 Water supply, wastewater generation and treatment capacities in Class I[a] and Class II[b] towns of India in the year 2008

Category	No. of towns	Population	Total water supply (MLD)	Wastewater generation (MLD)	Treatment capacity (MLD)
Class I towns	498	227,652,872	44,769	35,558	11,553
Class II towns	410	30,018,368	3324	2696	234
Total	908	257,671,240	48,093	38,254	11,787

Source Compiled from CPCB (2008)
Note MLD million litres per day
[a]Settlements having a population more than 0.1 million, popularly called cities in India
[b]Settlements having a population between 50,000 and 99,999

3 Defining Co-Benefits and Past Initiatives in Waste Sector

In general, with reference to climate change various agencies have explained the concept of Co-Benefits differently. Some of the definitions are summarized below for better understanding of the concept.

- Fourth Assessment Report of the Intergovernmental Panel on Climate Change (IPCC) on mitigation differentiates 'Co-Benefits' as benefits intended as the primary objective of certain actions or policies from those that are secondary or incidental to it are named simply as 'ancillary benefits' (IPCC 2007).
- The positive effects that a policy or measure aimed at one objective might have other objectives, irrespective of the net effect on overall social welfare. Co-Benefits are often subject to uncertainty and depend on local circumstances and implementation practices. Co-Benefits are also called ancillary benefits (IPCC 2013).
- Institute of Global Environmental Strategies (IGES) promotes a simple definition of Co-Benefits as potential benefits of climate change mitigation actions in other fields or areas not covered by climate change or UNFCCC (IGES 2007).
- The United States Environmental Protection Agency (USEPA) Integrated Environmental Strategies (IES) Program offers a less restrictive definition of 'Co-Benefits'. USEPA defines the 'Co-Benefits approach' as: 'Co-Benefits refer to multiple benefits in different fields resulting from one policy, strategy, or action plan. Co-beneficial approaches to climate change mitigation are those that also promote positive outcomes in other areas such as concerns relating to the environment (e.g., air quality management, health, agriculture, forestry and biodiversity), energy (e.g., renewable energy, alternative fuels, and energy efficiency) and economics (e.g., long-term economic sustainability, industrial competitiveness, income distribution)' (GoJ 2009).

The general Co-Benefits definition in the context of GHG emissions mitigation, pollutants reduction, health improvement and other aspects lies within the scope of sustainable development (Jiang et al. 2013). Though the Co-Benefits as a term has been used recently in the climate change literature, the efforts to minimize energy need and generation of bi-products have been documented for many years in the case of wastewater disposal. Shortage and the high cost of electricity, fuels, and chemicals have become important considerations while designing and operating wastewater treatment plants (Kapshe et al. 2013). Use of methane for electricity generation has been continuously discussed in the literature. The generation of electricity at the Palo Alto Sewage Treatment Plant in the USA with a gas engine was practised long ago in the 1930s, using digester gas as fuel, with one of the major objectives of providing the second source of power or captive power for operating the plant (Harvey 1936). Waste-to-energy power plants are in operation in 25 US states. They are fuelled by 26.3 million tonnes of MSW and have a generating capacity of 2700 MW of electricity. They also recover about 0.64 million tonnes of ferrous and non-ferrous metals annually. There are two main categories of WTE plants. In mass-burn plants, the MSW is fed as collected into large furnaces. In plants, the MSW is first shredded into small pieces and most of the metals are recovered before combustion (Psomopoulos and Themelis 2009).

Energy from Waste is regarded as a means for producing renewable energy, since approximately 70% of MSW is defined as being carbon neutral (Aeat 2005). Electricity generation is a primary concern of energy from waste developers, due to its favourable effect on project economics of capital-intensive schemes (Williams 2005), but in the opinion of the authors, this does not represent sound resource management when generation efficiencies of between 13 and 24% (Malkow 2004) result in a great amount of potentially useful energy being lost in the form of waste heat (Pepermans et al. 2005).

The government of India has prepared the National Action Plan on Climate Change (NAPCC) to address the urgent and critical concerns of the country (GoI 2008). It identifies measures that promote development objectives of the country while yielding Co-Benefits for addressing climate change effectively.

UNFCCC Conference of the Parties-21 (Paris Agreement) has given due importance to Co-Benefits in policy making by stating that all parties recognize the social, economic and environmental value of voluntary mitigation actions and their Co-Benefits for adaptation, health and sustainable development through enhanced action prior to 2020 (UNFCC COP-21 2015). It is well established taht role of Co-Benefits will continue to grow in importance following the launch of the INDC and Sustainable Development Goals (SDGs). INDC highlighted that the current development paradigm reiterates the focus on sustainable growth and aims to exploit the Co-Benefits of addressing climate change along with promoting economic growth. India recognizes the dual benefits by promoting waste to wealth and, accordingly, incentives are being granted to cities to take up waste to energy conversion projects (INDC 2015).

4 Enabling Legal and Policy Framework

Each of the state governments in India is under a constitutional mandate to take the necessary steps to improve public health and the quality of life. The 12th schedule of the 74th Constitutional Amendment Act of 1992 clearly assigns solid waste management as the primary function of ULBs. State laws also stipulate management of solid waste as an obligatory function of the ULBs (Planning Commission 2014). The Gazette notification by Government of India dated 25 September 2000 had made public the rules and regulations entitled 'The Municipal Solid Waste (Management and Handling) Rules 2000' [MSW (M&H) 2000]. There are various sections in these gazettes that enable the ULBs to regulate scavenging and ensuring removal and disposal of excrementitious and other filthy matter, all ashes and rubbish. They may also reclaim unhealthy localities, remove noxious vegetation and generally abate nuisances. The gazette also makes it clear that the solid waste must be deposited in public receptacles and depots which are the property of the ULBs. The municipal commissioner under this gazette is obliged to provide for convenient situations and places for the final disposal of municipal solid waste and is required to do so in a manner that causes the least practicable nuisance.

Before MSW (M&H) 2000 was in place, problems relating to solid waste were covered in the definition of 'pollution' as defined in the Water [Prevention and Control of Pollution] Act 1974. The Environmental Protection Act 1986, Section (3) sub-Sections 3 (2), 3, 4 and 6 provided the central government with standards for the quality of the environment in its various aspects, setting appropriate levels for emissions or discharges of environmental pollutants from various sources and finally lays down procedures and safeguards for the prevention of accidents which may cause environmental pollution (Ministry of Environment and Forests 1986: 2–3).The policy statement for abatement of pollution in 1992 identified the increasing trend of environmental pollution and mentioned traditional municipal waste as one of the pollutants (Ministry of Environment and Forests (MoEF) 1992: 2–4). The Acts of 1986 and 1992 are the most recent Acts. The National Environment Tribunal Act 1995 provides for strict liability for damages arising out of any accident occurring, focusing more on hazardous wastes and only superficially addresses municipal waste management issues. Apart from the MSW (M&H) 2000, specific legislations such as the Plastic Manufacturer and Use Rules 1993 amended in 2003 and the Batteries (Management and Handling) Rules 2001 are now in place that enables environmental friendly recycling in the country. The Plastic Waste (Management and Handling) Rules 2011 were issued in supersession of the Recycled Plastic Manufacture and Usage Rules, 1999 notified under the Environment (Protection) Act 1986. The National Urban Sanitation Policy pertains to management of human excreta and associated public health and environmental impacts. With a view to expeditiously improve solid waste management in the country, the MoEF in July 2013 notified the draft MSW Rules, 2013 in supersession of the MSW Rules 2000, inviting objections and suggestions from citizens. A timeline of environmental laws and regulation in India is provided in Table 3.

Table 3 Timeline of environmental laws and regulations in India

Year	Environmental acts/rules
1974	Water (Prevention and Control of Pollution) Act
1981	Air (Prevention and Control of Pollution) Act
1982	Air (Prevention and Control of Pollution) Rules
1986	Environmental (Protection) Act & Environmental (Protection) Rules
1989	Hazardous Waste Rules
1992	The 74th Constitutional Amendment of 1992
1995	National Environmental Tribunal Act
1998	Biomedical Waste Rules
2000	Municipal Solid Waste (Management and Handling) Rules
2001	Batteries Rules
2006	National Environment Policy
2011	Plastic Waste (Management and Handling) Rules
2011	National Urban Sanitation Policy
2013	Draft MSW Rules

Source Compiled by authors from various sources

5 Technology Options and Case Studies of Waste to Energy from Solid Waste

The World Health Organization (WHO) has observed that 22 types of diseases can be prevented/controlled by improving the MSW management system.[3] This will indirectly save huge financial resources currently spent on health and medical services. Energy recovery in the form of electricity, heat and fuel from the waste using different technologies is possible through a variety of processes, including incineration, gasification, pyrolysis and anaerobic digestion. These processes are often grouped under 'WTE technologies'. Two groups of technologies could be used for processing the fractions of wastes:

1. Bio-chemical waste to energy technologies
2. Thermo-chemical waste to energy technologies.

Bio-chemical conversion of biodegradable MSW can be categorized into composting, biomethanation and fermentation. Thermal processing of MSW can be accomplished in several ways including incineration, pyrolysis, gasification and mass burning. Only 22 states/UTs have set up processing and disposal facilities but the rest of the states/UTs made no effort until 2013. Of the 279 conventional composting and 138 vermicomposting facilities, 172 biomethanation, 29 and 8 WTE plants reported to have been established many are either closed or underperforming (Planning Commission 2014). Options available for MSW treatment

[3]Planning Commission (2014).

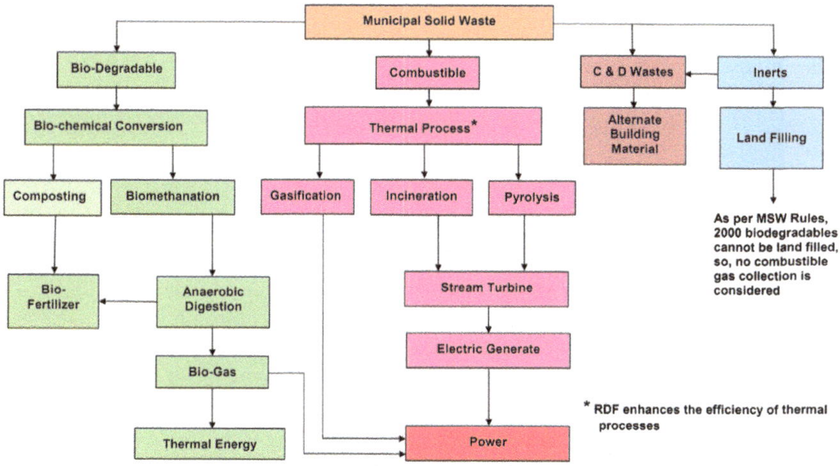

Fig. 1 Options available for MSW treatment and utilization. *Source* Planning Commission (2014)

and utilization are shown in Fig. 1 and MSW generation and treatment facilities are provided in Table 4. Most of the efforts are on a small scale and project-level initiatives are not integrated into national policy frameworks. Further, both technologies (bio-chemical waste to energy and thermo-chemical waste to energy) involve higher costs and a relatively higher degree of expertise.

India has off-grid renewable power capacities of 238 MW from biomass cogeneration, 125 MW from biogas, 53 MW from waste-to-energy, 3 MW from solar PV plants, and 1 MW from hybrid systems. The MNRE provides incentives for the recovery of energy from MSW and other urban waste streams under the Programme for Energy Recovery. It also provides INR 20 million (US$400,000) per MW installed or up to 20% of the project costs at INR 100 million (US$2 million) total, whichever is lower; and also provides 40% of the total project costs with a maximum incentive of INR 20 million (US$400,000) per MW for the generation of electricity from biogas at sewage treatment plants and 30% of the project costs with a maximum incentive of INR 30 million (US$600,000) per MW for power generation from biomethanation-based projects. Both private and public sector projects are eligible for these waste-to-energy incentives (NREL 2010). The MNRE took up a new initiative for bottling of biogas to demonstrate an integrated technology package in entrepreneurial mode on medium-size mixed-feed biogas-fertilizer plants (BGFP) for a generation, purification/enrichment, bottling and piped distribution of biogas. Installation of such plants aims at meeting stationary and motive power, cooling, refrigeration and electricity needs in addition to cooking and heating requirements. There could be a huge potential of installation of medium-size biogas-fertilizer plants in the country.

Keeping in mind the energy shortage in the country, there is a need to tap biomass resources such as cattle dung, kitchen waste, agricultural waste and so on, for the generation of biogas through the involvement of entrepreneurs and

Table 4 MSW generation and treatment facilities of India (2012–13)

Generation in TPD	133,760
Collection efficiency	68%
Total waste collected	90,957
Waste lost/littered	42,803
Composition of MSW	
Organic waste (51%)	68,218
Inert and non-organic (32%)	42,803
Recyclable (17%)	22,739
MSW treatment facilities	
MSW treated/processed in TPD	25,884
No. of ULBs having treatment/processing facilities	626
1. Composting	279
2. Vermi-compost	138
3. Biogas plant	172
4. Palletization (RDF)	29
5. Waste to energy	8
Landfilling	
Landfilled	65,489
Total waste to be landfilled	108,292

Source Compiled from Planning Commission (2014)

industries to set up a decentralized biogas-based energy infrastructure in the country, especially in rural areas at the potential sites where large quantities of biomass is available. Based on the availability of cattle dung alone from about 304 million cattle, there exists an estimated potential of about 18,240 million cubic metres of biogas generation annually.[4] There is huge potential for installation of medium-size biogas-fertilizer plants in various villages and other areas and in the agro/food processing industry of the country. During the demonstration phase, the Ministry was providing central financial assistance (excluding the cost of land) for a limited number of such projects for implementation following an entrepreneurial mode (Kumar and Sharma 2013). One such commissioned project was in the Talware Nasik district where energy was produced from biomass cow dung, agricultural waste and so on. The upgraded biogas is used for power generation, cooking and industrial applications, and the slurry of biogas plant being used as an organic fertilizer in the nearby agro fields is considered as a co-benefit from the project. Another co-benefit project studied is in Pune, Maharashtra. The purified biogas generated in the project is used for generating electricity and the fibre contents (the organic material) used as manure for the public gardens is considered as a co-benefit from the project. The two successful WTE pilots are documented in Boxes 1 and 2.

[4]Kumar and Sharma (2013).

Box 1: BGFP Project at Talwade, District- Nashik, Maharashtra

A biogas generation project for a generation, purification/enrichment, and bottling of biogas has sanctioned by MNRE during the year 2009–10 to Ashoka Biogreen Pvt. Ltd at Talwar in the Nasik district of Maharashtra. This first biogas bottling plant under the Technology Demonstration policy of the MNRE was commissioned in March 2011 after obtaining a licence for filling and storage of compressed biogas in CNG cylinders from the Petroleum & Explosives Safety Organization. The biogas generation capacity of the plant is 500 m^3 per day and based on NISARGRUNA (BARC) Technology. The purity of biogas is about 98% which is compressed to 150-bar pressure for filling in cylinders. A purity of 98.4% methane has been achieved through a test conducted by Shriram Institute for Industrial Research, Delhi. The purity of the enriched biogas is continuously monitored by the online analyyzing system along with caliberation of analyzers. The purified biogas is equivalent/similar to CNG. These upgraded biogas is used for power generation, cooking and industrial application. The slurry of biogas plant is being used as an organic fertilizer in the nearby agro fields. The field trials have indicated 150% growth in agro-production and substantial improvements in the quality.

Source MNRE (2013).

Box 2: 1 × 5 TPD Capacity Segregated Organic MSW-based Decentralized Biomethanation Plant at Ward No. 34, Model Colony, Pune, Maharashtra

The population of Pune city as per the 2011 Census is more than 3 million, a growth of more than six times in the last 60 years (CDP 2012). Pune city is generating about 1300 tonnes of municipal solid waste per day. The broad composition of solid waste generated in the city is from: commercial 25%, market area 5%, hotels and restaurants 25%, vegetable market 5%, household 40%. The organic and biodegradable portion of solid waste is 70% and therefore can contribute significantly to solid waste management (MEDA 2013). In the Model Colony of Pune city the solid wastes generated in urban areas from vegetable markets, hotels, hostels, kitchen wastes, and so on are collected and processed in organic MSW based biomethanation plant. The solid wastes generated in urban areas from vegetable markets, hotels, hostels, kitchen wastes and so on, are best suited for this process due to the high moisture content and organic fractions (up to 90%). The total solids in the organic waste decompose rapidly (i.e. are highly putrescible) and therefore these wastes can be treated by the biomethanation process, more commonly called anaerobic digestion. The biomethanation plant at the Model Colony is based on the two-stage anaerobic process. The sizes of the digesters for the first stage and the second stage are decided on the basis of the

suspended organic contents of the slurry to be treated. The first stage fermentation is the hydrolysis stage and the second is the methanation and polishing stage. The first stage is designed to give maximum solid retention time for the hydrolysis and the second stage for acidification and biomethanation process operate in the mesophillic range. In this process, the wet waste generated within the colony (from household kitchens, commercial complexes, hotels/restaurants, fruit and vegetable markets wet wastes, etc.) is collected and brought to the plant site by the Pune Municipal Corporation (PMC). Though it is segregated wet waste, it still contains 2–5% non-biodegradable material, such as plastics, glass, metal and so on. All such material is removed manually in the first stage, known as 'fine segregation'. Thereafter, the segregated wet waste is mixed with water in 1:1 proportion and crushed in the shredder to convert it into slurry, before being fed into the primary digester. The slurry is then treated in closed vessels called anaerobic digesters (primary and secondary digesters) where, in the absence of oxygen, microorganisms break down the organic matter into a stable residue, and generate a methane-rich biogas in the process. The generated biogas is cleaned with the help of scrubbers. In the scrubbing process, moisture, hydrogen sulfide and to a certain extent carbon dioxide are brought to an acceptable level and then the purified biogas is stored in a biogas balloon, made up of neoprene rubber. The purified biogas is then supplied to a 40 kVA indigenized biogas engine (run on 100% biogas) to generate electricity. The solid residue that remains after the biomethanation process, comprising of solid/fibrous material and liquid, is separated in the slurry drying beds. About 50% of the liquid manure is then recirculated into the system, as it contains nitrogen and some active anaerobic microorganisms. The fibre contains an organic material, which is being used as manure for the PMC's public gardens. The biomethanation plant has been in operation from November 2009.
Source MNRE (2012).

There are many such initiatives at project level. These projects have not been documented and included in policy guidelines and as a result, these initiatives have remained fragmented. Local bodies have so far not prepared a time-bound action plan to undertake collection, segregation, storage, transportation, processing and disposal of MSW. There is no agency which can assist local bodies technically, either at the state or national level to prepare the plans. Further, a detailed assessment does not exist as to how ULBs would meet targets as per MSW rules, including financial requirements. There is currently no guaranteed performance report of any waste-processing technology and under such circumstances, ULBs are not in a position to take decisions about technology to be adopted. Guidelines on optimization of technology corresponding to the quantum of wastes generation do not exist. Awareness amongst the states/ULBs about the benefits of integration of various technologies for MSW processing is lacking (Planning Commission 2014).

6 Waste to Energy from Wastewater: The Case of Surat, India

Surat city located on the west coast of India, in the southern part of Gujarat state, is one of the major economic centres in India particularly for diamonds, textiles and for diamond-studded gold jewellery manufacturing. The city with its continuing high rate of growth has reached a population of 4.4 million in 2011 from 2.8 million in 2001. It is likely to reach 5.5–6.0 million by the year 2020 (CDP 2008). Surat Municipal Corporation (SMC) is the first municipal corporation in India that is maintaining its Sewage Treatment Plants (STPs) with biogas energy. Utilization of wastewater for power generation is perhaps the most suitable mitigation option, since it not only generates clean energy resulting in a reduction of GHG emissions but also helps in the disposal of wastewater, water pollution reduction and employment generation as Co-Benefits (Box 3).

Box 3: Methane Recovery and Power Generation from Sewage Treatment Plant at Surat, Gujarat

SMC has installed a one megawatt captive power plant at each of the Bhatar, Karanj and Singanpore STPs. In addition, a fourth plant at Anjana has an installed capacity of 0.5 MW. Sewage gas is generated by the process of anaerobic digestion of sludge. The project is designed to generate the sewage gas by treating the sludge in an anaerobic processing system (digester) so as to restrict the atmospheric emission of methane gas. At the same time, the methane gas is recovered without releasing it in the atmosphere to utilize it for high efficiency power generation by a gas engine. The electricity generated is used by the STP thus mitigating the greenhouse gas emission that would have resulted from the use of a fossil-fuel based grid power supply (SMC 2012). The total installed capacity of the power generation at SMC from the four plants is 3.5 MW. All four plants have more or less the same biogas characteristics. However, the electricity generation performance of Karanj plant is better with the slightly higher amount of electricity generation and net availability of electricity. The study has tried to analyse how the intervention was able to improve the environmental situation with regard to reduction in CO_2 emission in the without and with project intervention scenario. CO_2 equivalent emission is estimated according to UNFCCC guidelines (CDM 2012a, b) and shown in Fig. 2. If the captive power plants were not established, these STPs would have taken power from the grid that would have added more carbon dioxide into the atmosphere because the grid electricity primarily comes from thermal power plants using coal or diesel. Thus, biogas production from sludge and electricity generation from this biogas has resulted in reduction in emissions. As the methane is used for power generation, CO_2 emissions due to the use of grid electricity have reduced by about 50%. Thus, there is substantial emission saving from the plants. Each plant is able to save between 15,000 and 22,000 tonnes of CO_2e amounting to a total of 80,000 tonnes of emissions in Surat every year.

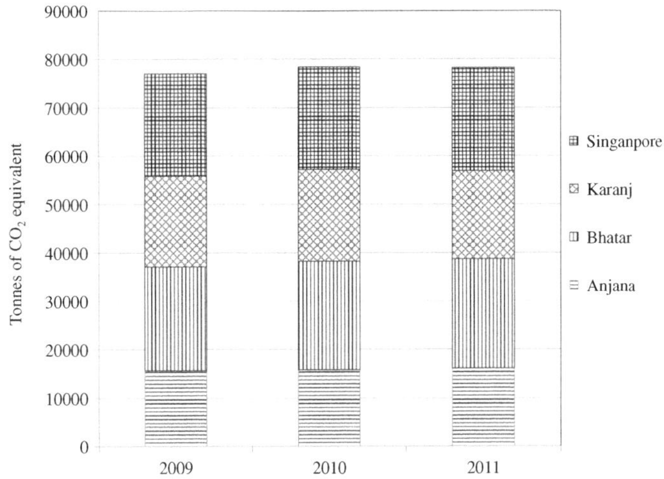

Fig. 2 Total CO_2e emission savings from the four STPs

7 Conclusions

Cities, mainly in developing countries, are facing more challenges as compared to the developed countries as competition for the limited resources, financial and natural endowments, is always a challange. Moreover lack of technical capacity also play an important role. As cities are growing fast the demand for energy is also growing rapidly especially in the industrial, transportation and agriculture sectors. Since these sectors consume a huge amount of fossil fuels, billions of tonnes of GHGs are emitted to the atmosphere every year. Indian cities are also experiencing a similar situation and can significantly benefit from the co-benefit approach for protecting the environment along with development. Waste to energy and associated Co-Benefits will continue to grow in importance following the recent launch of the SDGs, INDC and 'Swachh Bharat Mission' initiatives. There are many cases (three mentioned in this chapter) for initiating action on converting waste to energy but most of these efforts are small-scale, fragmented and project level initiatives that have not been integrated into national policy frameworks. In the case of Surat, with the establishment of the Anjana Sewage Treatment and Biogas based electricity generation, the Surat Municipal Corporation has taken good lessons from the first project (in 2003) and started three more projects in 2009. At present, three more projects are under commissioning stage and methane flaring from the plant is practised. One of the new sewage treatment plants at Bamroli that is ready to be commissioned has started the process of sewage treatment with the tertiary treatment process. In terms of technological functions, it can be concluded that SMC is currently at the stage of 'intensifying' the activity. Existing management for current systems, with continuous research and development in collaboration with external agencies, and enhancing a number of plants with the tertiary treatment process (promoting good initiatives that happen in one part of the city to other parts) can be a learning example for many other municipal authorities in India.

As evident from various studies, only a few states/UTs have set up processing and disposal facilities and as those plants are either closed or underperforming, this is a major cause of worry. The challenge includes the higher cost of technologies (bio-chemical waste to energy and thermo-chemical waste to energy), lack of awareness, lack of a focused national policy and not having expert manpower, particularly at ULB level, to make this kind of action more common in India. This requires serious attention from policy-makers to disseminate the learning and integrate this knowledge into national policy frameworks by which these practices could well expand in other regions and bring further positive Co-Benefits.

Acknowledgements The work presented here is a contribution towards a larger study on 'Urban Development with Co-Benefits Approach', carried out in collaboration with the United Nations University—Institute of Advanced Studies (UNU-IAS), Tokyo, Japan submitted to the Ministry of Environment Japan (MoEJ). The authors gratefully acknowledge the technical and financial support received from UNU-IAS and MoEJ. This chapter also benefits from our earlier published research namely, 'Analysing the Co-Benefits: Case of municipal sewage management at Surat, India' in the *Journal of Cleaner Production* (2013).

References

Aeat. (2005). *Renewable heat and heat from combined heat and power plants—Study and analysis*. London: DTI & DEFRA.

CDM. (2012a). *Clean Development Mechanism (CDM) methodology booklet* (4th ed.). United Nations Framework Convention on Climate Change (UNFCCC). http://cdm.unfccc.int/methodologies/documentation/meth_booklet.pdf. Accessed on March 20, 2013.

CDM. (2012b). *Indicative simplified baseline and monitoring methodologies for selected small-scale clean development mechanism (CDM) project activity categories, AMS-III.H.* https://cdm.unfccc.int/methodologies/SSCmethodologies/approved Accessed on March 20, 2013.

CDP. (2008). *Revised city development plan 2008–2013*. Surat Municipal Corporation. http://www.suratmunicipal.org/content/downloads/citydevelopmentplan/Surat_CDP.zip. Accessed on August 21, 2011.

CDP. (2012). *Revising/updating the city development plan (CDP) of Pune city—2041 under JNNURM*. http://www.punecorporation.org/pmcwebn/informpdf/CDP/Executive_Summary-Revised%20CDP.pdf. Accessed on August 20, 2014.

Census of India. (2011). *Rural urban distribution of population*. http://censusindia.gov.in/2011-prov-results/paper2/data_files/india/Rural_Urban_2011.pdf. Accessed on August 21, 2011.

CPCB-Central Pollution Control Board. (2008). *Status of water supply, waste water generation and treatment in class-I cities and class-II towns of India*. In Control of Urban Pollution Series. http://www.cpcb.nic.in/upload/NewItems/NewItem_153_Foreword.pdf Accessed on August 16, 2012.

CPCB-Central Pollution Control Board. (2013). *Status on municipal solid waste management*. http://www.cpcb.nic.in/divisionsofheadoffice/pcp/MSW_Report.pdf. Accessed on July 20, 2014.

GoI. (2008). *National action plan on climate change. Prime Minister's council on climate change.* Government of India. http://pmindia.gov.in/climate_change_english.pdf. Accessed on March 20, 2013.
GoJ. (2009). *Manual for qualitative evaluation of the co-benefits to climate change projects, version* 1, June 2009. http://www.env.go.jp/en/earth/cc/manual_qecba.pdf. Accessed on August 21, 2011.
Harvey, O. B. (1936). Generation of electricity at Palo Alto with a gas engine. *Sewage Works Journal, 8*(1), 68–80.
IGES. (2007). *The developmental co-benefits of climate policies in Asia.* http://www.iges.or.jp/en/cp/pdf/activity15/summary.pdf. Accessed on July 12, 2012.
INDC. (2015). *India's intended nationally determined contribution: Working towards climate justice.* http://www4.unfccc.int/submissions/INDC/Published%20Documents/India/1/INDIA%20INDC%20TO%20UNFCCC.pdf. Accessed on January 9, 2016.
IPCC. (2007). *Climate change 2007: Mitigation of climate change.* In Contribution of working group III to the fourth assessment report of the intergovernmental panel on climate change. Cambridge: Cambridge University Press.
IPCC. (2013). IPCC WGII AR5 glossary, final draft. http://ipcc-wg2.gov/AR5/images/uploads/WGIIAR5-Glossary_FGD.pdf. Accessed on January 05, 2014.
Jiang, P., Yihui, C., Geng, Y., Dong, W., Xu, B., Xue, B., et al. (2013). Analysis of the co-benefits of climate change mitigation and air pollution reduction in China. Journal of Cleaner Production, 58(1), 130–137.
Kapshe, M., Kuriakose, P. N., Srivastava, G., & Surjan, A. (2013). Analysing the co-benefits: Case of municipal sewage management at Surat, India. *Journal of Cleaner Production, 58*(1), 51–60. doi:10.1016/j.jclepro.2013.07.035.
Kumar, S., & Sharma, A. (2013). *Recent advances in bioenergy research* (Vol. 1). Kapurthala: Sardar Swaran Singh National Institute of Renewable Energy. http://www.academia.edu/5333560/Recent_Advances_in_Bioenergy_Research_Volume_I_Edited_by. Accessed on March 20, 2014.
Malkow, T. (2004). Novel and innovative pyrolysis and gasification technologies for energy efficient and environmentally sound MSW disposal. *Waste Management, 24,* 53–79.
MEDA. (2013). 1 × 5 TPD capacity segregated organic MSW based decentralized biomethanation plant at ward no 34, model colony, Pune. *Maharashtra Energy Development Agency.* http://www.mahaurja.com/PDF/BEPolicies/1.1%20Success%20Story%20on%20Biogas%20based%20Distributed%20%20Grid%20Power%20Generation%20Programme.pdf. Accessed on March 20, 2014.
MNRE. (2010). *Biogas generation, purification and bottling development in India—A case study.* New Delhi: Ministry of New and Renewable Energy. http://mnre.gov.in/file-manager/UserFiles/case-study-Biogas-Generation_Purification_and_Bottling-Development-In-India.pdf. Accessed on August 02, 2014.
MNRE. (2012). Creating wealth from organic waste. *Ministry of New and Renewable Energy, 5* (6), Success story. http://mnre.gov.in/file-manager/akshay-urja/may-June-2012/EN/38-40.pdf. Accessed on March 20, 2014.
MNRE. (2013). The first biogas bottling plant towards commercialization in India—A success story. *Ministry of New and Renewable Energy.* http://mnre.gov.in/file-manager/UserFiles/success-story_first_Biogas_Bottling_Plant_towards_commercialization-in-India.pdf. Accessed on March 20, 2014.
MoEF. (1986). *The environment [protection] act—1986.* New Delhi: Ministry of Environment and Forests, Government of India.
MoEF. (1992). *Policy statement on abatement of pollution.* New Delhi: Ministry of Environment and Forests, Government of India.
MoUD. (2012). *Proceedings of the consultation workshop on national mission on sustainable habitat.* New Delhi: National Institute of Urban Affairs. http://urbanindia.nic.in/programme/uwss/NSHS_workshop.pdf. Accessed on March 25, 2013.

NREL-National Renewable Energy Laboratory. (2010). *Indian renewable energy status report background report for DIREC 2010*. http://ren21.net/Portals/0/documents/Resources/Indian_RE_Status_Report.pdf. Accessed on March 20, 2014.

Pepermans, G., Driesen, J., Haeseldonckx, D., Belmans, R., & D'haeseleer, W. (2005). Distributed generation: definition, benefits and issues. *Energy Policy, 33*, 787–798.

Planning Commission. (2014). *Report of the task force on waste to energy (Volume I)—In the context of integrated MSW management*. http://planningcommission.nic.in/reports/genrep/rep_wte1205.pdf. Accessed on August 02, 2014.

Psomopoulos, C. S., Bourka. A., & Themelis, N. J. (2009). Waste-to-energy: A review of the status and benefits in USA. *Waste Management, 29*(5), 1718–1724. doi:10.1016/j.wasman.2008.11.020. Accessed on January 15, 2015.

Sharma, S., & Shah, K. W. (2005). Generation and disposal of solid waste in Hoshangabad. In *Book of proceedings of the second international congress of chemistry and environment* (pp. 749–751). India: Indore.

Shekdar, A. V., Krshnawamy, K. N., Tikekar, V. G., & Bhide, A. D. (1992). Indian urban solid waste management systems–jaded systems in need of resource augmentation. *Journal of Waste Management, 12*(4), 379–387.

SMC. (2012). *An unpublished report on sewage gas based power generation plants at sewage treatment plants*. Surat: Surat Municipal Corporation.

UNFCCC. (2015). *Conference of parties*. December 2015, FCCC/CP/2015/L.9. United Nations Framework Convention on Climate Change. http://unfccc.int/resource/docs/2015/cop21/eng/l09.pdf. Accessed on January 24, 2016.

Williams, P. (2005). *Waste treatment and disposal* (2nd ed.). Chichester: Wiley.

Zhu, D., Asnani, P. U., Zurbrügg, C., Anapolsky, S., & Mani, S. (2008). *Improving municipal solid waste management in India: A sourcebook for policy makers and practitioners*. https://openknowledge.worldbank.org/handle/10986/6916. Accessed on August 01, 2014.

Chapter 9
Co-benefits of Urban Biodiversity

Meenakshi Dhote and Debojyoti Mukherjee

Abstract Growth and development of an urban area modifies and creates biodiversity, which has a vital role in maintaining and improving the quality of the urban environment. The paper studies the growth and development of Delhi, the capital city of India, and outlines an approach to assess the biodiversity pattern of the metropolis. The co-relation between habitat scales and levels of planning to arrive at a biodiversity profile of an urban area, are explained. The Global Biodiversity Assessment and Convention on Biological Diversity has brought forth the need to conserve biodiversity at global and local levels. Because of their diminishing numbers, the role of species in urban areas has become critical. So far, in cities, open spaces are supposed to mainly cater to aesthetic demands and recreation needs. But if we examine the environmental role they play—pollution abatement, water recharge, indicator of pollution, climatic amelioration, flood control and so on —they are many. The strategy for conservation is related to the scale of habitat, thus a multiscale strategy for conservation is explained in detail. The legal, governance and policy tools in India relating to national, sub-national and local levels are put forth to give a holistic picture of the various aspects to be considered for mainstreaming biodiversity conservation in the urban planning process.

Keywords Convention on biological diversity · Urban biodiversity
New delhi · Master plan · Ecological and organismal diversity
Habitat scales · Land use planning · Strategic environmental assessment

M. Dhote
Department of Environmental Planning, School of Planning and Architecture, New Delhi, India
e-mail: m.dhote@spa.ac.in

D. Mukherjee (✉)
Consultant, IPE Global Ltd., New Delhi, India
e-mail: debojyoti1920@gmail.com

M. Sethi and J.A. Puppim de Oliveira (eds.), *Mainstreaming Climate Co-Benefits in Indian Cities*, Exploring Urban Change in South Asia,
https://doi.org/10.1007/978-981-10-5816-5_9

1 Background

Urban environments and biodiversity are often seen as mutually exclusive. In many parts of the world, especially in developed countries, the majority of the population lives in urban areas. This is now becoming the case in a growing number of developing countries as well. The current pace and scale of change—over 60 million people are added to the urban population each year—often strain the capacity of local and national governments to provide even the most basic services to urban residents. An estimated 25–50% of urban inhabitants in developing countries live in impoverished slums and squatter settlements, with little or no access to adequate water, sanitation, or refuse collection. In such situations, both environmental quality, human health and well-being are at risk. Urbanization is rapid in the fast-growing economies of the Asia-Pacific region, where the average annual urban growth rate is more than 4%. The level of urbanization in India as a whole increased from 27.7% in 2001 to 31.1% in 2011—an increase of 3.3% points during 2001–11 compared to an increase of 2.1% points during 1991–2001. The number of urban areas has increased from 5161 in 2001 to 7935 in 2011 (Census of India 2011); with 2774 towns having been added in ten years. With this, the densities of human settlements will rise, and put further stress on the urban environment. Thus, protection of natural habitats in urban settings is critical—not only are they facing the threat of being engulfed by the needs of development; their essential contribution for the betterment of environmental quality of urban areas is also being lost.

Human settlements are caught in the web of exhaustive and unsustainable consumption of resources and production of wastes and draw upon nature to offer a range of ecological services, biodiversity being one of them. Out of energy poverty, biomass fuels such as fuel wood, crop residues, and animal dung continue to be the dominant fuels used by rural households. These power generation units generally run on diesel. In the urban sector the important sources of energy are kerosene (10%), firewood and chips (22%) and liquefied petroleum gas (LPG) (57%). In 2007, the residential sector emitted 137.84 million tonnes of carbon dioxide (CO_2) equivalent, of which 69.43 million tonnes were in the form of CO_2 emissions, mainly from fossil fuel use in the residential sector. The methane (CH_4) and nitrous oxide (N_2O) emissions were 2.72 million tonnes and 0.036 million tonnes, respectively. The CH_4 emissions are driven by the biomass consumption in the residential sector. The commercial and institutional sector also sees extensive use of captive power generation across the country due to frequent power shortages in various seasons. In addition, crop residue is burnt in the fields in many Indian states such as Uttar Pradesh, Punjab, West Bengal, Haryana, Bihar, Madhya Pradesh, Himachal Pradesh, Maharashtra, Gujarat Chhattisgarh, Jharkhand, Tamil Nadu, Uttaranchal and Karnataka producing CO, CH_4, N_2O, NO_x, Non-Methane Hydrocarbon Compounds NMHCs, SO_2 and many other gases. As per the Government's estimates (MoEF 2010), 0.23 million tonnes of CH_4 and 0.006 million tonnes of N_2O was emitted from burning of crop residue in India in 2007.

Mainstreaming of biodiversity is defined by the Scientific Advisory Technical Panel of the Global Environmental Facility (GEF) as a process of embedding biodiversity considerations into policies, strategies and practices of key public and private actors that impact or rely on biodiversity, so that it is conserved and sustainably used both locally and globally. India was one of the first countries to have a proactive legislation and enacted a comprehensive Biological Diversity Act in 2002 to implement the provisions of Convention on Biological Diversity (CBD). The Biodiversity Rules were notified in 2004. The Act is being implemented through a three-tier structure, National Biodiversity Authority (NBA) at the national level, State Biodiversity Boards (SBBs) at the provisional level, and Biodiversity Management Committees (BMCs) at the local level. India's National Biodiversity Action Plan (NBAP 2008), formulated through a comprehensive inter-ministerial process and approved by the Government of India (GoI) in 2008, was developed prior to the CBD Strategic Plan for Biodiversity 2011–20. NBAP (2008) outlined broad actions and detailed actionable points, generally aligned with the targets laid down by the five Strategic Goals of Strategic Plan for Biodiversity (SP) 2011–20 and the corresponding 20 Aichi Biodiversity Targets. India is among the select countries in the world that have developed their own National Biodiversity Targets aligned with the Aichi Biodiversity Targets. A framework with indicators, agencies responsible for monitoring and reporting, and frequency of monitoring/reporting has been developed for the implementation of the National Biodiversity Targets, on the basis of consultations with a range of stakeholders and a review of the programmes and activities being undertaken by relevant ministries/departments in the GoI, SBBs and non-governmental organizations (NGOs)[1] (MoEF 2014).

Though, globally, several mainstreaming projects have been conceived and operationalized, in India incorporation of biodiversity concerns in urban land use planning and sectoral planning is not very common barring the LBSAP (Local Biodiversity Strategy and Action Plan) of Bhopal and application of Singapore Biodiversity Index to assess the Hyderabad City Biodiversity Index and MIRA Bhayender Biodiversity Index in 2012. Whereas instruments for mainstreaming biodiversity conservation in urban areas are available, their application, testing and wide usage in the country are just beginning. It is felt that with the appreciation of the co-benefits, the process would gain momentum.

Successful implementation of sustainable development requires action which is based on the understanding that biodiversity is more than natural resources, including food, fibre, fuel, and medicinal plants, which are essential for human development and are provided both from managed agricultural ecosystems and less managed 'natural' ecosystems. In 'The Future We Want' (2012), countries reaffirmed the importance of biodiversity and its critical role in maintaining ecosystems that provide essential services that serve as critical foundations for sustainable development and human well-being, poverty eradication, sustained economic

[1]Report on progress made to address biodiversity in poverty eradication and sustainable development, unep/cbd/wgri/5/6;7 May 2014.

development and the resilience of societies, allowing them to adapt to and cope with risk and change. It needs to be understood that there are fundamental interlinkages between biodiversity, ecosystem services, human health, and development. Ecosystem services provide food, water, and clean air, cultural and spiritual services, and they help regulate disease and climate. The loss of biodiversity threatens human health by undermining ecosystem resilience and reducing genetic diversity (Campbell 2012). It reduces the availability of traditional foods and medicines, directly impacting food and nutritional security as well as drug development. Land degradation and resulting impacts to ecosystems influence vector-borne disease, emerging infectious disease including pandemics, and water-related disease. Changes in land use often impact the water cycle, with implications for sanitation and access to clean drinking water. Furthermore, habitat loss undermines ecosystem resilience leading to increased vulnerabilities to disturbances and is an underlying cause of emerging infectious diseases.

In decision X/2, the tenth meeting of the Conference of the Parties (COP) held from 18 to 29 October 2010 in Nagoya, the Aichi Prefecture of Japan adopted the Strategic Plan for Biodiversity, including 20 Aichi Biodiversity Targets, for the 2011–2020 period. Both the vision of the Strategic Plan for Biodiversity and its mission emphasize the role that biodiversity and ecosystem services play for sustainable development, referring explicitly to essential elements such as poverty eradication and human well-being. The vision is that 'by 2050, biodiversity is valued, conserved, restored and wisely used, maintaining ecosystem services, sustaining a healthy planet and delivering benefits essential for all people'. The mission of the Strategic Plan is:

> to take effective and urgent action to halt the loss of biodiversity in order to ensure that by 2020 ecosystems are resilient and continue to provide essential services, thereby securing the planet's variety of life, and contributing to human well-being, and poverty eradication. To ensure this, pressures on biodiversity are reduced, ecosystems are restored, biological resources are sustainably used and benefits arising out of utilization of genetic resources are shared in a fair and equitable manner; adequate financial resources are provided, capacities are enhanced, biodiversity issues and values mainstreamed, appropriate policies are effectively implemented, and decision-making is based on sound science and the precautionary approach (Saxena et al. 2012).

The UNEP Global Biodiversity Assessment and Convention on Biological Diversity (CBD) has brought forth the need to conserve biodiversity at global and local levels. In order to understand this we should appreciate the role of living organisms, especially in urban settings. The American ecologist Robert Whittaker proposed that the life forms can be divided into five kingdoms—*Prokaryotae, Protoctista, Plantae, Fungi and Animalia*. Each organism has its own pattern of functioning and the habitat in which it can survive. Also, they have a role to play in the habitat they live in. When we examine this, with respect to human settlements and especially the case of urban settlement, we find that species from all the five kingdoms have either dwindled or vanished when areas change from natural areas to agricultural areas, to urban areas. Because of their diminishing numbers, the role of species in urban areas has become critical. So far, in cities, open spaces are supposed to cater to mainly aesthetic demands and recreational needs. But if we examine the environmental role they play—pollution abatement, water recharge, indicator of pollution, climatic amelioration, flood control and so on, there are many. In other words, urban biodiversity needs to be protected for sustainable urban growth.

2 Urban Biodiversity

Growth and development of an urban area modifies natural biodiversity and at times introduces newer species, which collectively play a vital role in maintaining and improving the quality of urban environment. The city, originally a natural system, is made artificial, the functioning of which requires the presence of a certain number of different species. The human species is not the only one living in cities. There are always either remnants of nature, which preceded the city, or biotic elements introduced by man. Accordingly urban biodiversity can be categorized as natural, semi-natural and artificial habitats, as conceptualized in Fig. 1.

An attempt to view human settlements as habitats or ecosystems identifies the critical role played by the ecosystems to sustain human populations. The various physical, biological, and chemical components (i.e. land, air, water, vegetation, animals, etc.) collectively contribute in regulating water availability, controlling climate, improving air quality, enhancing aesthetic potential, and so on. These resources can be viewed as the ecological infrastructure of the settlement. Various types of functions performed by habitats or ecosystems can be summarized in Table 1. In the case of urban areas most of these functions are carried out by open spaces, green areas, water bodies and unpaved areas.

Urban biodiversity conservation efforts in India date back to efforts of the rulers in ancient and medieval periods where location of groves and gardens were planned simultaneously with the city layouts. The towns and cities during the Mughal and

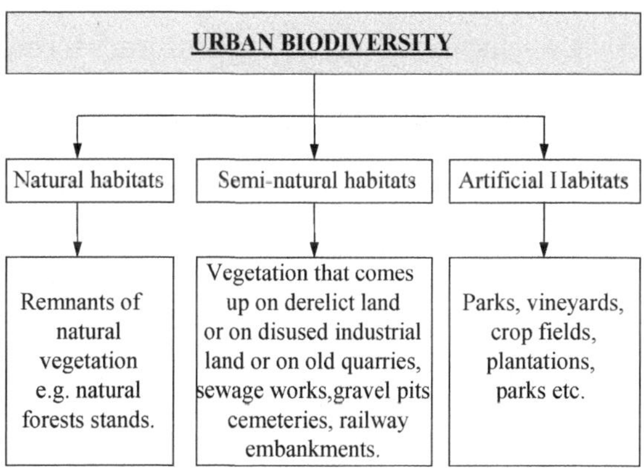

Fig. 1 Natural, semi-natural and artificial habitats within urban biodiversity

Table 1 Types of functions performed by habitats, ecosystems, open/greens paces in urban areas

Regulation functions	Capacity of the natural/semi-natural ecosystem to perform the essential ecological process and life supporting system (providing clean air, water and soil, etc.)	• Regulates the chemical composition of atmosphere and purifies the local air • Controls the run-off and flooding • Regulates the hydrologic cycles
Carrier functions	Capacity of the natural and semi-natural ecosystem by supporting many of the human activities by providing adequate space and medium	• Supports the biological diversity in the city; prevents soil erosion and sedimentation; regulates the local and global climate • Provides food and raw materials • Conserves the energy in the city through controlling the micro climatic variations • Provides recreation and tourism • Integrates urban man with nature
Production functions	This is the capacity of nature to produce food and raw materials for human consumption and industrial use and to function as an energy resource and genetic material	• Produces oxygen • Produces raw materials for some of the human activity • Recharges the ground water tables • Provides medical resources
Information functions	This is the capacity of nature to contribute to the spiritual enrichment, cognitive development and aesthetic experience	• Aesthetic information • Spiritual and religious information • Cultural and artistic inspiration • Scientific and educational information source

British periods propagated a variety of plants and trees for their fruits, affording shade in large bungalows and roadside plantations. Modern town planning in independent India also puts down a certain percentage of developed area to be mandatory for open spaces and recreation. An overview of planning and development of the National Capital Territory (NCT) Delhi brings forth an approach for addressing biodiversity conservation concerns alongside development goals.

3 Development of NCT Delhi and Urban Biodiversity

NCT Delhi stretches along the banks of the Yamuna river between 28°12′N–28°53′N and 76°50′E–77°23′E latitudes and longitudes respectively. Located in the semi-arid biogeographic zone; the natural vegetation of this zone consists primarily of tropical thorn forests. Delhi is also situated on the water divide (i.e. the Aravalis) dividing the two mighty river systems—the Ganga draining into the Bay of Bengal and the Indus falling in the Arabian Sea. It is therefore not surprising that Delhi has not only been physiographically important but also strategically important. Delhi, the capital city of India, makes a good example for a study of urban biodiversity, as in spite of rapid urban growth it still retains a large area (19% of developed area) under planned green and within the heart of its urban area lies a forest ecosystem—the Ridge, 7782 ha of arid scrub forest—and a river ecosystem—the Yamuna river, 51 km in length in NCT Delhi with 9700 ha of floodplain.

3.1 Growth of the City up to First Master Plan

The Aryans who entered India from the steppes of Southern Russia and Central Asia in the second millennium BC had a capital named Indraprastha, which tradition and archaeology have identified with Delhi. Dating from 1000 BC the epic *Mahabharata* recounts the capital city of Pandavas as *Indraprastha.* B.B. Dutt in his book *Town Planning in Ancient India* mentions the high level of importance given to choice and placement of flora and associated fauna in the cities of this era. The Mughals laid down a number of their beautiful gardens in Delhi, six of which exist even today—five in Old Delhi and one in New Delhi. The Jahanara Garden or Queen's Garden around Town Hall, Qudsia Bagh near the Interstate Bus Terminus (ISBT), Roshanara Gardens near Shakti Nagar, Shalimar Bagh and Beriwala Bagh near Azad Market. The Jahanara Garden is a typical specimen of a Mughal garden, replete with a pond and water channels. There are some ancient date palms growing along with other species of trees. The Talkatora Gardens in New Delhi, and Old Mughal Garden laid out by Shahjahan, have a few 300-year-old tamarind trees, in addition to many other beautiful trees.

When Lutyens was entrusted with the stupendous task of planning the Imperial city of Delhi; familiarization with the terrain was the first task. In one of his many sojourns north and south of Shahajahanabad and the east bank of the Yamuna. The subsequent shift of the capital to Delhi in 1912 changed the landscape of Delhi. The city was planned to accommodate 30,000 to 57,000 persons in ten square miles. A meticulous plan was made for lining the avenues and setting gardens in consonance with Lutyens' design of New Delhi (Fig. 2). Their tree-planting efforts road by road, which extended through the winter of 1919 up to 1924, consisted of a list of 121 trees, of which the British chose only a handful. The choice was based on what was indigenous, shade giving, stately and long lived, in that order of preference. Species planted included Neem, Tamarind, Jamun, Arjun, Banyan, Peepul, Siris, Pilkhan and Willow fig.

The British period spans the settlement development stage of industrial society and the neo-classical age. Here, both the ecosystem diversity and organismal diversity were modified. Also, emergence of the biodiversity which are vectors of diseases, pests started increasing in a large way. As part of classical planning, avenue plantation, large expanses of formal greens, manicured garden estates created protected areas for a few species of flora and fauna. These were modified ecosystems and species/organismal diversity compared to the original ecosystems.

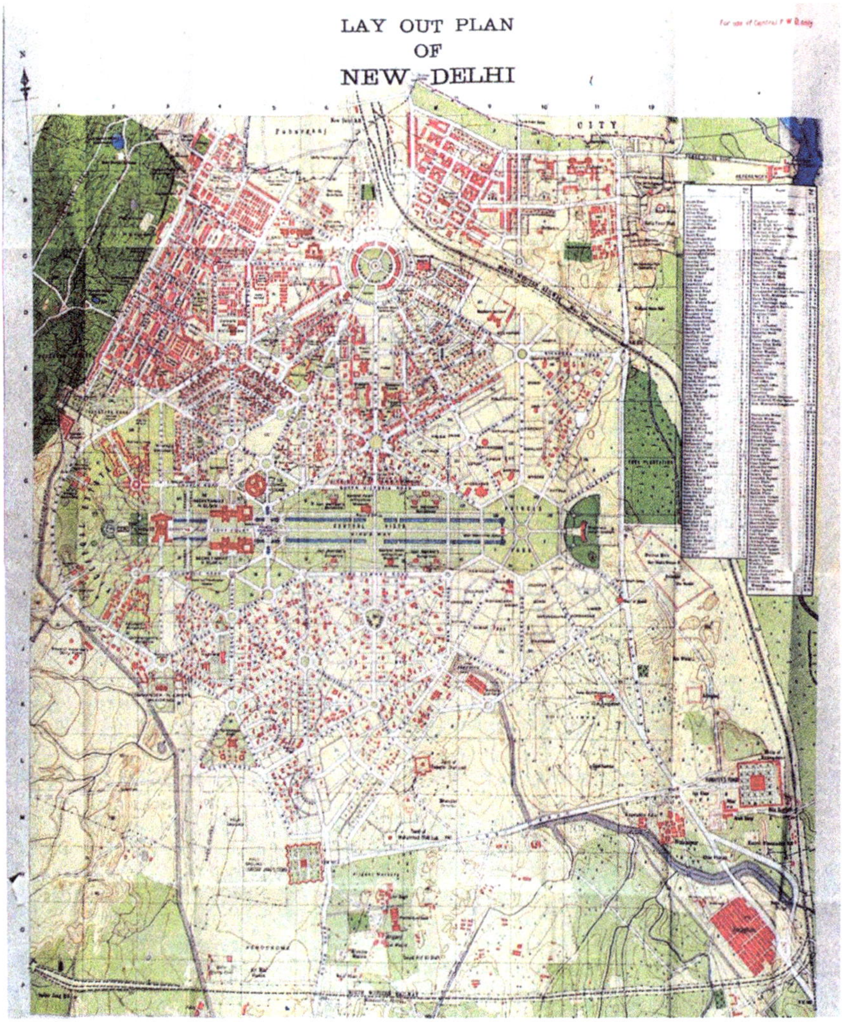

Fig. 2 Lutyens' Delhi layout plan. *Source* Work studies of first Master Plan 1962 (DDA 1962)

The number of exotic species and domesticated fauna grew steadily. Animals were seen frequently straying in human settlements that were becoming sources of food, which was increasingly unavailable due to modification of habitats in the surrounding hinterland.

Following the British planning of Delhi the next major spurt of development occurred during the partition. Growth towards the west of the Ridge occurred during this time. Prime agricultural land was taken up for urbanization. Many orchards were razed to accommodate residential areas. After independence the city grew steadily and in 1962 the first Master Plan for Delhi was notified. The work studies have recorded the land use and environs in 1958–1959. These indicate that the urban area of Delhi constituted about 11.7% of the total union territory. The major natural features included the ridge, Yamuna river, Najafgarh lake and drain. The northern and eastern parts were dotted with marshy areas. Most of the streams flowing from the ridge in south Delhi were also present. We can infer that most of the natural features were present until independence.

An overview of data sources indicates that at the level of ecosystems, there is representation of ecosystems that are characteristic of semi-arid biogeographic regions. The listing of species indicates that large to small fauna were present in the terrestrial and aquatic habitats. Introduction of exotic species in fruit gardens and gardens attached to large residences and public buildings had begun. Avenue plantation, consciously propagating native species is a gift of the planning exercise of the British. Proliferation of stray animals in dense settings also seems to be a fallout of the city's expansion.

3.2 Growth of Delhi from 1962 to Present

During the British Regime, Delhi was a District in Delhi Division of the North West Province of Punjab. After independence it was accorded the status of Union Territory, presently called the NCT of Delhi. Subsequently, the urban area has increased from 17,290 ha to 58,285 ha from 1958 (see Fig. 3) to 2001, that is, from 11.7% of the Union Territory to almost 40%. According to the estimates of the Delhi Development Authority (DDA), by 2021 the area under urban sprawl would account for 56.5% of the developed area and by the speculation trends it would be almost 66%, that is, 98,054 ha.

The 1958 land use classification indicates area under agriculture, urban, ridge and orchards, waterbody and flood plains of the river. In 1996 this classification constituted urban, agriculture, forest, river/waterbody, wasteland, farmhouses, open land, plantation, open scrub, quarry, wildlife sanctuary and roads. The major transformation has been the conversion of cropland to urban built-up area. Attempts have been made to maintain the forest cover quantitatively, however qualitatively the density has reduced. Some agricultural lands have been sold to make farm-houses, where a large variety of exotic species thrive. Documentation of flora and fauna during this period indicates introduction of exotic and hardy species. Proliferation of smaller and hardy mammals seem to have started.

3.3 Master Plans of Delhi

The Master Plan of Delhi (MPD), both 1962 and 1981, had proposed a network of green spaces within the urban limits to meet the recreation needs of the people and to function as lung spaces. These had their genesis in the London County Plan prepared

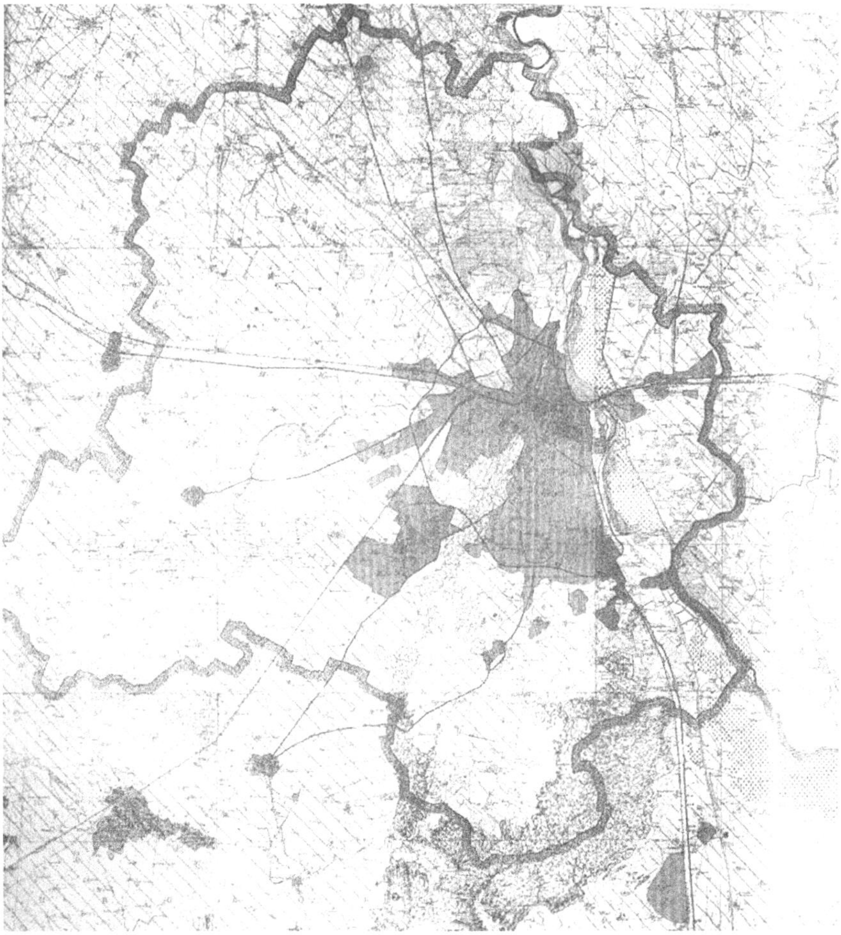

Fig. 3 Environs of Delhi, 1958–1959. *Source* Work studies of first Master Plan 1962 (DDA 1962)

by Sir Abercrombie and Haussmann's plan for Paris and Ebenezer Howard's garden city. The MPD 2001 had proposed 8722 ha as area under recreation which was at the rate of 9.7 m^2 per person (DDA 1990). Further conversion of recreational areas to other uses could only be carried out under special circumstances. Although the green area in Delhi is managed by different agencies, DDA has the largest role to play with over 5050 ha (approx.) under its jurisdiction. The green cover in the capital was 19% of the total area, much larger than other cities.

Vision-2021 is to make Delhi a global metropolis and a world-class city, where all the people would be engaged in productive work with a better quality of life, living in a sustainable environment. This will, amongst other things, necessitate planning and action to meet the challenge of population growth and in-migration into Delhi; provision of adequate housing, particularly for the weaker sections of the society; addressing the problems of small enterprises, particularly in the unorganized informal sector; dealing with the issue of slums, up-gradation of old and dilapidated areas of the city; provision of adequate infrastructure services; conservation of the environment; preservation of Delhi's heritage and blending it with the new and complex modern patterns of development; and doing all this within a framework of sustainable development, public private and community participation and a spirit of ownership and a sense of belonging among its citizens. For an overview of the existing and proposed urban area, see Fig. 4.

The hierarchy in the planning of greens by DDA are: regional parks, district parks, neighbourhood parks, city forests, historical landscapes, sport complexes, landmark greens, green belts, tot lots etc. In planning and development of regional parks and city parks care has been taken to reduce all artificial landscaping; therefore areas under the ridge and areas possessing remnants of natural growth come under this category. The other categories of greens work towards creating an overall socio-physical environment for the city—serene gardens, bustling picnic huts, lively musical fountains and joyous play fields along with children's parks.

3.4 *Methodology to Study Biodiversity of Delhi*

In Delhi, the growth pattern has fragmented the ridge, polluted the rivers and streams, reclaimed and polluted the wetlands and large agricultural areas have been converted to urban use. Apex species such as the leopard cat (Canis *lupus bengalensis*) of the terrestrial ecosystem, and the marsh crocodile (Crocodylus *palustris*) of the aquatic ecosystem have disappeared. Many species of plants which are indigenous to the ecosystem-*dhak* (Butea *monosperma*) are sparse. Presently a large variety of exotic species of plants and hardy species thrive. In order to study the biodiversity of the metropolitan city with a view to ascertain its environmental role and contribute to sustainable development, concepts applied in various studies in India and other countries need to be tested for their application in the context of Delhi. The following is an approach to understand the character of its biodiversity.

The existing biodiversity of Delhi could be understood in the spatial planning process and can be categorized at two levels—ecological diversity and organismal diversity. The ecosystem diversity has been based on the physiographic characteristic of Delhi and has been divided into six categories, detailed in Fig. 5.

1. Lowlands	– Flat agricultural/ horticulture, grazing lands
2. Hills	– Rocky outcrops with arid vegetation
3. Forest ecosystems	– Dry deciduous arid forest of the Ridge
4. Valleys	– Natural storm water drains
5. Freshwater wetlands	– Lakes and ponds
6. Riverine ecosystem	– River and flood plains

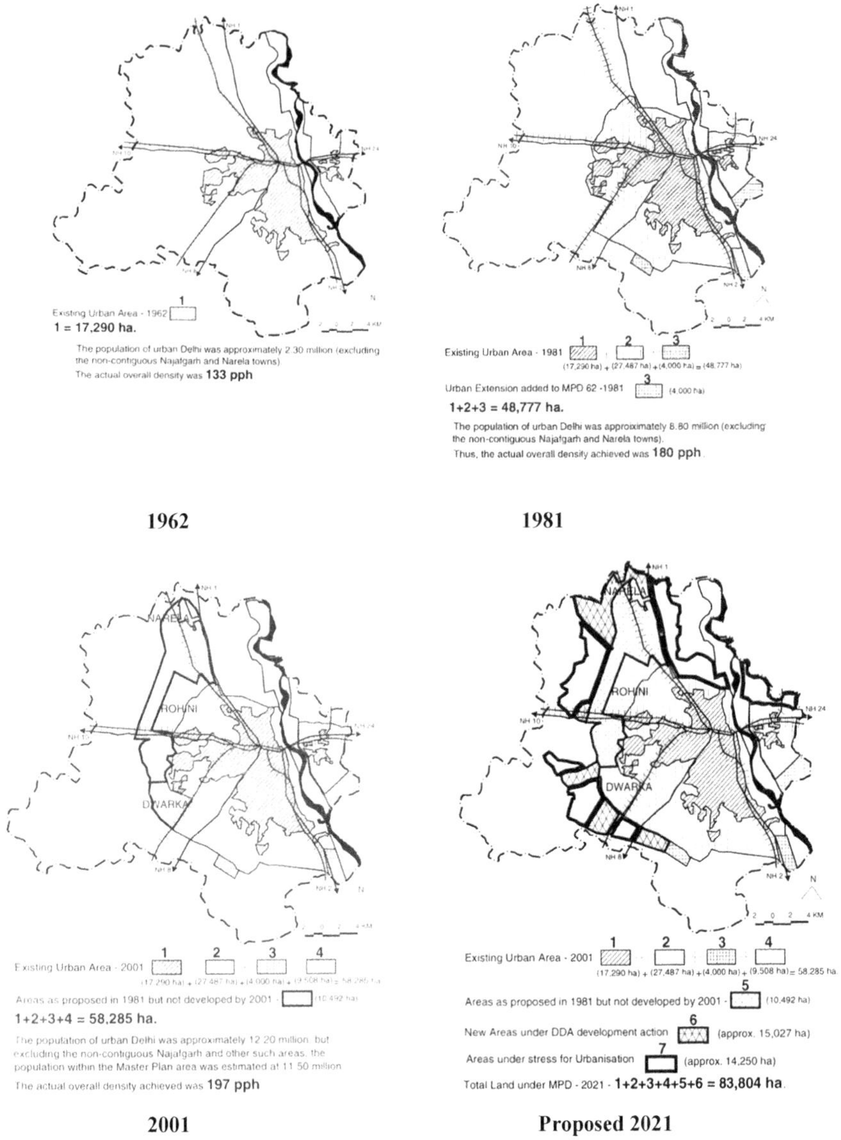

Fig. 4 Existing and proposed urban area, Delhi. *Source* DDA 2006

The organismal diversity is basically diversity at the species level. Comprehensive data is available for the flora of Delhi by Maheshwari and the fauna of Delhi by the Zoological Survey of India. A listing of recently introduced plants is also available from the Forest Department, Delhi Development Authority, Central Public Works Department, New Delhi Municipal Committee, Municipal

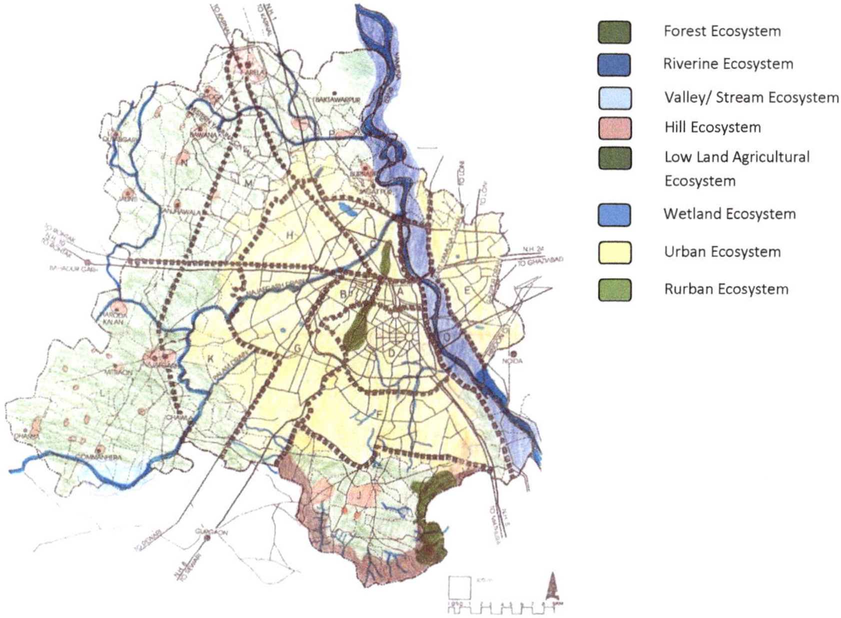

Fig. 5 Ecosystem diversity of Delhi

Corporation of Delhi and Cantonment boards. Many non-governmental organizations NGOs, and Eco-clubs record the fauna existing in various areas. The biodiversity status has to be related to the status of other resources for integrated resource conservation. An analysis of the natural resource profile of Delhi indicates:

- The temperature differences between city area and its fringe vary from 5 °C in December to about 7 °C from January to March
- High precipitation is recorded in areas of high population and larger greens
- When NW winds blow, the core of the city falls in the downwind direction of the large industrial areas (air polluting industries have thus been closed down). Air pollution is also due to traffic and high levels of Suspended Particulate Matter (SPM)
- Most of the area suitable for development has already been developed and the present growth is extending into conservation and transitional areas. Thus, there is a need to control haphazard and accelerated development
- On the basis of the water table fluctuation and trends, the Central Ground Water Board (CGWB 2007) has worked out the unsaturated zones available for water recharge in Delhi. These indicate areas where water recharge actions should be undertaken, which include many water bodies.

Based on land use and land cover data in 1991, the ecological diversity of Delhi is mapped in Fig. 5 and can be characterized by the following major features:

- The water bodies of Delhi are important not only for contributing to the water balance but also supporting a variety of flora and fauna. Their status is critical for the environmental sustainability of Delhi.
- Delhi has been capable of holding a wide variety of trees, crops, vegetables and fruits, only wherever the right cultivation practices taking into account water and soil behaviour, have been undertaken.
- The major natural forests in Delhi are generally restricted to the ridge, while there are 19 mixed forest. The actual forest area is only 22 km^2 (which covers about 1.5%) of Delhi's area.
- Delhi is well represented by most groups of fauna from the unicellular *Protozoa* to the highest group of *Mammalia.* Among the vertebrates, as mentioned earlier, birds (*aves*) formed the largest group, followed by the fish, *mammalia* and *reptilia.* An analysis of the occurrence pattern of the faunal species indicate that they inhabit certain habitat types and indicate the health of the habitat—quality of the soil, water to ensure availability of food—fruit and flower bearing species and other prey organisms.
- The role of flora in environmental protection is critical. The choice of species has to be considered taking into account the contribution of the species to environmental up-gradation—dust absorption, air purification, noise barrier, soil erosion control, soil conservation, windbreak, etc.
- Interestingly enough, it is observed that of the recorded Delhi fauna, there are 24 species belonging to Schedule I of the Wild Life (Protection) Act 1972 and 15 species of these also feature under Appendix I of CITES. All these belong to vertebrates except one, which is an invertebrate (*Lepidoptera*). Among the vertebrates six species were mammals, eleven species of birds and six species of reptiles, belonging to Schedule II. This itself is sufficient to be more conscious of Delhi's environment and conserve its biodiversity. Delhi has the distinction of possessing the largest number of bird species compared to other metropolitan cities in India. This could be plausible due to the presence of a major Asian flypath of birds along the River Yamuna.

3.4.1 Habitat Scales for Biodiversity Assessment

Habitat disaggregation would vary depending on the scale of the study. The area of NCT Delhi is 1483 km^2, which is a meso habitat. The urban area of Delhi is 582.85 km^2; also a meso habitat. At the scale of NCT Delhi, only major ecosystems can be identified and assessed for their biodiversity value. However, at the scale of urban Delhi and planning zone level, that is, Zone A to P (11.6–229.8 km^2), which are also meso habitats, landscape elements such as patch, corridor and matrix can be recorded.

3.4.2 Land Use and Habitats

At the NCT level (meso habitat) and urban area level (meso habitat) the land uses can be further classified into habitats. The land use of urban area is available from the master plans. A review of the Master Plan approach, in 1996, brought about the Urban Development Plans Formulation and Implementation (UDPFI) Guidelines (MUAE 1996). The land use classifications as suggested therein have been analysed to arrive at the habitat typology in Table 2.

It has been observed that habitat quality within similar land use zones varies with respect to:

- *Density of Development*—for example, single family residential (bungalows, plots) or multi-family residential (flats, group housing)
- *Ground Coverage of Buildings*—that is, the effective open soil/unpaved area available as habitat for flora and fauna
- *Age of the Locality*—pre-independence development has mature trees, native species, post-1962 Master Plan developments have quick growing, flowering trees and exotics
- *Size of the Land Use Zone*—since each land use is a 'Patch' as per landscape element classification, landscape metrics (such as patch size, perimeter to area ratio, and so on), would have an effect on the species composition therein.

Conservation of biodiversity has to be undertaken in an integrated manner, to ultimately lead to overall resource conservation. Any method applied has to understand the characteristics of the biodiversity specific to the settlement under study. For urban areas, the study would be limited to ecosystems and species diversity. The land use pattern of the settlement should be correlated to the habitats of the species it supports to accord protection/conservation measures as per the size of the area and the sensitivity of the area. Since land use classification is undertaken at various scales, the assessment of the biodiversity status and its role has to be documented at various scales for making the case for co-benefit. The strategy for conservation would thus emerge based on the type of species/habitats, scale of the area, status of the biodiversity and its environmental role.

3.5 Biodiversity Profile of the Urban Area

An urban area is an administrative unit and since it may not form an ecological unit; its setting within the ecological unit has to be understood. The biogeographic classification by Rodgers and Panwar has subdivided the Indian sub-continent into ten biogeographic regions. The classification identifies the ecosystem types present in the region. The biogeographic region within which the urban area is located and its major ecosystem types have to be recorded. Delhi lies in the semi-arid region, with scrub lands, rivers, wetlands, and forests forming the characteristic

Table 2 Land use and habitats (meso scale)

S.NO.	**Use zone**	**Habitat category**
1	*Residential*	
	Primary residential zone	Open scrub
	Mixed residential zone	
	Unplanned/informal residential zone	
2	*Commercial*	
	Retail shopping zone	
	General business commercial district/centres	Cliff and caves
	Wholesale, godown, warehousing, regulated markets.	
3	*Manufacturing*	
	Service and light industry	Open scrub
	Extensive and heavy industry	
	Special industrial zone Hazardous, chemical and noxious	
4	*Public and semi-public*	
	Govt./semi govt./public offices	Cliff and caves
	Govt. land (use undetermined)	Open scrub to woodland
	Educational and research	Open scrub to woodland
	Medical and health	Open scrub to woodland
	Social, cultural and religious	Open scrub
	Utilities and services	Open scrub
	Cremation and burial grounds	Open scrub to woodland
5	*Recreation*	
	Playgrounds, stadium, sports complex	Open scrub to woodland
	Parks and gardens (public open spaces)	Open scrub to woodland
	Multipurpose open space (*Maidan*)	Open scrub to woodland
6	*Transportation and communication*	
	Roads	
	Railways	
	Airport	Open scrub
	Sea port and dockyards	
	Bus depots truck terminal and freight complexes.	
	Transmission and communication	
7	*Agriculture*	
	Agriculture	Open scrub
	Forest	Woodland
	Brick Kilns and Extractive Area	
	Water bodies	Wetland
8	*Special area*	
	Old built-up areas	
	Heritage and conservation	

(continued)

Table 2 (continued)

S.NO.	Use zone	Habitat category
	Scenic value areas	Open scrub to woodland
	Village settlement	
	Other uses	

Source Use zone classification is derived from UDPFI Guidelines

ecosystems. Urban areas are dynamic and grow at a fast rate. The population size and area is recorded yearly by the department of statistics and by the census every decade. This growth results in expansion into agricultural areas, forests, wetlands and other ecosystems. A study and analysis of the settlement's growth and changes in biodiversity to understand the extent and nature of changes in ecosystems and type of species has to be appreciated. This would help to protect the ecosystems and species from further degradation. The biodiversity that presently exists—ecosystem, organismal and its environmental role—has to be recorded. The biodiversity character of the city can thus enhance the quality of the ecological infrastructure and therefore contribute to sustainability.

4 Co-relation of Land Use Planning and Biodiversity Conservation

Growth and development of an urban area is understood and spatially planned with the help of land use classification at various scales and at various points of time. As explained earlier, land uses are potential habitats for organisms and land use pattern can form the basis for recording the biodiversity character. Since land use is recorded at various scales, habitat scales can be correlated to land use planning as detailed in Table 3.

The scale of the habitat decides the level of detail of the study. At the macro scale—districts, regions, sub regions—the major ecosystems are studied to understand their role in maintaining the regional environment and as habitats for critical species. Meso regions relate more closely to the urban areas, as they include the notified planning area, urban area and planning zones. In fact most of the urban areas are meso habitats as their areal spread is between 1 and 10,000 km^2. At this scale the protected areas and ecosystems can be identified—reserved forests, sanctuaries, national parks and so on—and landscape elements—patch, corridor and matrix—can be disaggregated. The land use composition and its spatial pattern can be analysed for the biodiversity it supports. Micro regions relate to land use zones—residential, commercial, industrial, institutional, recreation, transport, agriculture and so on. At this scale detailed quantitative assessment is possible to study the contribution of biodiversity in improving the environment of an area. It

Table 3 Correlating biodiversity classification with spatial planning levels

Ecological diversity of planning regions						
Ecosystems at the global level	Ecosystems at the level of the Indian subcontinent	Habitat scales, planning, region and ecological diversity		Habitats and organismal diversity		
Biomes	Biogeographic regions	Landscapes	Ecological diversity/ ecosystems	Habitats	Niches (conservation status)	Organismal diversity
		Macro regions (District and regional planning levels—Sub-regions, Metropolitan regions etc.) 10,000–100,000 km^2 **Meso regions** (Planning area, developed area—Notified planning area) 1,000 –10,000 km^2 (Urban area and planning zones) 1 –1,000 km^2 **Micro regions** (Land-use zones) up to 1 km^2		Terrestrial	Apex species	Plant kingdom
						Animal kingdom
					Keystone species	Plant kingdom
						Animal kingdom
					Dominant species	Plant kingdom
						Animal kingdom
				Aquatic	Apex species	Plant kingdom
						Animal kingdom
					Keystone species	Plant kingdom
						Animal kingdom
					Dominant species	Plant kingdom
						Animal kingdom

follows, therefore, that a multi-scale strategy has to be adopted to conserve the biodiversity in an urban area.

4.1 Strategy for Conservation at the Macro Habitat Level

At the level of metropolitan regions and districts, the legally protected areas and representative ecosystems of the biogeographic region have to be spatially demarcated. Their buffer zones need to be identified and sensitive habitat units (areas which include home ranges of sensitive species) marked for protection. The conservation status of species as per the International Union for Conservation of Nature (IUCN) and the Wildlife Protection Act (WPA), needs to be listed while ensuring habitat protection. The biodiversity value of these in terms of contribution to environmental quality has to be understood. Listing of permitted and not permitted land uses and activities within protected areas and around protected areas would help in ensuring conservation of biodiversity at this scale. A schematic representation is presented in Table 4.

In the case of Delhi, the macro region/habitat is the National Capital Region. The legally protected areas within this region are environmentally fragile areas, that is, Aravalis, Sariska National Park/ Project Tiger area, Sultanpur National Park and Bird Sanctuary. Representative ecosystems include the flood plains of River Yamuna and Ganga, numerous wetlands such as Damdama lake and Badkal lake in the Haryana sub-region, and forested areas in the Uttar Pradesh sub-region near Meerut, for example. The spatial extent of protected areas is available but the spatial extent of the buffer area has not been marked. In the case of the Aravalli ecosystem, the extent is documented as names of areas to be included, but its spatial mapping has not been undertaken. In the absence of concrete boundaries, encroachments are possible. Detailed listing of species is available and their conservation status is available with the forest and wildlife authorities and not with the planning authorities. No comprehensive listing of permitted and non-permitted land uses and activities in the buffer zone has been done. Representative ecosystems have been mapped, however their species listing and area demarcation is not available. Listing of permitted and not permitted land uses has not been undertaken.

4.2 Strategy for Conservation at the Meso Habitat Level

The Meso scale represents the planning area, urban area and planning zones of a city. At the planning area level, the protected areas and ecosystem diversity have to be mapped. The buffer area needs to be demarcated, and listing of permitted and not permitted land uses and activities has to be undertaken. Detailing of sensitive habitat units within these areas to protect the species has to be mapped.

Table 4 A schematic representation of strategy for conservation at the macro habitat level

Scale of study	Land uses for conservation/protection	Special extent of protection	Assessment of biodiversity value	Measures within protected area	Measures around protected area
Macro metropolitan region	Legally protected areas —national parks, sanctuaries, reserved forests, notified fragile areas	– Demarcated boundary of protected area – Buffer area around national parks, sanctuaries. Areas whose buffer is not defined, efforts to be undertaken to define and demarcate to buffer area	**Qualitative** – Conservation status of species – Biodiversity value on basis of species richness and environmental role of the area **Quantitative** – Species area curve. Estimating impact of change in habitat size on species numbers – HEP (Habitat Evaluation Procedure) to analyse the status of habitat and critical habitat variables	– Identification of habitat units (HU) for species to ensure no encroachment into home range of critical species – Measures to ensure habitat quality is maintained by ensuring requisite vegetative cover, adequate water availability and adequate food supply	– Listing of permitted land uses within buffer area – Listing of activities permitted and not permitted around protected areas

Green land uses—low density residential/green neighbourhoods, institutions, heritage areas, large parks, critical urban gradients, and landscape elements (patch, corridor, matrix) have to be spatially identified and assessed for their biodiversity value. These areas have to be protected from unsuitable development as they serve as habitats for species as well as contribute to environmental protection. Conservation status of the species has to be available with the planning authorities with sensitive habitat locations.

Designated protected areas—studies suggest that a width of at least 8 km should be provided along River Yamuna to protect the habitat of bird species. Areas around the Ridge, River and Asola Sanctuary should have low density development—parks for passive recreation, single family residential, research institutions (uses that do not cause air, noise pollution, possess large green areas, approximately 75% of plot area, no high rise development, and low traffic flow). These ecosystems have to be mapped and their status in terms of activities and land uses, organismal diversity and environmental quality have to be recorded. Areas that can be brought under protection should be identified and appropriate measures of controlling development, afforestation and so on, should be undertaken.

Green land uses—old low density residential areas—civil lines, Lutyens' Bungalow Zone, cantonment areas—constitute large plots with dense vegetative growth, with varieties of species forming ideal habitats for many types of fauna. Since they also function as pollution sinks, water recharge areas and bring down the effect of urban heat island, they have to be protected from further densification. Institutional land uses (Jawaharlal Nehru University, Indian Agricultural Research Institute), parks, heritage areas (Qutub Minar, Siri Fort, Tuglaqabad etc.) which are larger than 100 ha in area and possess about 30% forested areas, and waterbodies function as habitats of a wide variety of fauna. It is easier to control development here as they are already identified as low development areas. Control in activities that attract traffic and noise or appropriate regulation of activities can protect the habitats here.

Critical urban gradients—certain transacts across the city indicate appropriate arrangement of land uses for species to find contiguous habitats that they can accommodate—food patches, home range, foraging areas and tracts in the case of birds. In other words the urban fabric shows a gradual increase in open spaces to accommodate many scales/types of habitats. These are species-rich areas and development should not degrade the habitats.

Landscape elements—the urban green structure that allows for linkages of green patches with green corridors within the land use matrix is most preferred as it allows species to travel to various food and water areas. The First Master Plan for Delhi had proposed a system of flowing greens. Even though it was not completely executed, the various water bodies, storm water drains, heritage areas, parks, green land uses can be interconnected to provide 'greenways' for the city, while protecting habitats for species. A detailed recording of existing species through scientific floral and faunal surveys has to be undertaken to accord protection of suitable species, removal of noxious species and to introduce species to enhance degraded habitats. A systematic inventory of the natural resource profile which is available at the level of the notified planning area (preferably at the scale of 1:10,000) is the first step for any resource conservation exercise. The resources can be categorized as below:

- Land resource—geology, soil, land use (the existing land use determines the resource quality in an urban area and needs to be documented in detail at all scales)
- Water resource—surface and ground resources
- Atmospheric resource—micro-climatic variations, ambient air quality
- Biotic resources—biodiversity profile

A schematic representation of strategy for conservation at the meso habitat level is presented in Table 5.

4.3 Strategy for Conservation at the Micro Habitat Level

The micro scale constitutes various land use zones—Residential, Commercial, Mixed Use, Institutional, Industrial, Transport Recreation, Water Bodies, Agriculture, Forest. Each land use represents a patch, corridor or matrix. In the rural zone and urban zone, the land uses and their landscape element classification are grouped in Table 6.

Each land use/patch can be classified as a habitat—terrestrial or aquatic or both—scrub, woodland scrub, woodland, undulating scrub, and so on, depending on species composition. The species composition is related to the age, size and landscape design of the land use patch and may vary for similar land use—area under trees, shrubs, climbers, ground covers, waterbodies, landforms. The age and type of floral species would determine the faunal species. The more the variety of floral species the larger the number of faunal species observed. It follows that within the development controls/by-laws, the type of landscape that supports biodiversity can be made mandatory while sanctioning plans, as is already being done for water harvesting systems. A combination of large trees, small trees, shrubs, climbers and wherever the size permits, the creation of a waterbody, would result in habitats for many species.

Since a metropolitan city has many environmental problems—water scarcity, air and noise pollution, water pollution, heat build-up, waste generation—the selection of species should also contribute to reducing the environmental stress, by fulfilling some of the functions, such as recharge water, abate pollution, treat waste and improve microclimate. The ambient air quality, temperature, humidity, soil conditions, existing vegetation, water recharge potential, existing development controls, activity pattern, waste type and generation specific to land uses has to be assessed to propose the modifications in landscape development to enhance habitat quality. A schematic representation of strategy for conservation at the micro-habitat level is presented in Table 7.

Table 5 A schematic representation of strategy for conservation at the meso habitat level

Scale of study	Land uses for conservation/protection	Special extent of protection	Assessment of biodiversity value	Measures within protected area	Measures around protected area
MESO notified planning areas Planning zones	Mapping of eco system diversity apart from those under legal protection. Green land uses, low density residential green neighbourhood, institutions, heritage areas, large parks Identification of critical urban gradients Identification of landscape elements patch corridors matrix	– Spatial extents of eco system to be demarcated to include habitat units, plot boundaries in case of green land uses, right of way in case of corridors, storm water channels, roads and railways	**Qualitative** – Conservation status of species. Biodiversity value on the basis of species richness and environmental role of the area **Quantitative** HEP to analyse the status of habitat and critical habitat variables Species richness in case of patches, green neighbourhoods	– Identification of habitat units (HU) for species to ensure no encroachment into home range of critical species – Measures to ensure habitat quality is maintained by ensuring requisite vegetative cover, adequate water availability and adequate food supply – Measures to maintain continuity of greens, especially maintaining green corridors along water channels	– Maintaining compatible land uses along green land uses, corridors, etc. – Adequate plantation measures to control air and noise pollution

Table 6 The land uses and their landscape element in the rural zone and urban zone

Rural zone	Urban zone
Patch	**Patch**
Residential, industrial, public utility, water bodies, forest etc.	Agriculture, commercial, recreational, industrial, public utility, water bodies, forest etc.
Corridor	**Corridor**
River, stream, roads	River, stream, roads
Matrix	**Matrix**
Agriculture	Residential use

5 Mainstreaming Biodiversity Conservation in Urban Planning in India: Way Forward

India is a signatory to the Convention on Biological Diversity 1992. As per the Biodiversity Act 2002 the country has to protect its biological resources in in situ and ex situ situations. Draft CBD decision for COP10 on cities, local authorities and biodiversity. Agenda item 4.9 (Cooperation with other conventions and international organizations and initiatives, engagement of stakeholders, including business and biodiversity, cities and biodiversity, and South/South cooperation).

The Conference of the Parties, considering decision IX/28, encourages Parties to the Convention on Biological Diversity to recognize the role of cities and local authorities in national strategies and plans, and invites Parties to support and assist cities and local authorities in implementing the Convention at local level. It recognizes that the significant contribution of local authorities to the implementation of the Convention and its 2011–20 Strategic Plan will be further enhanced and leveraged by a coherent plan of action which will coordinate efforts at local, sub-national, national, regional and global levels. It acknowledges the progress achieved by the Global Partnership on Cities and Biodiversity and by ICLEI (through its Local Action for Biodiversity programme), consolidated in events such as the Second Curitiba Meeting on Cities and Biodiversity, the 5th World Urban Forum in March 2010 in Rio de Janeiro, and Expo Shanghai 2010, and the contribution of Singapore in the development and testing of the City Biodiversity Index.

A presentation of Delhi's biodiversity was made by the Department of Environment, NCT Delhi at Aichi-Nagoya City Biodiversity Summit, on 24–26 October 2010, at the margins of COP 10 relating to the plan of action on cities, local authorities and biodiversity (2011–20) and addressing the above concerns that sustainability habitat parameters under the National Action Plan for Climate Change should also include 'urban biodiversity'. Accordingly, actions taken in India can be summarized as follows.

5.1 **Biodiversity profile of urban areas scientific studies**

Flora of urban areas—published document available for many urban areas, such as Delhi, Bhopal and so on.
Published documents on plants for landscaping—inventory of trees, shrubs, climbers, grasses, indoor plants and so on—indigenous, naturalized and exotics suggested/ propagated in urban areas.
Faunal survey of urban areas—Delhi, Avian species inventory available for many cities from Wildlife Institute of India.
Field survey and documentation of species of representative ecosystems—documents prepared by committees, environmental impacts assessment (EIA) project reports, eco-sensitive area studies, eco-clubs, NGOs, dissertations undertaken by courses on environmental sciences, management, planning, landscape architecture.

5.2 **Mainstreaming of biodiversity conservation in development process in urban settings**

Environmental impact assessment of development projects under the Environmental Protection Act:

- Classification of projects into categories based on significant impacts on environment
- Impact on biotic environment to be recorded
- Protection of flora within a 10 km radius of project site and within site
- Compensatory afforestation in case trees are removed.

Declaration of eco-sensitive areas under the Environmental Protection Act:

- Areas of significant ecology brought under protection through notification
- Development plans to record, floral, faunal and cultural resources
- Development plans for the area to ensure protection of natural and cultural diversity.

5.3 **Local biodiversity strategy action plan for urban area**

Global Studio organized by SPA Bhopal and UNU-IAS, Tokyo, CBD and ICLEI (May–June 2012) to conduct a study based upon biodiversity of the city of Bhopal and propose local strategies and an action plan for biodiversity conservation. Presented at the City Biodiversity Summit, COP 11, 2012.

5.4 **City biodiversity summit held at COP 11**

A landmark programme was launched on day two of the summit: the LAB India programme, which joins a global network of local governments committed to galvanizing local action for biodiversity. Notably, the cities of Hyderabad, Thane, Delhi-NCT, Guntur, Shimla, Anantapur District, Kurunegala, Matale and Varanasi signed the Durban Commitment. Also introduced at the summit was the Cities in Biodiversity Hotspots Programme, a ten-year global initiative seeking to provide

Table 7 A schematic representation of strategy for conservation at the micro habitat level

Scale of study	Land uses for conservation/protection	Special extent of protection	Assessment of biodiversity value	Measures within protected area	Measures around protected area
Micro land use zone	Land use greens – parks, open spaces, water bodies, urban forests, plantations, forests, agriculture – Green land uses—low density residential neighbourhood, institution, heritage areas – Identified green corridors	– Spatial extents of eco system to be demarcated to include habitat units, plot boundaries in case of green land uses, right of way in case of corridors, storm water channels, roads, railways	**Qualitative** – Conservation status of species – Biodiversity value on basis of species richness and environmental role of the area. **Quantitative** – Species richness in case of patches, green neighbourhood – Biodiversity index calculation for similar species richness areas	– Measures to ensure habitat quality is maintained by ensuring a variety of vegetative cover, trees, shrubs, herbs, climbers. Introduction of water features and waterbody – Measures to maintain continuity of greens, especially maintaining green corridors along water channels	– Adequate plantation measures to control air and noise pollution. Species that support faunal species should be introduced

over 250 cities located in and around the 35 biodiversity hotspots of the world with a platform to take action on biodiversity and aid city-to-city learning.

5.5 **Legal and governance/ management framework for protection, conservation and sustainable use of biotic resources at national, sub-national and local levels (relevant to urban biodiversity conservation)**

National—Biodiversity Act 2002 deals with protection of biodiversity in in situ and ex situ situations. The National Action Plan for Climate Change 2008—Mission on Sustainable Habitat, Mission on Greening India, National Biodiversity Action Plan and National Green Tribunal—advocates protection of environmental resources.

Sub-national—State Biodiversity Boards which prepare inventory of biotic resources in the Peoples Biodiversity Register and a detailed checklist developed to record urban biodiversity. Format 29 and 30 of PBR relate to urban biodiversity.

Local—master plans of urban areas have to provide areas for open spaces/recreation preferably including existing pockets of green. Climate action plans of cities include greening of urban areas in tune with the biodiversity.

5.6 **Strategic environmental assessment as a tool for urban biodiversity co-benefits**

Strategic Environmental Assessment (SEA) is a systematic, ongoing process for evaluating, at the earliest appropriate stage of publicly accountable decision making, the environmental quality, and consequences, of alternative visions and development intentions incorporated in policy, planning or programme initiatives, ensuring full integration of relevant bio-physical, economic, social and political consideration (Partidário 1999). SEA is important as it enables the integration of environmental and sustainability factors into policy-making by addressing the cause of the environmental problems at their policy source rather than treating the symptoms of impacts as in EIA (Sadler and Verheem 1996). The internationally accepted benefits of SEA are that it allows for a wider consideration of impacts and alternatives, is a pro-active tool that can be used to support strategic action formulation for sustainable development, can increase the efficiency of tiered decision-making (including strengthening of project EIA), it allows for a systematic and effective consideration of the environment at higher tiers of decision-making and allows a greater level of public participation and consultation.

Though there is no legislative provision for SEA, for the past 4–5 years the merits of applying SEA approaches at an earlier stage of the decision-making process have commanded importance in India because these assessments have proved useful in reducing time and costs and the need for an EIA. One of the strategies of the National Environmental Policy 2006 is to encourage regulatory authorities, central and state, to institutionalize regional and cumulative EIA (R/CEIAs) to ensure that environmental concerns are identified and addressed at the planning stage itself. It is evident from here that the National Environment Policy provides for the incorporation of SEA in the plan-making process in India (MoEF 2006). Even the 11th and 12th National Five Year Plans and the Regional Plan for NCR 2021 (NCRPB 1999) provides for SEA type assessments. One of the examples of SEA as a tool in biodiversity conservation in India is the 'Strategic Environmental Assessment of proposed Human River Irrigation Project, Maharashtra, India'. In this biodiversity driven SEA, greater importance was assigned to evaluation of impacts on ecosystem components valued as vital habitat links and movement corridors for tigers; ecosystem functions and features that characterize habitat suitability for conserving and managing a viable prey base for tigers; and diversity of rare and endangered species. Consideration of benefits of conserving forests and natural habitats in the project area for long-term assurances of useful goods and services (e.g., perpetuity of genetic resources and maintenance of hydrological balance for sustained availability of water) for local communities was also inherent in the approach.

Another case study of SEA in India, applied within the macro region of the NCT of Delhi, at macro level is the 'Strategic Environmental Assessment of the Gurgaon-Manesar Land Use Plan (GMP) 2031' (TCPD 2012). The total study area covers an area of 340 km^2 with an existing population of around 9 lakh. Analysis of the environmental data reveals that the ground water status of the area is extremely critical with levels of 40–50 mbgl and above 300% ground water draft. The ecosystems such as Aravali (ridge), Sultanpur wetland and the scrub lands are highly vulnerable to the proposed urbanization and industrialization. Once the issues of the area are identified, the GMP is subjected to SEA. It begins with the understanding of the need of the plan, followed by studying the policy framework under which the GMP is prepared. This includes all the international conventions, national policies, plans of upper tiers such as the regional plan of NCR, and so on. After this, the core assessment takes place. It starts with the identification of 'SEA parameters' for the study area such as air quality control, water management, protection of Aravali, Sultanpur wetland and scrub lands, and public transport efficiency. This leads to the formulation of 'SEA strategies', some of which are 'reducing fresh water demand in non-domestic sectors', protection and replenishment of ground water, and so on. Before the SEA strategies are finalized for further assessment, these are tested against the area specific issues and the SEA parameters respectively through the tool 'matrices'.

Next follows a series of assessments of the GMP proposals against the area specific issues, the SEA parameters, the SEA strategies and an internal assessment of the GMP proposals against each other. Here, the plan provision having a high positive score means that the provision is acceptable; whereas a provision showing a negative impact means that this particular proposal needs reconsideration or needs to be negated; and a proposal showing a low positive score means that it must be improved upon. In the study, the assessment of GMP proposals shows that the proposals pertaining to transportation and the commercial sector need to be reconsidered as it is having an anticipated negative impact on the area. The proposals regarding provision of open space, a biodiversity park and an artificial wetland need improvement.

5.7 Biodiversity conservation and poverty eradication

In the case on India, large metropolitan cities house almost 30% slum dwellers on an average and many of them pursue primary occupations, such as fishery and urban agriculture. This is largely treated as informal and has not been documented or studied in detail for its role in biodiversity protection and ensuring livelihood generation, thereby eradicating poverty. However, in India's Fifth National Report to the Convention on Biological Diversity (CBD) 2014 (MoEF 2014), which is now also being assessed under the BIOFIN programme for funding, the Aichi Target 14 is addressed under National Target 8 as a major sector. Also, National Biodiversity Targets 2, 3, 6, 7, 9, 10 and 11 address directly or indirectly extreme poverty and hunger.

The CBD has addressed the issue of biodiversity and poverty eradication and development since 2008. The linkages between biodiversity and poverty eradication are at the heart of the vision of the Strategic Plan for Biodiversity 2011–2020, 'Living in Harmony with Nature' where 'By 2050, biodiversity is valued, conserved, restored and wisely used, maintaining ecosystem services, sustaining a healthy planet and delivering benefits essential for all people.' Further, Aichi Biodiversity Target 14 specifically focuses on the needs of women, indigenous and local communities, the poor and vulnerable: 'By 2020, ecosystems that provide essential services, including services related to water, and contribute to health, livelihoods and well-being, are restored and safeguarded, taking into account the needs of women, indigenous and local communities and the poor and vulnerable.' The second meeting of the Expert Group on Biodiversity for Poverty Eradication, took place in Chennai, India, from 4 to 6 December 2013, supported by the Ministry of Environment and Forests and the National Biodiversity Authority of the Government of India. The Expert Group discussed, among others, issues related to mainstreaming biodiversity and ecosystems, national accounting systems, the funding situation and the benefits of investing in the Aichi Biodiversity Targets, best practices of integration of biodiversity and ecosystems, poverty eradication and sustainable development at regional and organizational/institutional levels including initiatives such as the UNEP/UNDP–led Poverty–Environment Initiative and country case studies in order to compile the elements proposed in the Dehradun/Chennai Recommendations and the related Guidance for Implementation of the Integration of Biodiversity and Poverty Eradication. One of the major recommendations was to identify linkages between biodiversity and poverty eradication for sustainable development, as well as drivers of biodiversity loss and poverty acceleration, inter alia, by using specific voluntary tools such as mapping of social and environmental vulnerability, regional poverty–environment profiling, and distributional studies assessing country- and region-specific links between biodiversity and poverty; and ensuring that the selected tools are gender sensitive and consider the diversity of views from indigenous and local communities, women, the poor, marginalized and vulnerable. Minimizing impacts, capacity building, and appropriate funding was also addressed (see Footnote 1).

5.8 Biodiversity and air pollution mitigation

Co-benefits from biodiversity and air pollution need to be quantified. Air pollution kills more than 5.9 million people annually, with more than 90% of these deaths in the capital city of India, Delhi. For improving the status of air pollution in Delhi, various policies and laws have been implemented. But even after the implementation of CNG, there was no significant change of pollutants (NO_x, O_3, SPM, RSPM (Respirable Suspended Particulate Matter) and CO) except SO_2. To carry out the analysis, daily ambient air quality secondary data (January 2002–December 2009; Source: CPCB 2009) of all the above discussed pollutants were used. For generating primary data, NO_X and O_3 monitoring were carried out at four different sites, namely Site I (Yamuna Biodiversity Park (YBP), away from traffic intersection), Site II (Traffic intersection at outside YBP, outer ring road, Gandhi vihar), Site III (Aravali Biodiversity Park, away from traffic intersection) and Site IV (traffic intersection at outside ABP, ring road, Vasant Vihar) during monsoon season (August–September 2009). The concentration of ozone was higher at sites which are at traffic intersections (Sites II and IV) than those which are away from traffic intersections (Sites I and III) (Saxena et al. 2012). One of the observations of the data were the levels of all pollutants were significantly below the permissible limits near both the biodiversity parks, whereas in other areas the values were either slightly below the permissible limits or above it. One of the main strategies of the Delhi Climate Action Plan, 2009–12, which is still underway, is greening of Delhi, through maintaining and creating a network of parks and roadside plantations.

5.9 Role of biodiversity conservation in India's INDC

India's contribution takes into account its commitment to conservation of nature as well as the imperatives of meeting the competing demand of resources for addressing the challenges of poverty eradication, food security and nutrition, universal access to education and health, gender equality and women empowerment, water and sanitation, energy, employment, sustainable urbanization and new human settlements and the means of implementation for enhanced action for achieving among others, the sustainable development goals for its 1.2 billion people (MoEF 2015). India is one of the few countries where forest and tree cover has increased in recent years transforming the country's forests into a net sink owing to national policies aimed at conservation and sustainable management of forests. As per the latest assessment, forests and tree cover has increased from 23.4% in 2005 to 24% of the geographical area in 2013. Government of India's long-term goal is to bring 33% of its geographical area under forest cover eventually. (1) With its focus on sustainable forest management, afforestation and regulating diversion of forest land for non-forest purpose, India has been successful in improving carbon stock in its forest by about 5%, from 6621.5 million tonnes in 2005 to 6941 million tonnes in 2013. (2) Initiatives like Green India Mission (GIM) aim to further increase the forest/tree cover to the extent of 5 million hectares (mha) and improve quality of forest/tree cover another 5 mha of forest/non-forest lands along with providing livelihood support. It is expected to enhance carbon sequestration by about 100 million tonnes CO_2 equivalent annually. (3) These efforts have been further augmented by policies like the National Agro-forestry Policy (NAP), REDD-Plus policy, Joint Forest Management; National Afforestation Programme and proposed devolution of about US$6 billion under Compensatory Afforestation.

References

CBD Website https://www.cbd.int/sp/targets/rationale/target-14/

Campbell, M. (2012). Biodiversity and endangered species: Issues of significance. *J Biodivers endanger species, 1,* e102. doi:10.4172/2332-2543.1000e102.

Census of India. (2011). *Provisional population totals -2011, Paper–II, Vol–II* (p. 1). New Delhi: Census of India.

CGWB-Central Ground Water Board. (2007). *Ground water report–Gurgaon District, Haryana.* New Delhi: Central Ground Water Board.

CPCB-Central Pollution Control Board. (2009). *Air Quality in Delhi.* New Delhi: Central Pollution Control Board.

DDA-Delhi Development Authority. (1962). *Master Plan of Delhi.* New Delhi: Delhi Development Authority.

DDA-Delhi Development Authority. (1990). *Perspective Development Plan.* New Delhi: Delhi Development Authority.

DDA-Delhi Development Authority. (2006). *Master Plan of Delhi 2021.* New Delhi: Delhi Development Authority.

MoEF, (2006). *National environment policy.* New Delhi: Ministry of Environment and Forests, Government of India.

MoEF. (2010). *India: Greenhouse gas emissions 2007* (pp. 1–54). New Delhi: Indian Network for Climate Change Assessment, Ministry of Environment & Forests, Government of India.

MoEF. (2014). *India's fifth national report (NR5) to the convention on biological diversity.* New Delhi: Ministry of Environment, Forests and Climate Change.

MoEF. (2015). *India's intended nationally determined contribution: Working towards climate justice* (pp. 15–16). New Delhi: Ministry of Environment & Forests, Government of India.

MUAE-Ministry of Urban Affairs and Employment. (1996). *Urban development plans formulation & implementation guidelines.* New Delhi: Ministry of Urban Affairs and Employment, Government of India.

NBAP. (2008). *National Biodiversity Action Plan.* New Delhi: Ministry of Environment and Forests, Government of India.

NCRPB–National Capital Region Planning Board. (1999). *Delhi 1999 A Fact Sheet.* New Delhi: National Capital Region Planning Board.

Partidario, M. (1999). Strategic environmental assessment—principles and potential. In J. Petts (Ed.), *Handbook of Environmental Impact Assessment* (Vol. 1, pp. 60–73). London: Blackwell Science Ltd.

Sadler, B., & Verheem, R. (1996). *Strategic environmental assessment: Status, challenges and future directions.* The Hague: Ministry of Housing, Spatial Planning and the Environment.

Saxena, P., Bharadwaj, R., & Ghosh, C. (2012). Status of air pollutants after implementation of CNG in Delhi. *Current World Environment, 7*(1), 109–115.

TCPD-Town and Country Planning Department. (2012). *Final development plan—2031A.D. of Gurgaon- Manesar Urban Complex.* Chandigarh: Government of Haryana.

United Nations Development Programme. (2012). *The Future We Want: Biodiversity and ecosystems— Driving sustainable development.* United Nations Development Programme Biodiversity and Ecosystems Global Framework 2012-2020, New York.

https://www.cbd.int/2011-2020/

Part IV
Promoting Co-benefits in the Urban Context: Innovations and Reforms

Chapter 10
Smart and Livable Cities: Opportunities to Enhance Quality of Life and Realize Multiple Co-benefits

Shilpi Mittal and Mahendra Sethi

Abstract Throughout human history, cities have been centres of prosperity, ideas and innovation. These days, smart cities are creating a new buzzword across the world. Examples boom in Japan, Europe, UAE and Singapore while several others are shaping up on the drawing board. With the recent announcement of 100 new Smart Cities, the Government of India has strategically responded to both the international call for innovation and transformative sustainability as well as growing domestic pressure in cities. Interestingly, there is neither an internationally accepted definition of a Smart City, nor does India have any national policy on urbanization. Within this science-policy vacuum, there is a fair degree of consensus on what a smart city looks like, but no understanding on what are the inputs and strategies to achieve one. With numerous expectations, inhibitions and euphoria around this theme, this paper attempts to systematically investigate what is a smart city, how it is different from similar prototypes like a sustainable, green and low-carbon city and what are the global best practices. The article addresses some of the ideological, technical, societal, governance and financial challenges that India faces to attain the '100 Smart Cities' goal, and what would be its policy implications. In the process, the research proposes a new idiom for SMART—Sustainability, Metrics, Adaptiveness, Reporting, Technology for Inclusiveness.

Keywords Smart cities · Best practices · Issues and challenges
Quality of life · SMART framework

S. Mittal (✉)
Indian Society for Applied Research & Development, New Delhi, India
e-mail: shilmittal@gmail.com

S. Mittal
GD Goenka University, Gurgaon, India

M. Sethi
National Institute of Urban Affairs, New Delhi, India

M. Sethi and J.A. Puppim de Oliveira (eds.), *Mainstreaming Climate Co-Benefits in Indian Cities*, Exploring Urban Change in South Asia,
https://doi.org/10.1007/978-981-10-5816-5_10

1 Introduction

As the saying goes, the road to the city of emeralds is paved with yellow bricks. Throughout human history, urban settlements have been centres of business, talent, entrepreneurship, wealth, ideas, art and culture. In 1800 only 3% of people lived in a city of 1 million or more. In 1950 there were only 83 cities worldwide with populations over 1 million; by 2007 there were 468. In April 2008, the world passed the 50% urbanization mark (UNDESA 2011). With more than half of the world's humanity now living in urban areas, some of it in most abject poverty, destitute and vulnerable circumstances, it is evident that the path to sustainable development must pass through substantially improving quality of life in cities. The role and responsibility of cities is undoubtedly increasing in international debates of economy, environmental governance and sustainability. They are estimated to contribute 70–80% of global gross domestic product and likewise accountable for 70–80% of greenhouse gas (GHG) emissions that cause global warming (World Bank 2010; UN Habitat 2011). The global urbanization trend is creating an urgency to find smarter ways to manage the accompanying challenges (Nam and Pardo 2011), and make cities more competitive and livable. UN Habitat forecasts that by 2050, the global urban population would comprise 75% of the planet's humankind and seeks a larger role of cities in Sustainable Development Goals (SDG), that replace the Millennium Development Goals and post 2015, steer the world's future growth trajectory. The SDG has set a goal to 'Build inclusive, safe and sustainable cities and human settlements' (under Goal 11: Sustainable cities and human settlements) in one of the recent Working Group meetings. The Seventh World Urban Forum (WUF) in Medellin pledged to convert cities into 'inclusive, safe, prosperous and harmonious spaces for all', which was eventually concurred in the New Urban Agenda adopted at the UN Conference on Housing and Sustainable Urban Development (Habitat III) in Quito, Ecuador. Meanwhile, India has focused upon developing a climate resilient infrastructure and urban centres in its Intended Nationally Determined Contribution submitted before the recently convened UNFCCC COP21 at Paris in 2015 (MoEF 2015). The need for smart and livable cities is becoming clearer and louder.

India on its development pathway faces an uphill urban challenge. According to the 2011 census, about 32% (377 million) of the country's population lived in urban areas as against 28% in 2001 and 17% in 1991. Projections of Census of India 2011 data suggest that the urban population is about to grow at a pace of 2.83% from 340 million in 2008 to 590 million in 2030, living in at least 60 cities with a population of more than one million (Mckinsey 2010; MoUD 2011). By 2039, most estimates consider India to be 50% urbanized. To keep pace with that, India would have to spend $1.2 trillion in its urban areas. A critical challenge for cities is that while these are meant to constantly guzzle economic growth engine from fossil fuels and absorb labour migrating from rural areas, these are simultaneously expected to reduce their emission footprints by becoming smart. With the recent announcement of 100 new Smart Cities, the Government of India has strategically responded to both the international and the domestic audience. Worldwide there is a call to battle global challenges like climate change, poverty, inequality and sporadic development in developing societies through transformative sustainability, innovations, livability and low-carbon societies. Meanwhile, there is a smoldering demand from Indian urbanites for enhanced economic growth, job creation, systems efficiency, urban renewal, municipal reform and international level of quality of life.

The Union Budget earmarked Rs. 7060 crore (70.6 billion) during the 2013–14 financial year for developing 100 'Smart Cities' in the country. Announcing this in his budget speech, the finance minister said that the prime minister has a vision of developing 100 satellite towns of larger cities as 'Smart Cities'. The government will also modernize existing mid-sized cities under the programme. 'With development reaching an increasingly large number of people, the pace of migration from rural areas to cities is increasing. The new cities should be developed to accommodate the burgeoning number of people. Otherwise, existing cities will soon become unlivable' (PTI 2014). Very soon, these advanced, intelligent and livable human-ecosystems are set to find their foot in India. But in the initial one year there were little details divulged on what the Government perceives as a Smart City and what would be the means and methods to achieve 100 of them in the next five years. The long silence was interrupted by the Union Urban Development (UD) minister himself, saying that, "Smart Cities should bring quality of life with inclusivity". The minister discussed future urban development plans and further suggested that there would be two different schemes, one for renewal of 500 urban habitations (now known as the Atal Mission for Rejuvenation and Urban Transformation or AMRUT) and the other for 'Smart Cities' (Zee News 2014). Meanwhile, the overall allocation for the sector was hiked by a whopping 133% in the 2014–15 budget. The UD ministry was allocated Rs. 17,628 crore (176.28 billion) in 2014–15 as against Rs. 7548 (75.48 billion) during 2013–14. In a bid to bring in more foreign investors and help them develop projects in smaller cities, the finance minister announced the government's decision to not only reduce the minimum carpet area criteria for construction to 20,000 m^2 from the existing 50,000 m^2 but also to reduce the minimum capital to US$5 million from US$10 million with a three-year post lock-in. All the above indicated that the Government perceives 100 Smart Cities more as a massive Greenfield project that is set to change the landscape of this country. In the initial media reports, it was reported that 7 out of 25 Smart Cities are to be planned in the first phase of Delhi Mumbai Industrial Corridor. There would be one each in Uttar Pradesh, Haryana, Rajasthan, Maharashtra, Madhya Pradesh and two in Gujarat. The leading one in the pack being Gujarat International Financial Tec-City in about 886 acres, while Dadri-Noida-Ghaziabad Investment Region extending over 50,000 acres in Uttar Pradesh is also being commissioned (Indian Express 2014). The contours of this scheme have become clearer with the launch of the Smart Cities Mission (SCM) in mid-2015 that offers a blueprint for planned urban development (for provisions and features of the SCM, see Box 1).

But there seem to be certain unsettled issues and challenges in this upcoming venture. There is neither an internationally accepted technical definition of a 'Smart City', nor does India have any national policy on urbanization. Within this science-policy vacuum, there is a fair degree of understanding on how some of the Smart Cities will look, but no consensus on what are the inputs and strategies to achieve one. With numerous expectations, inhibitions and euphoria around this theme, this chapter attempts in Sect. 2 to systematically investigate what constitutes a Smart City, and how it furthers the concepts of a livable, sustainable, green and low-carbon city. Global best practices are explored in Sect. 3. Section 4 addresses some of the major challenges—ideological, technical, societal, governance and financial—that India faces to attain the '100 Smart Cities' goal and its policy implications. Consequently, the research explores new possibilities in Sect. 5 that could serve as a new idiom for Smart City, while Sect. 6 presents the research conclusion.

Box 1: Smart Cities Mission: Provisions and Features

The Government of India launched the Smart Cities Mission (SCM) on 25 June 2015. The mission acknowledges that cities are engines of growth for the economy of every nation, including India. The SCM accepts that there is no universally accepted definition of a Smart City. It asserts that a Smart City would have a different connotation in India than, say, Europe. Even in India, there is no one way of defining a Smart City. In the approach to the Smart Cities Mission, the objective is to promote cities that provide core infrastructure and give a decent quality of life to its citizens, a clean and sustainable environment and application of 'Smart' Solutions. The focus is on sustainable and inclusive development and the idea is to look at compact areas, create a replicable model which will act like a lighthouse to other aspiring cities.

The Mission will cover 100 cities and its duration will be five years (FY 2015–16 to FY 2019–20). The SCM lists ten core infrastructure elements including water, electricity, waste, mobility, housing, environment, health and education. It also illustrates over 20 technology supported 'smart solutions'. In addition, the SCM offers some typical features of comprehensive development in Smart Cities, that include promoting mixed land use in area-based developments, housing and inclusiveness, creating walk-able localities, preserving and developing open spaces, promoting a variety of transport options, making governance citizen-friendly and cost effective, giving an identity to the city; and applying Smart Solutions to infrastructure and services. With no further checklists, parameters and indicators defined, the SCM features are empirically difficult to measure, verify and report.

The strategic components of area-based development in the Smart Cities Mission are city improvement (retrofitting), city renewal (redevelopment) and city extension (Greenfield development) plus a Pan-city initiative in which Smart Solutions are applied covering larger parts of the city. The Government does not prescribe any particular model to be adopted by the Smart Cities. The approach is not 'one-size-fits-all'; each city has to formulate its own concept, vision, mission and plan [Smart City Plan (SCP)] for a Smart City that is appropriate to its local context, resources and levels of ambition.

The SCP may include assured electricity supply with at least 10% of the Smart City's energy requirement coming from solar, adequate water supply including waste water recycling and storm water reuse, sanitation including solid waste management, rain water harvesting, smart metering, robust IT connectivity and digitalization, pedestrian friendly pathways, encouragement of non-motorized transport (e.g. walking and cycling), intelligent traffic management, non-vehicle streets/zones, smart parking, energy efficient street lighting, innovative use of open spaces, visible improvement in the area (e.g. replacing overhead electric wiring with underground wiring, encroachment-free public areas, and ensuring safety of citizens especially children, women and elderly).

The criteria for selecting Smart Cities is based on nomination by States/UTs (Union Territory) cum competitive selection. The total number of 100 Smart Cities have been distributed among the States and UTs on the basis of an equitable criteria. The formula gives equal weightage (50:50) to urban population of the State/UT and the number of statutory towns in the State/UT. Based on this formula, each State/UT will, therefore, have a certain number of potential Smart Cities, with each State/UT having at least one.

The implementation of the Mission at the City level will be done by a Special Purpose Vehicle (SPV) created for the purpose. The SPV will plan, appraise, approve, release funds, implement, manage, operate, monitor and evaluate the Smart City development projects. Each Smart City will have an SPV which will be headed by a full-time CEO and have nominees of Central Government, State Government and urban local bodies (ULBs) on its Board. The States/ULBs shall ensure that: (a) a dedicated and substantial revenue stream is made available to the SPV so as to make it self-sustainable and could evolve its own creditworthiness for raising additional resources from the market; and (b) Government contribution for Smart City is used only to create infrastructure that has public benefit outcomes. The execution of projects may be done through joint ventures, subsidiaries, public-private partnership (PPP), turnkey contracts, and so forth, suitably dovetailed with revenue streams.

The SPV will be a limited company incorporated under the Companies Act, 2013 at the city-level, in which the State/UT and the ULB will be the promoters having a 50:50 equity shareholding. The private sector or financial institutions could be considered for taking an equity stake in the SPV, provided the shareholding pattern of 50:50 of the State/UT and the ULB is maintained and the State/UT and the ULB together have majority shareholding and control of the SPV. Funds provided by the Government of India in the Smart Cities Mission to the SPV will be in the form of tied grant and kept in a separate Grant Fund. These funds will be utilized only for the purposes for which the grants have been given and subject to the conditions laid down by the Ministry of Urban Development. The SPV may appoint

Project Management Consultants (PMC) for designing, developing, managing and implementing area-based projects.

The Smart City Mission will be operated as a Centrally Sponsored Scheme (CSS) and the Central Government proposes to give financial support to the Mission to the extent of Rs. 48,000 crores (480 billion) over five years, that is, on an average Rs. 100 crore (1 billion) per city per year. An equal amount, on a matching basis, will have to be contributed by the State/ULB; therefore, nearly Rupees one lakh crore (1000 billion) of Government/ULB funds will be available for Smart Cities development. The Mission would be monitored by an Apex Committee at the national level, a High Powered Steering Committee at the State level and a Smart City Advisory Forum at the city level.

Source: MoUD (2015).

2 Deciphering Smart Cities in Literature

With the emergence of sustainable development on the global consciousness (Brundland Commission: Our Common Future in 1987; World Summit on Sustainable Development, Rio in 1992), Sustainable Cities became a favourable but challenging goal for future human development. These cities converge economic growth, environmental conservation and social well-being while maintaining inter-generational equity. With the growing importance of climate change and GHG emissions dawning upon the global environmental discourse, this trend was followed by a much nuanced approach to aim for a low-carbon city which could essentially reduce the carbon footprint of its citizens by mitigating harmful impacts from industries, transportation, thermal power plants and municipal waste. Meanwhile, the concept of ‘quality of life’ and livable cities aims to represent how well human beings’ needs are met or the extent to which individuals or community perceive satisfaction in various domains of urban life (Costanza et al. 2007). Since the early 2000s, the concept of Smart City, understood as the new process of urbanization, is being seen as a tool to realize some of these concepts on the ground and has become quite fashionable in the policy, entrepreneurial and academic arena.

According to recent reports, Stanford University’s research shows that there are currently around 150 Smart City projects ongoing or completed. Most of them are found in Europe (47), Asia (40) and North America (35). Smart City is also used as a marketing label by companies and the cities themselves to guide their urbanization and enhance global competitiveness. Top IT-based leading companies have targeted Smart Cities as its main markets and blue oceans of business development. Furthermore, academia is also increasingly embracing the topic of Smart Cities as one of the hottest emerging research areas launching postgraduate courses and research lines centred exclusively on the theme (CIDOB 2014). Surprisingly, in spite of so much of the buzz around Smart Cities, with numerous agencies promoting, evaluating and developing them worldwide, there is a great deal of ambiguity even on its normative definition, parameters and international authority for standardization. While some argue ISO 37120: 2014 to be the standard indicator for Smart Cities, the International Organization for Standardization itself designates them as 'Sustainable Development of Communities—Indicators for City Services and Quality of Life'. They are basically an extension of the Global City Indicators Facility, initially sponsored by the World Bank (World Bank 2008).

For the scope of this investigation, the scholarly definition of Smart Cities is, "Cities where investments in human and social capital and traditional (transport) and modern (ICT) communication infrastructure fuel sustainable economic growth and a high quality of life, with a wise management of natural resources, through participatory governance" (Caragliu and Nijkamp 2011, p. 6). Smart Cities highlight important aspects of sustainability, such as the need for responsible resource management, energy efficiency, and citizen engagement (Colldahl et al. 2013). Furthermore, they could be understood with six characteristics, namely: smart economy, smart people, smart governance, smart mobility, smart environment and smart living (Giffinger et al. 2007, p. 11). The concept is intricately related to intelligent, innovative or knowledge-based settlements that mobilize information and communication technologies (ICT) to deliver better services, reduce carbon footprint, create sustainable environments and improve living conditions. The core idea behind Smart Cities could be better interpreted through their best practices evident across the globe.

3 A Global Review of Best Practices in Smart Cities

Famous examples of Smart Cities include cities like Tokyo, New York, London, Shanghai, Yokohama, Kawasaki, Singapore, Barcelona, Amsterdam, Vienna, Rio de Janeiro, Brisbane, Stockholm, Copenhagen, Toronto, and so forth, while relatively non-descript ones include Kitakyushu (Japan), Songdo (South Korea), Masdar (Abu Dhabi), Skolkovo (Russia), Dubuque (US) and several in China like Meixi, Tianjin knowledge city, Suxhou, Guangzhou and Szechuan. A global review of Smart Cities reveals a great deal of variety in their scale, economic structures, technological innovations and sectoral priorities. For instance, while Tokyo and New York are mega city-regions and international financial hubs, cities like Songdo is confined to 1500 acres near the Incheon international airport while Dubuque, that proclaims itself to be the first Smart City in the US, has a population of only 60,000. By Indian standards, that may only translate into a modestly dense middle-income locality or a portion of *Dharavi* slum in Mumbai.

The winners' podium of the world's 'smartest' cities is occupied by Tokyo, London and New York, respectively, for the third consecutive year (IESE 2014), among 135 cities on 50 indicators along 10 different dimensions that are (with the city topping that dimension in parentheses): *Governance* (*includes citizen and business engagement,* Auckland is the top city in this area), *Public Management*

(Tokyo), *Urban Planning* (Berlin), *Technology* (London), *Environment* (Zurich, Geneva and Basel), *International Outreach* (London), *Social Cohesion* (Eindhoven), *Mobility and Transportation* (Berlin), *Human Capital* (Tokyo), and *Economy* (New York). Like IESE there are rankings by several other agencies. Based on their own valued reasoning, some top rated Smart Cities include Vienna, Toronto and Paris (Cohen 2014) and Singapore, Hong Kong and Curitiba (Kotkin 2014, featured in Forbes).

The global review demonstrates that cities have either targeted different sectors or in fact intermittent co-benefits to become smart. Japanese cities like Yokohama, Kawasaki, Kitakyushu, and Toyama have pioneered the multi-sectoral co-benefits, covering renewables like solar energy, the waste sector including segregation, load and emission reductions, smart grids with community and household energy management systems in place, and e-mobility that prioritizes light rapid transit and electronic vehicles (IGES/YCU 2011). Certain cities have started with one major sector. Singapore with heavy investment in its public transportation infrastructure succeeded in creating one of the most modern, affordable and highly used public transport networks in the world. London has similarly worked to ease traffic congestion leading to variable pricing for road tolls. Sacramento in California has focused on the power sector, installing 615,000 smart meters at customer premises through home area networks, which in turn are connected to an advanced metering infrastructure network. These smart meters enable adjusting electricity consumption within houses in tune with grid supply so that black or brown outs are proactively avoided. European cities are increasingly moving towards e-governance related co-benefits using automated revenue collection systems. In Amsterdam, the Digital Road Authority mines different types of traffic data to provide services such as on-demand parking space, and expected travel time to users, thus reducing congestion, waiting time and the associated air pollution, thereby improving road safety and quality of living of its citizens (Vembu and Sridhar 2014). Similarly, Copenhagen moved to smart travel cards for seamless travelling between various transport modes, facilitating shopping and payment of bills. Madrid plans to invest about US$20 million in a technology platform to manage a range of public services such as street maintenance, lighting and waste management. There are other cities that have initiated particular 'smart' solutions like Dubuque in Iowa, which operating on a fairly limited budget recently completed a three-year pilot project to install smart electricity and water meters. Smartness is involved in data collection, often in real time, leading to efficient decision-making. Rio de Janeiro, for example, monitors real time data from 30 agencies streaming into an operation centre from where responses to emergencies and accidents are efficiently coordinated. Similar measures in other cities include using Global Positioning System-assisted waste removal, helium gas-aided water pipeline leakage detection and a smart card facility for citizens for various services.

Meanwhile, there is a growing evidence of cities in middle- and lower-income countries taking advantage of technology to create co-benefits. While identifying Smart Cities for an award for the period 2007–2011 in the Asia region (on the basis of broadband connectivity; a knowledge-based workforce; digital inclusion;

innovation; and marketing and advocacy), three Indian cities—namely Bangalore, Hyderabad and Jaipur—figured in a list of 20 cities from the region (Ramachandran 2014). Modest beginnings made by other Indian cities could not be overlooked; for instance in Delhi, the municipal corporations have an online system to disburse birth/death certificates and enable people to pay some of their bills and property taxes. In Bangalore, real time technology relays information of bus timings, congested routes and so on; and in Indore, the traffic police have installed infrared devices to nab rule violators (Narayan 2014). Meanwhile, a city like Lavasa near Pune, has recently woken up and repackaged its technical prowess for the coveted 'smart' tag. The investigation of global best practices in Smart Cities shows the growing use of ICT in energy and mobility sectors or specific solutions in others. This has come about as a natural evolution for the developed countries that have already addressed issues of physical and social infrastructure and now need to gain immediate results in resource consumption, energy efficiency, cost recovery, business and civic engagement to ensure global competence and internal sustainability. India, which is increasingly becoming connected to the global systems will have to take this challenge to leapfrog into Smart Cities.

4 Issues and Challenges to Generate 'Smart' Co-benefits

In practice, actualizing Smart Cities to generate several co-benefits poses a formidable test. The SCM identifies some challenges that arise in this process that include competition among cities, support and leadership roles played by States and ULBs, capacity assistance required by stakeholders, prior investments of time and resources in the planning phase and pro-active participation of people (MoUD 2015). Indeed, Smart Cities not only seek technological leapfrogging but a giant leap of faith. Unlike smart phones and smart card technology, Smart Cities could neither be imported nor transplanted; they rather need to be adapted to the local context and assessed for their value to sustainability. This poses numerous intervening issues and challenges as discussed below.

4.1 Ideological Challenges

India in the past has had a sort of love–hate relationship with urbanization. It primarily started with the pressing need to resettle in-migrants after partition of India in 1947. This was followed by a perception of cities as a 'necessary evil' while modern industrial and capital towns came into being (believing India lives in its villages, cities were regarded as a social threat). Post-1970s drew a gradual attitude of acceptance (as the Census noted the realities of rural-urban migration, growing urban slums that led to commissioning of several development authorities, counter-magnets and *awas yojna* projects). The economic liberalization era from

1991 onwards has viewed urbanization as a challenge. While the private sector capitalized upon it through real estate, the government made attempts with the National Urban Renewal Mission (JNNURM), *Rajiv Awas Yojna,* Provision of Urban Infrastructure in Rural Areas, and so forth, with varying interventions and results. Today, India stands at a crossroads as the new government has taken urbanization head on as an opportunity, with a dream to improve quality of life (QoL) by making the cities smart. In this regard, it becomes vital for the government to clarify its vision and strategy for Smart Cities, considering the fact that there is no universally accepted definition and metrics. To make the matters complex, India does not have a national urbanization policy that could provide a blueprint to how many, what kind of cities should come where and when. This raises doubts of whether 100 new Smart Cities would put conventional urban-regional theories, scientific, integrated and multi-level planning head over heels? Indeed, this has been a practice with previous public policies, more as a norm than an exception while commissioning special economic zones, industrial areas, development authorities, transport corridors that had little or no relevance to regional needs, resource base, land use, district and local-level planning whatsoever. The present ravaged landscape of the country today is a mute witness to indiscriminate abuse and misuse of agricultural lands, non-compliance of regulations by authorities, intermittent and patchy urbanization, urban fringes, sprawls, squatter settlements that the entire country's urban arena seemingly portrays a slum and filth in continuity. The government will have to convince the citizens that if '100 Smart Cities' are to be seen as an opportunity, how will it collectively and effectively contribute to India's aim of sustainable urbanization and enhancing QoL? Will this scheme provide a paradigm shift from a prevailing pattern of unchecked migration, urban growth and rural-urban divide? After all, inequity of energy, resources, income, infrastructure, socio-economic facilities between urban and rural areas is a direct consequence of poor planning, industrialization and urbanization patterns in the past.

Even amongst and within the cities there is a stark disparity in their QoL and availability of municipal services. In 2011, the urban development ministry surveyed 1405 cities in 12 Indian states and found that more than half of them do not have full access to either piped water supply or sewage systems; that 80% of the households there get water for less than five hours a day; and, more gravely, over 70% do not have access to toilets. Between 2001 and 2011, India's urban population grew from 27 to 31% of the national population, but the urban infrastructure hopelessly lagged behind (HPEC 2011; Narayan 2014). Some Indian cities or specific localities within them desperately call for a basic level of urban amenities and quality of life. While certain facilities like water and electricity are being actually wasted by some rich and upper middle-class people and definitely need to be plugged, but comparing against the developed countries, these are still being consumed at one-third to one-fourth of the level on a per capita basis. With such global and local inequities amongst the population at large, the ideological challenge is to justify how 100 new IT-savvy cities would balance the prevailing norms.

4.2 Technical Challenges

The ideological challenges discussed above would require exploring tough possibilities and decisions, which could be made relatively easy if the technical challenges in this process are meticulously addressed. Smart Cities would invariably require a network of sensors, cameras, wifi, data centres, e-meters, hand held monitors, smartphone apps, and the biggest of all 'the internet of things'; but what would one do with the data, privacy and security issues? The government would have to face the challenge of managing this big data. Who is going to process it, how and to what purpose? It is generally felt that city councils are in an increasingly complex position to set up and manage open data processes with their own capabilities. Thereby, most find themselves forced to outsource these services to the higher expertise of private sector. But at this point, too much dependency on private initiative and know-how can create a technological loop that compromises the security of data. It risks the system to fail; it diminishes the ownership of data storage; and it deals with the issues about its ethical use (CIDOB 2014). Hence, creating technical capacity at the local body level to handhold the entire process will be an utmost requirement, which could be possible if they are actively engaged as a stakeholder right from the beginning.

In addition to the much-debated ICT component, there are technical issues associated with traditional infrastructure, planning and construction. Indian cities rank very poorly when compared to their international counterparts. New Delhi and Mumbai rank 46 and 52, respectively, on the Spatially Adjusted Liveability Index and 52 and 53 on the Economist Intelligence Unit Liveability Index in 2013 (Financial Express 2014). Joshi-Ghani, urban sector manager at the World Bank, notes that many of the needs of developing countries centre on providing and maintaining modern infrastructure (roads, power plants, water treatment plants, sewage systems, transit systems). Since this built environment lasts a long time, getting the right infrastructure in place shapes a city for decades to come. Planning a city with the right infrastructure, and not merely replicating past practices that have often been haphazard, means relying more on evidence and analysis about how sustainable cities can and should grow (UN/World Bank 2012). In this regard, there is a strong technical basis that as India re-imagines its urban landscape, it needs to focus on making its top 100 cities livable rather than creating 100 new cities. Isn't the idea of creating new cities utopian—belonging to the era of manufacturing-led growth—as we know that India rides heavily on the tertiary wave that is not necessarily location based. This prompts to evaluate each project proposal on its technical merit for the additional time, money and efforts invested in a new city or scheme, given that in the present world an individual can work smartly from the luxury of one's home.

But the biggest challenge that the Government would have to face as it walks deep into this initiative is perhaps not on account of the inherent technicalities of greenfield projects (against the brownfield ones as pointed above), but associated more so with their associated timelines. The ICT technologies are installed at the end of the construction phase, which involves initial planning, site selection, land procurement, making detailed project reports, issuing tenders, developing land, civil construction, installing services, and so forth. Even if the government is able to manoeuvre time overruns in land acquisition, creating a special purpose vehicle for projects and working on public-private partnership (PPP) mode, it would still be an arduous task for the Smart City to see the light of day in less than 4–5 years. Greenfields appear to be relatively easier and fancier, but very soon the agencies involved would realize that in order to show tangible results to the public, it would have to put its money where its mouth is, that is, treading the challenge of working in a filthy old city. The bigger the risks, the greater the benefits. In practice, leaving aside a few sectors, with part-infrastructure already given, brownfields would turn out to be rather encouraging and affordable giving plausible and immediate relief to their citizens. Best practices across the world show that smart mobility and smart energy are low hanging fruits that ensure multiple benefits and public appeal, while the complex and expensive projects may follow suit. Smart after all, is not just about advanced technology, but being innovative, thinking out of the hat, and enterprising to experiment with little resources at hand. Common sense techniques —walk to work, shaded pathways, car pools, mixed land use, lively street designs and use of natural features and materials in cities—need to be genuinely brought back.

4.3 Societal Challenges

'What is a city, but the people?' said Shakespeare. This necessitates the question whether smart solutions would descend from the heavens or based on societal needs. The global discourse like the New Urban Agenda lays emphasis on social priorities in addressing future needs, advancing towards greater social cohesion, breaking down social divides, promoting participatory and inclusive local governance, and fostering sustainable development. Smart Cities cannot only be about displaying technology and delivering services; fundamentally, they have to be inclusive and equitable places to live in. A strong pursuit of technological innovation without well-defined and targeted inclusive policies can be risky. It is likely to leave lower income and vulnerable populations behind the opportunity path. In the process, the income inequality gap increases and citizen participation can be undermined (CIDOB 2014). If the policy-makers and the practitioners prefer to view Smart Cities as small enclaves on the urban periphery probably out of convenience, these would have the tendency of turning into expensive and exclusive gated communities. It would be the most preferable pathway if given a choice. India has had a long colonial and modern history of exclusive estates like the princely estates, civil lines, bungalow zone, urban farmhouses, rich-suburbs to the recent special economic zones, that only harboured servitude, inferiority and no respectable spaces to the commoners. Is Smart City movement scaling up the gated community phenomenon to the city level? If the state overlooks the existing city and privileges new enclaves, the urban fabric will split into two unequal halves.

There are some counter arguments to this apprehension. Joshi-Ghani argues that cities are growth escalators, but Smart Cities are more than that. They make urbanization more inclusive, bringing together formal and informal sectors, connecting urban cores with peripheries, delivering services for the rich and the poor alike, and integrating the migrants and the poor into the city. Promoting Smart Cities is about rethinking cities as inclusive, integrated, and livable (UN/World Bank 2012). But is there evidence in support of this assertion. Practitioners who have worked on creating Smart Cities have expressed difficulties in holding effective engagement processes. Further, prioritization processes were often determined by political ambitions and available budgets (Colldahl et al. 2013). For inclusiveness, they made recommendations with respect to sustainable urban development, effective planning, measuring success, engaging stakeholders, and developing actions plans through prioritization processes. As Mischa Dohler, chair professor in Wireless Communication at King's College London, points out, 'big data is not enough, it's just half the way'. Ensuring a pro-poor access to smarter public services is therefore crucial (CIDOB 2014). The SCM must go a long way ahead of merely delivering technical solutions, if they wish to socially walk the talk.

4.4 Governance Challenges

There is a general perception as to whether a Smart City is being passed as the new bottle with e-governance being the old wine. Well, just as a foreign technology has to undergo the test of society, so does it need to find concurrence with the prevailing legislative, policy and administrative framework. In order to explore co-benefits, there are several challenges and unanswered questions; for instance, is prevailing legal and governing environment conducive for 100 new Smart Cities? What role and initiative will the state governments, local government and the private sector have? Theoretically, there could be several mechanism for execution, namely: (i) a flagship programme as had been done previously with schemes like JNNURM, RAY, *Indira Awas Yojna*, Basic Services to Urban Poor under the urban portfolios with nodal agencies in respective states; or (ii) through a separate statutory body created to plan and develop all the 100 Smart Cities in the country; or (iii) through separate special purpose vehicle in every state with Centre, State, ULB as the stakeholders; or perhaps more likely (iv) a combined but modified approach of (i) and (iii) above where both Centre and State oversee fiscal planning and mobilization and at the same time develop Smart Cities on a project mode through either line departments, private agencies or preferably on PPP mode, while ULB is also a stakeholder within a state level constituted authority. The SCM has chosen to adopt the fourth approach in practice, which could work for both greenfield and brownfield projects.

The government would need to clarify how it would treat a Smart City. Is it merely a functional entity or would it have a statutory mandate akin to Exclusive Economic Zones and Special Economic Zones? Like industrial areas, corridor zones and development corporations they could also be governed by separate statutes. Its ideological, technical and societal implications are discussed above. In terms of good governance, there is an already overlapping complex of jurisdiction and functions between authorities created for special purposes and the constitutional agencies at the local level, namely the urban local bodies (ULBs) and the gram panchayats. It needs to be seen with keen interest how the government treads this tightrope. Apparently there is a greater ease in doing business by bringing Smart Cities under a corporate SPV or a new act, considering that manufacturing, foreign direct investment and greenfield development seem to be the key generator; rather than taking a more nuanced, demanding and unprecedented approach of building capacities within the ULBs to plan and manage Smart Cities by themselves. The latter could have led to better devolution of powers to the local bodies as mandated by the 74th Constitutional Amendment Act 1992. It needs to be underscored that municipal governments can play a vital role during implementation and creating synergies with the existing city. In fact a common authority will command citizen participation, ownership and global brand equity for both the parent city and the Smart City and could lead to a new sunrise for the urban bodies in the modern history of India.

It needs to be appreciated that Smart Cities should not turn out as centre's edifices or white elephants at the local level. There needs to be a fine balance of roles and responsibilities of the centre, state, local governments and the private sector for integration of several systems. The role of the private sector also needs to explore which ICT provider has the expertise of basic services and infrastructure. Probably none. In fact execution of ICT services comes at the far end of the project, though their technical inputs would be significant during the planning phase (in case of new projects). Resting the entire project onto an individual technology provider could not just turn out to be technically unwise, financially risky but a major managerial challenge. PPP should rather support a much wider technical participation from the private sector on multiple criteria: allow consortiums, joint bidding, and possibly sub-contracts to local enterprises. The Government also needs to devise a system that promotes healthy competition between cities and at the same time be flexible to offer opportunity for late comers, unlike in JNNURM.

4.5 Financial Challenges

India's steady urbanization has the potential to bring in investments of Rs. 120–150 trillion over the course of the next 11 years until 2025, says the Kotak Institutional Equities report entitled 'Multiplicities'. Though several aspects of finance have been discussed in previous sections, a few challenges in particular seek attention. Going by the commitment of the SCM, Rs. 200 crore per Smart City serves as a stepping stone as the physical infrastructure saps high investment. Co-benefits of Smart Cities are normatively undoubted but the most critical question is at what price? And accordingly, what are the cost recovery mechanisms? In the words of the Songdo's International Business District CEO, 'the city itself is just a normal city with state-of-the art technology that struggles like any other city to attract citizens and firms to settle down' (CIDOB 2014). The Smart Cities would have to come up with their own non-traditional and innovative ways of self-sustenance, in addition to the financial provisions within the SCM.

This prompts to ponder the role of the state and local government in question, and would it also involve a similar sharing of liabilities and profit? The above questions become important because India has not had a spectacular track record in inviting foreign investments in the manufacturing sector while investments from urban programmes like JNNURM in the past have ended up as an asset creating black hole. Secondly, if cities are developed on PPP mode as increasingly finding favour in government policy, amassing of land and its monetization is inevitable as it is the most vital cost-recovery component (as well as the most controversial). But the government would have to ensure that the SCM does not turn into another land grabbing real estate venture that anticipates higher returns as time over-runs. There need to be sufficient checks and balances in its financial planning, implementation, contracts, risks and liabilities. Moreover, the initial commitment should be used as a seed fund, and every effort should be explored to make Smart Cities self-sustaining. Cities ought to be smart when it comes to investments, project costs, gauging demand and supply, return on investments, for example. There should also be due financial consideration for maintenance and overheads during the implementation phase, capacity building of stakeholders, and so on.

5 Possibilities for 100 Smart Cities

In view of the global challenges, cities need to serve as living laboratories to deal with complex intervening issues like job creation, poverty and inclusion, freedom and democracy, safety and security, innovation, economic growth, global competitiveness, climate change, and so forth. In order to address the above discussed challenges, this research puts forward a series of possibilities to shape 100 Smart Cities in India, keeping in view Sustainability, Metrics, Adaptiveness, Reporting, Technology for Inclusiveness (collectively forming a new acronym for SMART):

S—Sustainability: Smart Cities should be taken as a logical opportunity to realize this challenge, with sustainable urbanization as the overarching goal. The government's decision to formulate a new urbanization policy with a vision for the next 25–30 years would go a long way in creating this. Land and urban planning are a state subject, but urbanization is a regional and national phenomenon with increasing global associations like globalization, climate change, and so on. The government should consider what should be its priority, to make '100 Smart Cities', 100 'Smart' cities or 100 'Smarter' cities. The basic idea should be to refrain from being over-ambitious, splurging settlements, as happened in the past—governments either abetting unplanned, squatter settlements that were then regularized later or else creating ill-planned economic zones, industrial and real estate townships sprawling over the landscape with no genuine takers. Sustainability also mandates to be flexible and adaptive with the laws of nature. Given the fact that the existing cities which host a bulk of the population in old localities, waste a lot of resources, are highly energy-inefficient and the ones that urgently require smart solutions. At times redeveloping dilapidated parts of a city may bring greater co-benefits than sub-urbanization. Global cities have time and again proved this through urban revival. Similarly, decentralized action in municipal services, traffic, waste and wastewater management and participation of people could create greater co-benefits, in spite of having a centralized control over data. In fact 'smart', 'livable' and 'sustainable' goes hand in hand.

M—Metrics: Smart Cities are all about finding and implementing suitable metrics, which forms the bedrock for transparency, objectivity and rational decision-making. The absence of a standard definition in this regard should rather be taken as an advantage. The state needs to adopt a normative basis for its Smart City concept considering the Indian context, decide the main objectives, its components and devise suitable indicators like quality of life, inclusiveness, level of services/performance and so forth, to be adopted during planning and implementation. Drafting of guidelines, toolkits, laying procedures for public and stakeholder participation needs attention. Incite specific ideas and problems from experts and stakeholders. Employ back-casting to do scenario analysis and find solutions. Make a checklist of doable actions, monitoring plans and reporting structures.

A—Adaptiveness: Smart Cities would need to adapt according to India's federal governance frame—which cuts across jurisdictions, sectors and spatial scale. Since land development is a State subject, their enthusiastic participation is crucial. The Government has to put in place a detailed framework to guide investment and demarcate responsibilities. The policy alternative (iv), a combined but modified approach, discussed in Sect. 4.4 seems more credible, where centre and state oversee fiscal planning and mobilization and simultaneously develop Smart Cities on a project mode or PPP mode, including ULB as a stakeholder. This could work for both greenfield and brownfield projects. Funds are only one part of the problem. The key challenge would be to overhaul urban governance and infrastructure, both physical and digital and explore the possibility of a cross-subsidy between one another at various project stages. Before initiating a robust scheme that could give excessive powers to technology providers of private origin, who essentially fall short of adequate skills and experience associated with non-ICT infrastructure at the city scale, it may be worth developing few prototypes in upcoming cities where basic infrastructure is already in place.

R—Reporting: Cities around the world are using big data on a range of urban issues including transport, employment, migration, housing and education to generate savings, promote innovation and make cities a better place to live and work (Centre for Cities 2014). As such the projects focus on how big data can be presented in an easy to interpret and/or visually compelling way. In the USA, for example, cities such as New York and Boston have seen the development of mobile apps and service improvements through making data available to all. Since there is no precedent of a planned Smart City in India, there is a need to simultaneously create and share knowledge through documentation, reporting and creating knowledge networks, while actualizing projects on the ground. Tim Campbell in his book *Beyond Smart Cities* acknowledges that learning networks tend to form between cities that share similar levels of development, socio-economic context, and challenges (CIDOB 2014). This would not only serve into the feedback loop but also lead to horizontal expansion of the Smart Cities movement.

T—Technology for Inclusiveness: Technology should work with the motto 'citizen first'. Directly linked to the empowerment of citizens and local institutions, technological change is driving urban transformation. For instance, opening the 'black box' of urban data stored in the hard disks of public computers has been another major breakthrough in promoting local level participatory planning. Concepts such as 'open data' and 'big data' refer to the facilitation of public data to public consultation and its processing for a social service-based use (CIDOB 2014). Technology serves as a medium to attain a higher ethical, socio-economic and public good. Holland (2008) views this internalization as the Smarter Cities. Existing educational and technical institutes should be pressed into service to handle urban data for innovation and practical use. Joshi-Ghani sees in this a key to adapting 'Smart City' precepts. The concept of 'Smart Cities', she says, is 'really about good governance. It's about giving basic services to our citizens. It's about livability. It's about how we are using our resources. It is how a city functions on a day-to-day basis. I think smartness is about doing more with less' (UN/World Bank 2012). In this regard, the importance of amenities like potable water, clean and noise free surroundings, space to walk and commune, safety and security, health and hygiene, reliable power supply, hassle free civic services for the most simplest of things like bill payments, issuance of birth/death, marriage certificates, forms attestation, police verification, and so on, could not be undermined. It is the provision of these petty small solutions that technology could bring immense value to the lives of common people.

6 Conclusion

This research started with some fundamental inquiries about Smart Cities like what they really stand for—international best practices, probable challenges and possibilities to achieve 100 new Smart Cities in India. The research reveals that, in spite of having no internationally accepted definition or a national urbanization policy in India, nonetheless the Smart Cities Mission holds immense potential to achieve multiple co-benefits of sustainability, systems efficiency, economic growth, participatory governance and better quality of life. While techno-financially, building a Smart City does not seem to be much of a complex issue, the biggest challenge is to fuse it with the existing urban, governance and social fabric of the country. This could be fulfilled effectively if SCM maintains focus on sustainable urbanization, good governance, transparent metrics and reporting. As evident, most measures aimed at enhancing QoL and co-benefits are realizable in transport, energy, and so on, though social sectors like education, health, basic amenities also need to be simultaneously prioritized. Thus the mission should keep people in the spotlight because what the large and burgeoning urban populace aspires to embrace is a safe, inclusive and sustainable fold of urbanization, to which 100 new Smart Cities is the yellow brick paver.

References

Caragliu, A., Del Bo, C., & Nijkamp, P. (2011). *Smart cities in Europe.* (Executive Summary) Amsterdam: VU University of Amsterdam.

Centre for Cities. (2014, February 24). *Big data can lead to big results for cities.* Centre for Cities, London.http://www.centreforcities.org/blog/2014/02/24/big-data-projects-david-willetts-cities/ . Accessed on August 16, 2015.

CIDOB. (2014). *Beyond smart cities: It's time for urban sustainable development.* Paper No. 92: 06/2014. Barcelona: Barcelona Centre for International Affairs (CIDOB).

Cohen, B. (2014). *The top 10 smart cities on the planet.* http://www.fastcoexist.com/1679127/the-top-10-smart-cities-on-the-planet. Accessed on August 22, 2014.

Colldahl, C., Frey, S., & Kelemen, J. E. (2013). *Smart cities: Strategic sustainable development for an urban world. Thesis submitted for completion of Master of strategic leadership towards sustainability.* Karlskrona, Sweden: Blekinge Institute of Technology.

Costanza, R., Fisher, B., Ali, S., Beer, C., Bond, L., Boumans, R., et al. (2007). Quality of life: An approach integrating opportunities, human needs, and subjective well-being. *Ecological Economics, 61,* 267–276.

Financial Express. (2014, August 20). Growing Cities. In *The financial express.* New Delhi.

Giffinger, R., Fertner, C., Kramar, H., Kalasek, R., Pichler-Milanović, N., & Meijers, E. (2007). *Smart cities: Ranking of european medium-sized cities.* Vienna, Austria: Centre of Regional Science, Vienna University of Technology. http://www.smart-cities.eu/download/smart_cities_final_report.pdf. Accessed on August 13, 2015.

Habitat, U. N. (2011). *Global report on human settlements 2011: Cities and climate change.* London/Washington DC: United Nations Human Settlements Programme & Earthscan.

Hollands, R. G. (2008). Will the real smart city please stand up? City: Analysis of Urban Trends. *Culture, Theory, Policy, Action, 12,* 303–320.

HPEC. (2011). *Report on Indian urban infrastructure and services.* New Delhi: The High Powered Expert Committee on Urban Infrastructure, Government of India.

IESE. (2014). *Which are the world's "smartest" cities?* IESE insight cities in motion index 2014. http://www.ieseinsight.com/doc.aspx?id=1582&ar=5. Accessed on August 18, 2015.

IGES/YCU. (2011). IGES—YCU joint seminar on low-carbon and smart cities. In *IGES Conference Report-KUC-2011-01,* July 26–27, 2011.

(The) Indian Express. (2014, August 18). At Greater Noida, Uttar Pradesh Government begins work on creation of industrial township. In *Indian express.* New Delhi.

Kotkin, J. (2014). *The world's smartest cities.* http://www.forbes.com/2009/12/03/infrastructure-economy-urban-opinions-columnists-smart-cities-09-joel-kotkin.html. Accessed on August 25, 2015.

Mckinsey. (2010). *India's urban awakening: Building inclusive cities sustaining economic growth.* New Delhi: Mckinsey Global Institute.

MoEF. (2015). *India's intended nationally determined contribution: Working towards climate justice*. New Delhi: Ministry of Environment & Forests, Government of India.

MoUD. (2011). *India's urban demographic transition: The 2011 census results-provisional*. New Delhi: JNNURM Directorate and National Institute of Urban Affairs.

MoUD. (2015). *Smart cities: Mission statement & guidelines*. New Delhi: Ministry of urban Development, Government of India.

Nam, T., & Pardo, T. A. (2011). Conceptualizing smart city with dimensions of technology, people, and institutions. *Proceedings of the 12th Annual International Conference on Digital Government Research.*

Narayan, S. (2014, July 19). First smarten our cities, then build smart ones. *Hindustan Times*, July 19, 2014.

PTI. (2014). *Rs. 7060 crore earmarked for developing 100 smart cities*. Press Trust of India. http://timesofindia.indiatimes.com/budget-2015/union-budget-2015/Rs-7060-crore-earmarked-for-developing-100-smart-cities/articleshow/38140103.cms. Accessed on October 21, 2015.

Ramachandran, M. (2014, August 9). Decoding smart cities. *Business Standard.* http://www.business-standard.com/article/opinion/m-ramachandran-decoding-smart-cities-114080900706_1.html. Accessed on October 29, 2015.

UNDESA. (2011). *Urban population development and the environment wallchart*. United Nations, Department of Economic and Social Affairs, Population Division. http://www.un.org/en/development/desa/population/publications/urbanization/urbanPopDevEnv_2011.shtml . Accessed on December 16, 2015.

UN/World Bank. (2012). Who needs smart cities for sustainable development? In *Sustainable development*. World Bank. http://web.worldbank.org/WBSITE/EXTERNAL/TOPICS/EXTSDNET/0,,contentMDK:23146568 ~ menuPK:64885113 ~ pagePK:7278667 ~ piPK:64911824 ~ theSitePK:5929282,00.html. Accessed September 16, 2015.

Vembu, D., & Sridhar, V. (2014). Requirements and future of smart cities. *Express Computer.* http://computer.expressbpd.com/columns/icts-and-smart-cities/. Accessed on September 19, 2015.

World Bank. (2008). *Global city indicators program report: A program to assist cities in developing an integrated approach for measuring city performance*. New York: World Bank.

World Bank. (2010). *Cities and climate change: An urgent agenda*. New York: The World Bank.

Zee News. (2014, August 25). Smart Cities' should bring quality of life with inclusivity: Naidu. *Zee News.* http://zeenews.india.com/news/nation/smart-cities-should-bring-quality-of-life-with-inclusivity-naidu_951059.html. Accessed on April 11, 2015.

Chapter 11
Social Entrepreneurship, Energy and Urban Innovations

Rama Krishna Reddy Kummitha

Abstract Energy has become a central focus of human existence and lack of access to energy is an indication of lack of access to many basic needs. Especially people in developing countries face a greater hurdle to access electricity. Given the constraints in the capacities of developing countries, it may be difficult to cater for the escalating energy needs in urban areas unless innovative methods are explored. However, what becomes very significant in this mileu in developing countries is the fact that the third sector is emerging very strong—in particular, social enterprises with their innovative strategies, processes and methods to provide access to energy and protect the environment from over-usage of natural resources. With this background, the current chapter, through a case study approach, attempts to explore how social entrepreneurs in India adopt better ways of providing energy to the excluded and marginalized social sections.

Keywords Social entrepreneurship · Energy access · Urban innovations
Renewable · Solar · Social inclusion

1 Introduction

The relationship between energy and poverty is long proven (Khandker et al. 2010). People without access to energy have been largely stuck in a vicious cycle of exclusion. It is also understood that the world poor spend a significant amount of their limited incomes on expensive and unhealthy forms of energy which result in their own hazardous health conditions (Karekezi et al. 2012). The importance of access to energy is long argued by vast literature field. For example, Sovacool (2012), while referring to the Millennium Development Goals (MDGs), claimed that it is impossible to achieve any one of the MDGs as declared in the Millennium Declaration without having access to adequate and affordable energy. It is apparent

R.K.R. Kummitha (✉)
School of Management, Politecnico di Milano, Milano, Italy
e-mail: mailtokrkr@gmail.com

M. Sethi and J.A. Puppim de Oliveira (eds.), *Mainstreaming Climate Co-Benefits in Indian Cities*, Exploring Urban Change in South Asia,
https://doi.org/10.1007/978-981-10-5816-5_11

that energy has become the central focus of human existence and lack of access to energy is an indication of lack of access to many basic needs. A significant portion of the world poor and excluded who live in disadvantaged areas continue to struggle in order to get access to energy. It is estimated that about 3 billion people, which is almost 40% of the total world population, cook their food using solid fuels and about 1.4 billion people, which is about 20% of the total population, do not have access to basic electricity (IIASA 2012). IIASA's report on *Global Energy Assessment: Towards a Sustainable Future* emphasizes that without any question, "a radical transformation of the present energy system will be required over the coming decades". For this change to happen, large and sustained investments, attractive policies and innovative institutional mechanisms need to be deployed (IIASA 2012, p. xiii).

Further, the rural versus urban divide in access to energy is quite disturbing. For instance, World Energy Outlook data for developing countries reveals that while urban areas have electrification rates of 90.6%, rural areas lag behind at 63.2% (WEO 2008). While a majority of urban residents have privileged access to electricity, a large number of people from rural areas that live in disadvantaged neighbourhoods continue to experience energy poverty. Even if the energy becomes accessible, there is a great disparity in urban and rural energy-use patterns in developing countries, where urban dwellers have substantial amount of final energy per capita than their rural counterparts, as high as two to three times (Sethi 2015). It is estimated that by 2030 two-thirds of the total population on the planet is expected to live in urban areas (Kummitha and Crutzen 2017). That means transformative migration from rural areas to urban areas will result in two-thirds of people settling in urban areas. Given the constraints in the capacities of developing countries, it may be difficult for them to cater for the escalating energy needs in urban areas unless new and innovative methods are explored. The poor capacities and investment levels in Third World countries may hamper their capacities to address the growing needs in urban areas. For example, India has recently expressed its concerns in COP 21 that such a massive increase in the rural urban transition may create troubles for the 'unfinished' development agenda and increase the development gap in the country On the other hand, the global debate to reduce the environment impact from traditional methods of electricity generation like fossil fuels has emerged as a concern. This has positioned nation states and parties involved in electricity facilitation to embark on innovative ways of promoting renewable energy.

In this regard, this research intends to understand the grassroots innovations that aim at providing affordable energy access (to the very same migrants who end up living in urban slums) and could also create multiple co-benefits, particularly to improve the local environment and socio-economic conditions.

This chapter argues that renewable energy, which is being promoted by such local initiatives apart from contributing to the global energy needs, is capable of offering many environment and socio-economic benefits including job creation, increased energy security, improved human health, environmental protection, and mitigation of climate change. IIASA (2012) opined that there have already been sufficient efforts carried out to make sure that 50% of the global energy supply by 2050 will come from renewable energy sources including biomass, hydro, wind, solar and geothermal. However, what becomes significant in the developing countries is the fact that the third sector is growing and emerging very strong; in particular, the social enterprises with their innovative strategies, processes and methods provide access to energy and protect the environment from over-usage of natural resources. With this background, the current chapter attempts to explore the process involved in creating and promoting innovative attempts by social entrepreneurs in India.

2 Background: Relevance of Expanding Clean Energy Access

In 2011, India became the fourth largest energy consumer in the world after China, the United States and Russia. For India, being the second largest populated nation, this is quite understandable. The current per capita energy consumption in India is one-third of the global average, but this may increase drastically as India is expected to embark on a potential developmental path way. The International Energy Outlook 2013, predicts that by 2040 both India and China will require half of the total global energy demand, while India's demand alone is projected to increase 2.8% per annum (http://www.eia.gov/countries/cab.cfm?fips=in). Meanwhile International Energy Agency (2012) believes that "India's energy sector is increasingly unable to deliver secure supply of energy amid growing demand and fuel supply" (p. 7). Further, it notes that the lack of required timely and adequate investments may push India towards a disastrous energy crises. There is a strong need for the country to plan much in advance for the expected economic development and urbanization in the near future. It requires, apart from planning, a strong investment pipeline into energy industry, which would need to expect long-term slow returns on the investment. Even currently, the shortage of energy is quite visible at times. Due to the shortage of energy production, India suffers from a major shortage of power, especially in the cities which leaves an array of activities to halt including the basic human survival. For instance, in July 2012, there was an unexpected electricity blackout in northern India for a couple of days, which resulted in an estimated 700 million people suffering. It is understandable that the people in urban areas are prone to experience harsher consequences from such conditions than people living in rural areas where energy access is limited.

Figure 1 denotes that India's energy needs are largely met by coal. Both Oil and Biomass represents about a quarter of the total requirement. India has recently started venturing into nuclear energy production which stands at present at 1% of total production while gas and hydro electricity offer 7 and 1% respectively (GWEC 2011). Electricity sector planning in the country has long noted the importance of renewable energy in order to attain energy security (Prayas 2010). The Energy

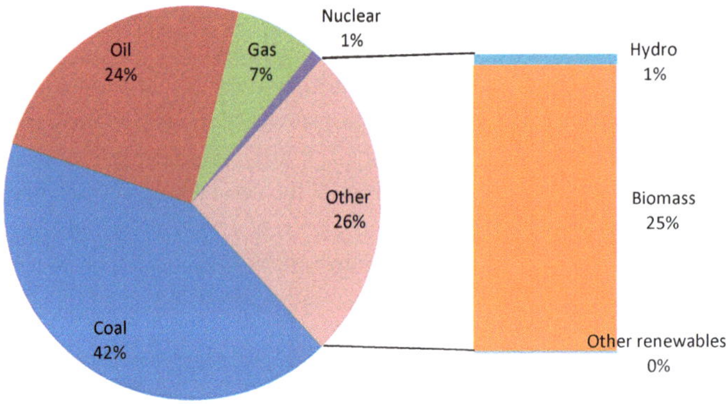

Fig. 1 Energy mix in India

Conservation Act 2001 talks about energy efficiency while the Electricity Act of 2003 describes about promotion of clean energy. Prayas (2010) opines that the two Acts enable strong participation of civil society in clean energy development. India is the first country to have a separate Ministry for New and Renewable Energy. The National Action Plan on Climate Change, which was commissioned in 2008, also presses the need for promoting renewable energy as an important attributor to attain energy efficiency. It is quite crucial to talk about utilization of domestic resources including institutions, techniques and approaches, in addition to promoting indigenous energy technologies to enhance energy based security in the long run (IEA 2012).

Solar energy, among all other renewable energy sources, is considered as crucial and is capable of liberating the nation from its dependency on fossil fuels in the long run (Planning Commission of India 2006). Further renewable energy could be instrumental to reach those in the far-flung hinterlands of the country, where more than 65% of its people live. The other reason for its promotion by the nation state is being upheld by its commitment to contribute for reduction of climate change. By 2009 the share of renewable energy was about 26% of total energy production. This includes biomass, hydro and other renewable energies, in which biomass occupies the largest share. Although the State has committed its complete cooperation, its role in promoting renewable energy is dominated by the private players who made investments in the field.

Private investors have started pouring investments into clean energy. For example, the investments grew from $851 million to $2.1 billion in 2008 (Prayas 2010). Renewable energy would play a critical role to solve some of India's energy problems and to improve the quality of life for citizens. Accordingly, ambitious policy targets would help build confidence among investors. However, effective implementation requires timely regulatory clearance and available human and manufacturing resources. Moreover, an adequate and meticulous learning process should be in place so that India can internalize the necessary experience and expertise. As one Indian solar entrepreneur puts it, 'let's walk before we can run'

(WSJ 2009). At present, India requires about 80,000 million units of power capacity in order to meet it's demand. Renewable energy has been identified as a capable instrument to meet part of the energy demand (MNRE 2011). Although the government has liberated several policies to initiate solar based mini-grids it continues to take a few weeks to months to get commencement approval, which would require visiting different government departments. In addition, the high import duty on solar products, and the need to satisfy compliance tests for hardware in order to acquire subsidies while there are generous subsidies for kerosene, are a few concerns (ADB 2012).

In cities, certain disadvantaged neighbourhoods slums have access to neither the local energy grid nor to the legalized electricity supply, leaving them vulnerable to be exploited by illegal entities. Even where the supply is available, people living in extreme poverty are unable to access it. In addition, the local household business owners have to undergo depressive health conditions and poor profits due to either lack of or insufficient supply of electricity. The growing urbanization showcases daunting circumstances where many poor end up in such deprived neighbourhoods in which minimum standards of living are disturbed by lack of solid basic services including electricity. There are always questions about the legality of slum dwellers' residential status in the cities and, thus, the State has not taken a keen interest to provide electricity supply for such areas. Whereas when it comes to India's plan to achieve universal electricity access, there must be efforts placed to address the energy needs of the slum dwellers irrespective of their residential status.

However, there are concerns whether such needs be met by customized innovations which are driven by local practices or strong top-down approaches, influenced by international players. The obvious answer is that both methods are plausible. In fact both methods should complement each other and the global community should facilitate such a process, given the fact that the developing countries are often restricted by their limited capacities, whereas on the other hand economically developed nations contribute to the larger carbon emissions and maintain distance from creating a balanced path. It is at this juncture that Narendra Modi, the Indian prime minister, emphasized in the recent United Nations climate talks that international partnerships must create a conducive environment where knowledge and technology transfer could easily take place so that both national and local level initiatives be cultivated in order to reach out to the most excluded and change the climate regimes. In fact, Article 11 in the COP21 Paris agreement asserts that, it is the responsibility of the nation states to build capacities at national, subnational and local levels by lessons learned and success achieved. However, it also re-emphasizes that technology and knowledge transfer from developed countries remains key for developing countries to build their capacities. While the talks are at nascent stage at global level about the possible connect between the developed and developing countries for technology transfer and capacity building at national levels, it is quite necessary to enrich our understanding about social innovations at grassroots, especially in the third sector, which uses technologies to meet the unmet energy needs of the poor and deprived, and combat climate change. It is also necessary to learn how such innovations handle technologies at grassroots level. Such learning will be useful once international technology transfer comes into effect, as the successful grassroots level technology usage mechanisms will be more handy to achieve actual results as specified in the COP 21 agreement.

Especially the role of social entrepreneurs to address the energy problems is quite significant. Various innovations they adopt to reach the most excluded while facilitating them with basic electricity supply is quite challenging. Innovation is not only about the production (supply side), but also includes innovative ways of providing energy access to the end users. The best practices in the field will teach the way forward and guide us how they can be further utilized, extended and strengthened to extend the services to the unserved population. Social enterprises adopt innovation in energy generation and enhance efficient methods to reach out to the needy. The social entrepreneurship space in India has both players ranging from niches such as SELCO, Husk Power and *Desi* Power, to large corporations such as Tata BP Power and Philips (ADB 2012).

Social enterprises in the energy sector in India challenge orthodox thinking and employ new ideas to serve the poor with affordable, clean and sustainable energy. Social entrepreneurs are strictly linked to the underserved markets and excluded population. So, it is basically not only about providing clean and sustainable energy but also promoting equity, and adopting innovative and effective ways in service delivery which play a key role. An Intellecap (2012) study demonstrates that a significant portion of social enterprises in India work in the energy sector. A total of 25% of the social enterprises studied as part of its research claimed to work in the energy sector, leaving it as the second most attractive sector next to agriculture which is at about 28%. Most of the emerging social entrepreneurs in India promote renewable energy with a special emphasis on the solar and biomass energy sector. Energy social entrepreneurs largely operate in areas where electricity grids are absent or in those areas which have no reliable access to electricity. They mostly focus on solar energy and biomass when it comes to renewable energy sources. The products or services developed by social entrepreneurs can be classified into three main categories including: (a) meeting basic energy needs of households including lighting and cooking; (b) household energy systems; and (c) energy extraction from waste and delivery to off-grid or underserved communities in the context. This research attempts to discuss the best innovations practised by three such social entrepreneurs to facilitate the poor and the excuded with access to energy in the Indian context. The remainder of the chapter is divided into two sections. The first section offers a rich understanding about social entrepreneurship, whereas the second section discusses about three social entrepreneurs in India who have unleashed innovative mechanisms to create new institutions and practices to address various energy problems faced by marginalized sections.

3 Theoretical Framework: Role of Social Entrepreneurship

Dees (1998) describes social entrepreneurship as a process which unlocks innovative solutions for the unaddressed social problems, whereas social enterprise as an institution catalyses such innovations through adoption of a dual mission (Austin et al. 2006; Seelos and Mair 2007). Dual mission or blended value creation in social enterprises refers to creation of both social and financial value simultaneously (Battilana and Durado 2010; Smith et al. 2013; Besharov and Smith 2014; Lee and Battilana 2013). While social value creation aims to promote a much robust social order where significant social problems are addressed and inequalities are reduced among different social groups, the financial value aims to facilitate the social value being created (Kummitha 2016b).

Literature in the field claims that adoption of innovations by social entrepreneurs often faces restrictions from several factors ranging from cultural aspects which are rooted in communities, to organizational dynamism and so forth (Mair and Marti 2006). The blended value creation principle in social enterprises also often reduces the organisational capabilities to innovate as the existing systems do not allow them to perform roles beyond the conventional institutional arrangements. For example, the conventional enterprises may not be expected to create social value beyond their corporate social entrepreneurship initiatives or interventions aiming at bottom of the pyramid or known as BoP initiatives whereas the non-government organizations are not expected to make profits, even for complementing their social goals (Newth and Woods 2014). While, hybrid ventures which are a recent entry in the field are relatively flexible to create blended value, a system to support such ventures is yet to be developed. Thus, it is argued that in the event of resource scarce environment, social entrepreneurs recombine resources in order to maximize the social value (Gundry et al. 2011). The overall intention of social entrepreneurship is to sustain the social value being created by adoption of income generation models (Kummitha 2017).

Such models may include: (i) strategies to create earned income stream; (ii) the adoption of business like skills; (iii) redefining stakeholder participation through greater democracy and accountability; and (iv) creation of multiple-bottom line approach (Wallace 1999; Lasprogata and Cotten 2003); whereas Hill et al. (2010) hold the opinion that social entrepreneurship as a phenomenon is widely circulated ranging from: (a) non-profits adopting earned income streams in order to supplement shrinking public support; (b) enhancing the efficiency of non-profits through application of business skills and discipline; (c) widening the purview of organizations to enhance ownership and governance structures; and (d) leveraging social innovation to create vibrant businesses that create robust social value.

The presence of entrepreneurship in social entrepreneurship raised several reservations against the basic existence of the concept, as social and entrepreneurship never felt to exist at the core of any one organisation. Accordingly, a considerable body of literature emerged which argues that social entrepreneurship, while taking some of the key strategies from entrepreneurship practice, continues to operate in social lines (Meyskens et al. 2010; Austin et al. 2006). Especially when social organizations are expected to use innovative methods to generate incomes, there are often questions raised about their social commitments. It becomes more intense when non-profit organizations, which have an obligation to promote public interest, are involved in such activities. In order to guard public interest, non-profit organizations should not only aim at following what is legal but also aim at rationalizing what is right. The legal and good intentions would create trust among the larger communities; however, there is a danger that entrepreneurship would compromise the non-profit trust-building process and further influence the legitimacy which they have earned among the communities (Newth and Woods 2014). In fact, legitimacy is the key for grassroots organizations to maintain their support base among the communities. When organizations succeed to retain their trust and legitimacy among communities, in several instances, the latter contributes to the former even in such conditions when they fail to get financial returns on their investments (Kummitha 2016b).

The role social enterprises plao address growing exclusions is well appreciated (Kummitha 2016a). It is at this juncture that innovations play a key role in terms of balancing the tensions that arise between blended value creation (Nicholls and Cho 2006). In fact, it is argued that non-profits gain a number of benefits from the marketization trend, including gaining access to reliable resource streams, which opens space to promote innovation, enhance efficiency and accountability (Aspen Institute 2001). It is against this background that social entrepreneurship in the energy sector takes a key role in addressing the energy needs of the poor and deprived who live in disadvantaged areas in cities. They employ innovations and entrepreneurial practices quite often to reach the unreached and to facilitate the value creation process.

4 Research Methodology

The methodology of content analysis is adopted to draw understandings, policy suggestions and necessary conclusions in this study. Berelson (1952) defines content analysis as 'a research technique for the objective, systematic and quantitative description of the manifest content of communication' (p. 18). Here content of communication refers to book chapters, articles, Ashoka web profiles, books, reports, formal and informal discussions, interviews, newspaper, advertising and historical documents (Lukenbill 2012). Content analysis delivers valid inferences or analytical constructs from the existing texts. It further offers a new insight, enhances researchers' understanding of the phenomenon being studied, and attributes to gain knowledge related to practical actions. Analytical constructs have been derived from: (i) existing understanding on various theories and practice; (ii) the experience or knowledge of experts; and (iii) previous research on the subjects being studied (see Krippendorff 2004).

Based on the premises discussed about social entrepreneurship, the study would further take up three social entreprises in reaching the unreached. They include: (i) SELCO initiated by Harish Hande; (ii) Sustaintech by Svati Bhogle; and (iii) The National Confederation of Dalit Organizations by Ashok Bharti. All three social enterprises studied in this research are Ashoka fellows, whose fellowship is considered significant in social entrepreneurship circles.

4.1 The Solar Electric Light Company

The Solar Electric Light Company (SELCO) is a social enterprise that aims to provide sustainable energy solutions and services to marginalized and excluded households and small and marginal entrepreneurs. It was initiated by Harish Hande in 1995 in Bangalore, Karnataka, a southern state in India. SELCO facilitates energy access to the poor and deprived. It promotes solar energy in addition to adopting various innovative methods to reach out to the poor and excluded.

After tired working hours, due to lack of electricity, people had to return to dark homes, which are generally subjected to toxins and heat from burning kerosene for lighting and cooking. Due to the failure of existing systems, the downtrodden and excluded were forced to use substandard forms of fuels. SELCO, as a social enterprise, disproves three established myths about poor and poverty: (i) poor people cannot afford sustainable technologies; (ii) poor people cannot maintain sustainable technologies; and (iii) social ventures cannot be run as commercial entities. It has not taken much time for SELCO to disprove these propositions and maintain steady growth to serve about 400,000 individuals, most of them earning less than $4 a day. Today, in addition to Karnataka, it is also operational in four other states such as Andhra Pradesh, Kerala, Telangana and Gujarat. It serves customers through its energy service centres located in 28 places across the four states.

4.1.1 Strategy

The idea for initiating SELCO came from Harish's PhD work at the University of Massachusetts. He was inspired when he visited the Dominican Republic where a mechanism was developed to use microcredit provision to facilitate energy services for poor and downtrodden . Then, he met Dr Nevile Williams, the founder of Solar Electric Light Fund (SELF) a US-based non-profit organization that aims to provide electrification worldwide. With the help from SELF, Harish initiated SELCO. The key innovation SELCO carries out is based on the need-based approach. Unlike many traditional solar technology providers, SELCO assesses the need of every customer and offers tailor-made energy solutions. The need assessment includes the capacity of light necessary to carry out the relevant activities in a household, based on their incomes and the willingness of the customers to own the solar systems.

The main priority Harish put forward was the viability of the technology among the poorest of the poor whom SELCO wanted to attract. It did not take much time for him to realize that the technologies were not available affront in the market which suits the need of the target customers . Thus, SELCO initially ventured into manufacturing the products using a design thinking procedure. Design thinking is a mechanism which requires the products and technology to be manufactured after deliberate discussion, testing and retesting among the target clients (Chen and Venkatesh 2013; Clark and Smith 2008; Kim and Ryu 2014; Bauer and Eagen 2008; Kummitha 2017b). Although SELCO's standard procedure has been threatened by many similar initiatives that came up in the due course of time, they have all faded away to create real time change as they largely depended on grants. Once the grant money is over, there is no means to sustain their efforts. So, the poor clients were in real risk. Thus, SELCO believes in adopting innovative methods of financing and elevating the capabilities of the poor and deprived to bring services up front, while not compromising their own marketing techniques.

Accessibility to a reliable and safer electricity source has improved the people's quality of life. It has a direct impact on earnings, health and education while contributing to the sustainable development discourse through reducing carbon emission. The lesser dependency on fossil fuels has created a healthy environment because of which families are able to engage in their work environment for a longer time and earn more.

4.1.2 Resource Mobilization and Participation

SELCO closely works with rural banks to facilitate the poor to gain necessary finances to access the customized solar technologies developed. Although the Grameen Bank initiative in Bangladesh has taught the world that the poor and marginalized are most likely willing to return or repay the amount they have borrowed, momentum still continues in certain banking segments where they have no faith in the capability or willingness of the poor to repay. The same trend was prevalent in Karnataka where SELCO initiated its operations. Even in cases where customers approach banks, they are required to deposit a quarter of the cost as down payment, which at times the poor and excluded customers are unable to afford. Thus, SELCO not only needed to demonstrate the importance of solar technologies, but it had to explore possible funding options to enable deprived customers to access renewable technologies. In general, the upfront costs involved in installing solar technology is too high and the poor may not be able to access them unless an innovative mechanism is derived. SELCO has created its system in accordance the need. They have negotiated with banks and stood as guaranty on behalf of the customers. SELCO not only guarantees but also deposits a quarter of the total amount on behalf of the customers in the bank.

The provision of solar devices is not only limited to low-income households, but also addresses the needs of micro and small-scale entrepreneurs, street vendors and those dependent on household/cottage industry. The solar technologies in this connection helped the street vendors, small and micro entrepreneurs to generate greater production which in turn enhanced their earnings and the well-being of their families. Generally, street vendors' work sees peak footfall after sunset, that forces them to rely on kerosene lamps which results in degenerating health of the owners, customers and further leads to environmental pollution. Kerosene lamps cost a fortune compared to the normal electricity bill, in excess of 3–4 times, more than those who have access to electricity. The government does not show any interest in street venders and they have no access to gain electricity legally. As a result of SELCO's intervention, several street vendors have gained access to lighting. SELCO's multidimensional support, especially networking and partnering with banks, has been the key.

The lobbying and networking with banks and microfinance institutions has ensured credit accessiblity to the poor and excluded, who were once considered unbankable. The interactions with investors has helped it to gain access over new markets and serve the poorest market segments. Investors during the early 2000s pressurized SELCO to scale up quickly. This led SELCO to operate its service centres using franchising model. Since franchise dealers had no interest in the mission of SELCO, their motive just revolved around maximizing profits. As a result, SELCO's profits started to fall during 2005–2006. This helped SELCO to learn a hard lesson and later a strong revival effort from internal mechanisms improved its presence in the market. However, SELCO has learned that it is worthless to accept investments from all those who show interest and it is necessary to carry out a proper review before accepting investors. SELCO within no time was able to get rid of the investors who forced it to scale up and identified new investors whose intentions fitted with their mission. Altogether SELCO has partnered with about 20 local financial institutions to focus on its mission and address the social problem.

4.2 *Sustaintech*

In India, street food vendors, who operate in an unorganized sector without having any need to maintain quality or safety standards, use inefficient stoves which require firewood for their operation. Sustained usage of firewood based stoves generates health risks for the street vendors, customers and has negative impacts on the environment and the business alike. Depending on their operations, street vendors use approximately 5–100 kg of firewood per hour for 10 hours which continues for 6–12 months in a year. On the other hand, an increase in the cost of firewood has direct implications on the profits of a small-scale business. It is estimated that a vendor who earns an approximate profit of US$1500 per year, spends about US $500 for firewood. In addition, it is estimated that about 20–80 tons of CO_2 is released by each establishment based on its size and operation levels.

Svati Bhogle, who completed her Master's in Chemical Engineering from the Indian Institute of Technology, Bombay, established a social enterprise called

Sustaintech to reduce the usage of firewood by street vendors. While as discussed earlier, SELCO targeted light to create a social change, here Svati, after identifying the problems that creates health risks and hampers growth of business, realized that the solution would require appropriate technology to promote stoves, and she especially realised the need of adopting design thining.

4.2.1 Strategy

Over a period of time, Svati realized that technology is necessary but not a sufficient condition to implement an effective system. She opined that in order to make sure that the technologies reach the target beneficiaries, it would require to incentivize all the stakeholders involved in the value chain. Similar to the SELCO model, Svati had to work on dual tasks. On one hand, creating customized technological inventions and on the other hand bringing banks on board which are expected to offer loans to the unorganized street vendors. Svati argues that though using firewood is certainly a wrong proposition, it continues to be the cheapest option for street vendors. Thus, with the technology that they can afford, she has invented three biomass-based stoves which are energy effective and require less wood. The three stoves include: (i) a flat plate to make pancakes; (ii) tea for road side tea shop; and (iii) multi-purpose cooking including frying. The stoves invented reduce fuel consumption by 40% which would now require 3 kg against the 5 kg required per installation earlier.

The experiment helped vendors to save about 40% on their energy bills which in turn helped them to repay the loans within 15 months. Moreover, it has contributed enormously to curbing a significant portion of CO_2 from entering the atmosphere. Though the innovation does not provide an ultimate alternative, given the circumstances of technology availability, finances available and the capacity of street vendors to pay, this particular technology offers enhanced energy efficiency and benefits compared to existing ones. However, another full-fledged technology which can offer a durable solution to the problem is required in the long run.

4.2.2 Resource Mobilization and Participation

The technologically advanced stoves facilitated by Svati encourage street vendors to shift from fuel and firewood-based production to biomass-based production. The stove is smokeless with an attached chimney that directs the smoke away from human activity. Thus, it not only provides a healthy, better and safer environment for those who work in the shops, but also customers feel relieved.

The fabrication of the stoves is outsourced from a private company while assembly of various parts is carried out locally by a designated entrepreneur, who provides employment in the vicinity. The local entrepreneur not only assembles the parts, but also takes care of the maintenance. Sustaintech looks after the quality checks and frequently organizes workshops and capacity building programmes for the local entrepreneurs. Negotiations with local banks enabled Sustaintech to directly get funds transferred by the banks while the product is delivered to the customer. The customer then repays the bank in monthly installments. The process also helps the street vendor to learn basic banking issues as they hardly carried out any banking transactions earlier. Once banks start trusting the vendors, the credit line accordingly goes up and the vendors probably have a chance to apply for another loan and further develop their enterprise. If it were not for the negotiations carried out by Sustaintech, it would have been very difficult for the street vendors to own the technology.

Each stove life is projected as five years and it has enormous implications for environmental protection. Sustaintech is estimated to sell about 65,000 such stoves in the next five years and it is estimated that at least 6.5 million people will be saved from eating in polluted areas, while it also enables 65,000 vendor families to get rid of traditional stoves and enhance fuel efficiency. It altogether saves about 3.7 million tons of carbon dioxide to be released by each street vendor in a period of five years (each kilogram of firewood is equal to one and half kilograms of CO_2). Based on the design thinking procedure, Sustaintech is venturing to scale it in other states in India.

4.3 The National Confederation of Dalit Organizations

The National Confederation of Dalit Organizations (NACDOR) aims to emancipate the most excluded sections in Indian society—scheduled caste, scheduled tribes and other backward classes. Historically, Indian society was divided on the basis of the caste system, which is into four sections namely priests, warriors, traders and labourers. The classification was hierarchical, which limited the participation of those who are placed in the bottom of the hierarchy in the mainstream. However, there is another section of people who were never considered part of the social hierarchy and were placed below hierarchy as untouchables, that is, the Scheduled Caste (Olivelle 2004; Radhakrishnan 2009). Scheduled Tribes were not acknowledged in the mainstream caste system as they preferred to inhabit in forests and mountains, away from the mainstream social life. The developmental approaches that modern society adopted are partly considered as reasons for exclusion of tribals. One could easily locate that the question of rights never become a question for those who are placed below the caste hierarchy. The more privileges the upper strata enjoys, the more vulnerability that downtrodden sections face. Those who are placed on top of the ladder, command power and gain benefits at the cost of those who are at the bottom. Thus, these sections are unable to take an active part in social life, because of which they never become an integral component of society. Hence, attaining social inclusion in India is largely about attaining equality for those groups which are socially placed at the lower end of the ladder in addition to addressing poverty and discrimination faced by other social sections (Kummitha 2015). The NACDOR initiated by Ashok Bharti aims to address among many other, the energy needs of the excluded and deprived sections who live in slums in Delhi.

4.3.1 Strategy

Bharti as a social entrepreneur attempts to bridge the gap between the excluded population, government and electricity companies. About 30% of the total population in Indian cities lives in slums, where the minimum living is hampered by lack of basic services. It is a common scenario in India where slum dwellers are not granted legitimate access to basic services such as water and electricity. The government's rejection to accommodate the given sections of people in mainstream society let them depend on different methods to access basic facilities. One such familiar and well-known method in Delhi slums is to depend on the black market for their energy needs. The local mafia steals electricity from a legitimate source and then sells it to the slum dwellers who in turn pay much more than market prices to get access. In spite of paying higher market prices, the quality of the electricity is always compromised, which results in frequent short circuits causing hazardous and fatal situations. At times, the entire residential area is at risk due to excessive use of temporary and flammable materials for constructing houses in slums.

Electricity companies and government which are aware about the black mafia, maintain distance from the scenario. Bharti argues that the mafia prevalence has influence over many other social issues such as (a) they recruit young people and make them directionless; (b) the growing mafia disturbs law and order; and (c) the crime rate rises. Bharti, who himself was born in Delhi slums and knew all the nitty-gritty about the problems related to urban slums, felt that government failures should not be supplemented by socially corrupt and illegal systems.

In Delhi, unlike many other states in India, while government manages power generation and bulk transmission, private companies take over local distribution. However, the private companies continue to rely on the age-old systems and infrastructure placed by the government to distribute the energy. This has facilitated the provision for contractors to look after the power supply in slums. The contractors quite often come with political affiliations and are well known as local level politicians, who get a range of commissions from the mafia. Although it is well known that private companies work to enhance their financial value, in this case, due to the large-scale corruption involved, the firms' financial value maximization is bypassed by private gains of a few individuals involved in the system. The existing structure thus has negative implications for all genuine and legal stakeholders including slum dwellers, private companies and the government, while a few individuals in the system benefit from illegal means.

Bharti who was fed up with the existing system decided to introduce an alternative system which benefits all legal stakeholders. Being an advocate of the participatory model, he found creating a legitimate electricity cooperative as an alternative which can facilitate the dream of direct electricity supply to the slums. However, it was not an easy task for him to carry it out smoothly. Especially dealing with the mafia, which was quite active, and mobilizing private companies and government to concentrate on the slums, while the government never had any interest to intervene in slums, and so on, were a few of the tough challenges. However, Bharti made the difference with his active involvement with all stakeholders. The proposed model is a win-win situation for everyone involved as communities stand to benefit from the direct and quality supply of electricity from the distribution companies while the private companies get bills paid directly from their customers.

4.3.2 Resource Mobilization and Participation

Initially Ashok carried out a pilot project in a sprawling slum called Haiderpur in Delhi to create a model cooperative. The cooperative which partnered with the North Delhi Power Limited (NDPL) works as a regulatory body and represents community interests during the meetings and discussions with the power supply officials. Initially Ashok's role was basically to bring community members together who do not generally cooperate with each other due to various cultural reasons. Thus, he had to pursue each and every household in order to bring them on board. After a few meetings and interventions, the community finally came together. Further, a working group of a few active members was created who can represent the community. They were trained in capacity building on how to manage the group and how to negotiate in different situations. Later the group was further strengthened by another women's group based on self-help spirit. The main intention of the latter group was to supplement the interventions initiated by the main group. In particular, women are responsible for the electricity usage at houses and they can well represent the electricity need. This activity has brought a sea change in the way women see themselves and how the community perceives them. Women were otherwise expected only to restrict themselves to their families, and households, whereas the current approach has brought them into social action and encouraged them to opt leadership roles.

An assessment of the electricity consumption is carried out using a participatory appraisal approach to estimate the need and to propose the necessary charges each family needs to pay. The assessment also includes identifying a safety audit of each household and training volunteers in slums to help during the leakages and breakdowns. The participatory approach is crucial because it allows the communities to assess and estimate their own energy needs and to plan their monthly expenditure accordingly. With a participatory approach, meeting concerned people in the government, electricity board and distribution companies resulted in convincing the government. In addition, his proposition was profitable to companies as well, which made them think and act accordingly. As a result, this small initiative by Ashok has helped the slum to access clean electricity for a lesser price.

After successful implementation of the pilot in the slum, Ashok has now plans to replicate the model in all the remaining slums of Delhi. The entire process, as discussed, has multiple co-benefits. The private companies especially will directly benefit from doing business with the end users. As the customers grow, revenues will also grow. In addition, it controls the mafia and its negative implications on society. The end users also benefit by paying less money than they used to pay to mafia—a fixed amount of INR 175 (US$3) every month to access a safe and reliable supply of about 500 V electricity.

5 Conclusion

The best practices discussed in this study not only contribute to the growing green and clean energy demand, but also addresses concerns for basic human survival and social integration. The innovations adopted from time to time enhance the prospects of the poor and excluded to gain access to the mainstream market and improve their participation. The underserved population are targeted by SELCO to see clean energy that aims to enhance their quality of life. The provision of solar energy includes servicing and financing. Sometimes solutions can be as small as mobilizing the stakeholders, which is really a challenging task. Although bringing the heterogeneous communities together was a difficult task in Ashok's case, with strong dedication and commitment it has been achieved. The case of Sustaintech has shown innovative ways of promoting alternative methods to save the local environment, offset carbon emissions, protect the health of street vendors and customers while also maximizing profits.

Here a lesson is learned that it is necessary to have social entrepreneurs and their innovative thinking to balance different needs and facilitate social functioning. Availability of technology or skills alone would not make any difference, unless it is made accessible to the poor and downtrodden communities. Thus, the multiple roles played by social entrepreneurs to accrue several co-benefits are worth applause. The most significant and crucial task is to bring governing public authorities and funding agencies or banks on board in order to enable the process.

Although the cases studied in this chapter are indicative, which are used to explain the range of activities being undertaken by social enterprises at ground level to address the basic and clean energy issues for the poor and marginalized, as a matter of fact all the cases have been well supported by government and its institutions. However, there are many such initiatives which require capacity building, technologies, training, awareness, education support and encouragement. Unless many such initiatives are identified and promoted, it would be difficult to address the growing demand for access to clean energy especially in poor and deprived neighbourhoods. In addition, such attempts help us to incrementally step up from existing fossil fuels-based lifestyles of energy consumption behaviour to green technologies meant for global sustainability.

It is to agree that the local level innovations and commitment by civil society organizations, are key to reach the unreached and to contribute to their well-being. In particular, the three cases discussed in this chapter strengthen this argument. While the COP21 agrees that these innovative initiatives driven by civil society are a matter for addressing the climate change issues, in response the Government of India has not laid any importance to the role of grassroots innovations in its nationally policies. Thus, it is to conclude that once the actual technology and knowledge transfer takes place from developed to developing countries as promised in the COP agreement, the Government of India will have to explicitly come up with a plan of how such technologies and knowledge will be instrumental at grassroots level to achieve actual results. The plan of action requires more than just commitments.

References

ADB. (2012). *India social enterprise landscape report*. Manila: Asian Development Bank.

Aspen Institute. (2001). *The nonprofit sector and the market opportunities and challenges*. Publications 01-13. Washington, DC: Aspen Institute.

Austin, J., Gutierrez, R., Ogliastri, E., & Reficco, E. (2006). *Effective management of social enterprises: Lessons from business and civil society organisations in Iberoamerica*. Cambridge, MA: Harvard University Press.

Battilana, J., & Durado, S. (2010). Building sustainable hybrid organizations: The case of commercial microfinance organisations. *Academy of Management Journal, 53*, 1419–1440.

Bauer, R. M., & Eagen, W. M. (2008). Design thinking—Epistemic plurality in management and organization. *Aesthesis, 2*(3), 64–74.

Berelson, B. (1952). *Content analysis in communication research*. New York: Free Press.

Besharov, M. L., & Smith, W. K. (2014). Multiple institutional logics in organisations: Explaining their varied nature and implications. *Academy of Management Review, 39*(3), 364–381.

Black, R., Natali, C., & Skinner, J. (2005). *Migration and inequality, background paper for world development report 2006*. http://siteresources.worldbank.org/INTWDR2006/Resources/477383-1118673432908/Migration_and_Inequality.pdf. Accessed January 7, 2016.

Chen, S., & Venkatesh, A. (2013). An investigation of how design-oriented organisations implement design thinking. *Journal of Marketing Management, 29*(15–16), 1680–1700.

Clark, K., & Smith, R. (2008). Unleashing the power of design thinking. *Design Management Review, 19*(3), 8–15.

Dees, J. G. (1998). *The meaning of 'social entrepreneurship*. http://the-ef-org/resources-Dees103198.html. Accessed May 18, 2011.

GWEC—Global Wind Energy Council. (2011). *Global wind report*. Annual Market Update. http://gwec.net/wp-content/uploads/2012/06/Annual_report_2011_lowres.pdf. Accessed January 14, 2015.

Gundry, L. K., Kickul, J. R., Griffiths, M. D., & Bacq, S. C. (2011). Creating social change out of nothing: The role of entrepreneurial bricolage in social entrepreneurs' catalytic innovations. In G. T. Lumpkin & J. A. Katz (Eds.), *Social and sustainable entrepreneurship* (pp. 1–24). Bingley: Emerald.

Hill, T. L., Kothari, T. H., & Shea, M. (2010). Patterns of Meaning in the social entrepreneurship literature: A research platform. *Journal of Social Entrepreneurship, 1*(1), 5–31.

IEA—International Energy Agency. (2012). *Understanding energy challenges in India: Policies, players and issues*. http://www.iea.org/publications/freepublications/publication/India_study_FINAL_WEB.pdf. Accessed December 9, 2014.

IIASA. (2012). *Global energy assessment—Toward a sustainable future*. Cambridge, UK: Cambridge University Press.

Intellecap. (2012). *On the path to sustainability and scale: A study of India's social enterprise landscape.* http://intellecap.com/sites/default/files/publications/intellecap_landscape_report_web.pdf. Accessed March 18, 2013.

Karekezi, S., McDade, S., Boardman, B., & Kimani, J. (2012). Energy, poverty and development. In *Global energy assessment – toward a sustainable future* (pp. 151–190). Cambridge, UK and New York: Cambridge University Press.

Khandker, S. R., Barnes, D. F., & Samad, H. A. (2010). *Energy poverty in rural and urban India: Are the energy poor also income poor?* Policy Research Working Paper 5463. Washington, DC: The World Bank.

Kim, J., & Ryu, H. (2014). A design thinking rationality framework: Framing and solving design problems in early concept generation. *Human–Computer Interaction, 29*(5–6), 516–553.

Krippendorff, K. (2004). *Content analysis: An introduction to its methodology*. Thousand Oaks, CA: Sage.

Kummitha, R. K. R. (2015). Social exclusion: The European concept for Indian social realities. *Social Change, 45*(1), 1–24.

Kummitha, R. K. R. (2016a). Social entrepreneurship as a tool to remedy social exclusion: A win-win scenario? *South Asia Research, 36*(1), 1–19.

Kummitha, R. K. R. (2016b). *Social entrepreneurship: Working towards greater inclusiveness*. New Delhi: Sage.

Kummitha, R. K. R. (2017). *Social entrepreneurship and social inclusion: Processes, practices and prospects*. London: Palgrave Macmillan.

Kummitha, R. K. R., (2017b). Institutionalising design thinking in social entrepreneurship: A contextual analysis into social and organizational processes. *Social Enterprise Journal* (in press).

Kummitha, R. K. R., & Crutzen, N. (2017) How do we understand smart cities? An evolutionary perspective. *Cities, 67*, 43–52.

Lasprogata, G. A., & Cotten, M. N. (2003). Contemplating 'enterprise': The business and legal challenges of social entrepreneurship. *American Business Law Journal, 41*(1), 67–113.

Lee, M., & Battilana, J. (2013). *How the zebra got its stripes: Imprinting of individuals and hybrid social ventures*. Harvard Business School Working Paper 14-005. http://hbswk.hbs.edu/item/how-the-zebra-got-its-stripes-imprinting-of-individuals-and-hybrid-social-ventures. Accessed September 29, 2015.

Lukenbill, W. B. (2012). *Research in information studies: A cultural and social approach*. Bloomington, IN: Xilbris.

Mair, J., & Marti, I. (2006). Social entrepreneurship research: A source of explanation, prediction, and delight. *Journal of World Business, 41*(1), 36–44.

Meyskens, M., Robb-Post, C., Stamp, J. A., Carsrud, A. L., & Reynolds, P. D. (2010). Social ventures from a resource-based perspective: An exploratory study assessing global Ashoka fellows. *Entrepreneurship Theory and Practice, 34*(4), 661–680.

MNRE—Ministry of New and Renewable Energy. (2011). *Strategic plan for new and renewable energy sector for the period 2011–17*. http://mnre.gov.in/file-manager/UserFiles/strategic_plan_mnre_2011_17.pdf. Accessed November 4, 2014.

Newth, J., & Woods, C. (2014). Resistance to social entrepreneurship: How context shapes innovation. *Journal of Social Entrepreneurship, 5*(2), 192–213.

Nicholls, A., & Cho, (2006). Social entrepreneurship: The structuration of a field. In A. Nicholls (Ed.), *Social entrepreneurship: New models of sustainable social change* (pp. 99–118). New York: Oxford University Press.

Olivelle, P. (2004). *The law code of Manu*. New York: Oxford University Press.

Planning Commission of India. (2006). *Integrated energy policy: Report of the expert committee*. https://cleanenergysolutions.org/content/integrated-energy-policy-report-expert-committee. Accessed November 19, 2014.

Prayas. (2010). *Clean energy regulation and civil society in India: Need and challenge to effective participation report*. http://electricitygovernance.wri.org/files/egi/Clean_energy_regulation_csos_india_peg_oct10.pdf. Accessed November 29, 2014.

Radhakrishnan, P. (2009). Globalisation and exclusion: The Indian context. *Global Asia, 4*(1), 64–73.

Seelos, C., & Mair, J. (2007) Profitable business models and market creation in the context of deep poverty: A strategic view. *Academy of Management Perspectives, 21*(4), 49–63.

Sethi, M. (2015). Mapping global to local carbon inequities in climate governance: A conceptual framework. *International Journal of Development Research, 5*(8), 5387–5397.

Sovacool, B. K. (2012). The political economy of energy poverty: A review of key challenges. *Energy for Sustainable Development, 16*(3), 272–282.

Wallace, S. L. (1999). Social entrepreneurship: The role of social purpose enterprises in facilitating community economic development. *Journal of Developmental Entrepreneurship, 4*(2), 153–174.

WEO. (2008). *World Energy Outlook 2008*. Paris: International Energy Agency.

WSJ—Wall Street Journal. (2009, December 16). India makes push for solar power. http://online.wsj.com/articles/SB126090748287292467. Accessed November 29, 2014.

Chapter 12
A State-level Framework for Integrated Land Use and Transport

Shabana Charaniya

Abstract According to the International Energy Agency, India was the fourth largest carbon dioxide emitter in the world after China, the United States and the European Union in 2011. Considering the accelerated economic growth dynamics in the Indian cities, the contribution of urban agglomerations to the overall carbon emissions is only going to increase in the coming years. The Government of India has formulated the National Mission on Sustainable Habitat and the National Urban Transport Policy as a part of its efforts to induce low carbon and sustainable urban growth. These policies together advocate integrated land use and transportation planning as a tool to reduce greenhouse gas (GHG) emissions. However, there exists no policy to guide integrated land use and transportation in the cities or assist in future decision-making in similar matters. In this paper, a framework is devised for such a state-level integrated land use and transportation policy aimed at reducing GHGs and improving sustainable accessibility. The framework draws policy pointers from similar efforts globally to explore the barriers in implementing such a policy in the Indian context. The research also identifies existing schemes/programmes that can support the implementation for such a policy. The chapter concludes with a list of actions that would facilitate the implementation of an integrated land use and transportation policy in India. Whereas, such a framework is rather befitting new developments, travel demand management can be better utilized to render existing developments less carbon intensive. However, the scope of this paper pertains to new development/growth.

Keywords Integrated land use and transportation
National urban transport policy · Comprehensive mobility plans
Co-benefits · UMTA—Unified metropolitan transport authority

S. Charaniya (✉)
London School of Economics, London, UK
e-mail: shabanac@gmail.com

S. Charaniya
NIUA, New Delhi, India

M. Sethi and J.A. Puppim de Oliveira (eds.), *Mainstreaming Climate Co-Benefits in Indian Cities*, Exploring Urban Change in South Asia,
https://doi.org/10.1007/978-981-10-5816-5_12

1 Introduction

The 12th Five Year Plan (2012–17), Sustainable Development Goal No. 11 (Urban SDG) and most recently, India's Intended Nationally Determined Contributions (INDCs) provide a perfect backdrop for Indian cities to pursue opportunities for creating inclusive, safe, resilient, sustainable and low-emission cities (Planning Commission of India 2013; Woodbridge 2015; MoEF 2015). The Indian INDCs focus on making India's future growth less emission intensive. Given that rapid urbanization is becoming the most dominant trend, decreasing the emission intensity of urban development should be of priority along with rendering its inclusivity, safety, resilience and sustainability. As was recently witnessed in Chennai, the disastrous floods were not only a function of extreme weather events due to climate change, but also that of faulty urban planning (Misra 2015). This points in the direction of pursuing strategies for combating climate change that cuts across functional and administrative boundaries in local and sub-national contexts. Transportation is one such functional category which often plays a key role in climate change mitigation strategies, especially in cities.

Integrated land use and transportation (ILU&T) policies have gained recognition as a consistent co-benefits approach in cities to achieve low-carbon growth by reducing travel distances or the need to travel (Macario et al. 2005; Bongardt et al. 2010). Combined with an augmented public transit network ILU&T is also globally recognized as an essential precondition of sustainable development (Meyer and Miller 2000; Priemus, Nijkamp, and Banister 2001; Bertolini et al. 2005; Swamy and Bhakuni 2014). However, lesser known are the development-related co-benefits accrued on account of easily accessible, low-cost and safe modes of physical transportation made possible by ILU&T policies. Improvements in low-cost physical mobility options provide better and inclusive access to economic opportunities leading to upward social mobility (Gakenheimer 2006; Adeel et al. 2016). Thus, ILU&T policies are uniquely placed in the co-benefits framework by providing a platform for the convergence of climate change action and development benefits.

In India, the National Urban Transport Policy (NUTP) and National Mission on Sustainable Habitat (NMSH) suggest linking land use and transport planning through preparation of comprehensive mobility plans (CMPs), and integrated land use and transport plans. However, there is no guidance available on how to realize their co-benefits. Thus, there is a need to identify a governing policy framework at an appropriate level that can guide the local level efforts towards achieving land use and transportation co-benefits. This chapter is an effort towards building this framework.

Research has very well-established interactions and correlations between land use and transport. Whereas land use planning and decision-making influence travel behaviour (Taylor and Sloman 2011), much of transport decisions have direct and indirect impacts on land use (Litman 1995). Land use factors like residential/employment density, neighbourhood design, location, and so on, affect the trip length, trip frequency, and mode choice, whereas transport factors like accessibility have an impact on location of residence and business, for example (Paulley and Peddler 2000). Thus, integrating the two at a policy level, covering planning and decision-making aspects, is essential. In order to derive the policy framework, the co-benefits within land use and transport planning in the Indian context are examined and an appropriate level at which these need to be integrated is established. Policy pointers are obtained by studying critical aspects like aim, approach, measures and mechanisms of global efforts/initiatives to integrate land use and transport planning. Once a policy framework has been identified, an exploration of implementation conditions is done by first recognizing the barriers in implementing such a policy in the Indian context, then highlighting existing schemes/programmes that can support implementation of such a policy. Finally, a list of specific actions that would facilitate the implementation of an integrated and co-benefitting land use and transportation policy in India is illustrated.

2 Land Use and Transport Planning in the Indian Context

According to various acts (Town and Country Planning Act, Urban Development Authority/Municipal Act), the function of land use planning including transport legally rests with the Town and Country Planning Organization (TCPO) in the state. Whereas, land use planning is being performed by an increasing number of local urban development (UD) authorities, planning for transport infrastructure is still essentially a state function in India. The City Master Plans, Traffic and Transportation Plans and the Comprehensive Mobility Plans (CMPs) guide land use and transport planning respectively. Development Plans, generally referred to as Master Plans, are the instruments through which different land uses are prescribed for achieving the desired pattern of spatial and economic development in urban settlements. These spell out the land uses according to the required level of social and economic growth (Khurana 2011). A CMP on the other hand is a long-term policy document for mobility plan framework and urban transport projects in the city (Chotani 2010). Table 1 explains the current realm of land use and transport planning through Master Plans and CMPs respectively and the extent of integration (land use and transport) in the two.

Thus, there is a partial link between land use and transport by virtue of the Master Plan and CMPs. It is important to note that Master Plans fix the land use for 20 years and Transport Plans are made to fit into it. In order to achieve minimization in transport demand, it is inevitable to plan land use and transport in an integrated way.

Investments in improving the transportation infrastructure induce growth that is not confined to local jurisdictional boundaries (ICF Consulting 2005). There is evidence that local land use practices influence regional climate, vegetation, and stream flow patterns in adjacent natural areas (Stohlgren et al. 1998). A policy at state level can thus analyse the regional impacts and articulate efficient response to

Table 1 Characteristic of master plans and CMPs

Description	Master plans	CMPs
Scale	City level	City level
Guiding document/policy	State Town and Country Planning Acts/Urban and Regional Development Plans Formulation and Implementation (URDPFI) guidelines/UD authority acts	NUTP, urban development authority acts in some cases
Prepared by	TCPO/local UD authority	Consultants usually appointed by the state
Implementation body	Urban local body, local development authority	Urban local body, state/para statal institutions, public transport authority
Land use-transport linkages	Some UD authorities are entrusted responsibility of both land use and transport planning, master plans of these authorities are required to have transport considerations	Takes cognizance of the existing and future land use plan as laid out in the master plan

the same. In India, State governments independently control local land use policies, motor vehicle and sales tax rates, bus transport systems, and policies for private sector participation (Singh 2005). The scale of transportation investments is fairly large and hence local governments are dependent on centre/state for the same. Considering the above, it can be said that a policy at state level can easily leverage the regional perspective, technical expertise, fiscal and procedural decision-making with respect to land use and transportation of the state to formulate an integrated land use and transport policy.

3 Global Efforts in Realizing Integrated Land Use and Transport Co-benefits

Although there are not many ILU&T policies globally, there are several instances of ILU&T planning where concrete measures have been taken by concerned agencies to integrate the two. The aim, approach, implementation mechanism and measures for these ILU&T instances/practices are studied to derive pointers for the framework, the same is summarized below.

3.1 Aim

It is observed that global efforts on ILU&T aim to avoid or reduce negative land use impacts of transportation projects (ICF Consulting 2005), achieve reductions in vehicle miles travelled (VMT) (Planning Institute Australia 2008; Jones 2009), manage and guide growth, encourage more efficient and effective transportation investment and land use, improve livability and maintain the character of surrounding communities, improve access to new transportation assets created (Robinovitch 1996; ICF International 2009) and achieve sustainable urban development (Paulley and Peddler 2000). In certain cases, the objective of land use and transport decisions is to efficiently use public infrastructure investments to maintain the mobility and safety of the highway system, create compact livable cities while minimizing negative effects on the natural environment and fight sprawl and automobile reliance (Hrólfsdóttir 2008; Oregon Department of Transportation 2011).

3.2 Approach

Studies show that efforts to integrate land use and transport in the American context follow the 'predict and provide' model (ICF Consulting 2005; Ward et al. 2007) wherein the aim is to utilize planned or projected future land uses to calculate transportation demand and investment or development induced by transport project is utilized for land use projections. Most European efforts are local manifestations of national policies towards sustainable development, where the approach followed is to introduce aspects of sustainable accessibility by progressive land use and transport policies that encourage compact development, use of non-motorized vehicles and discourage sprawl and use of personalized vehicles (Paulley and Peddler 2000; European Commission 2003).

3.3 Implementation Mechanism

Many state-led agencies have sought integration by getting engaged in local land use planning through statewide growth efforts (designating growth areas and providing financial and technical assistance in planning for the same), through engagement in the local planning process (help in growth planning and understanding travel demands of land use decisions) and through recognizing local goals in transportation decision making (ICF Consulting 2005). Curitiba achieved integration through its Master Plan that was in line with the urban priorities identified through a participative and consultative process (Robinovitch 1996). The European context identifies efforts towards ILU&T that range from national level rule, state/regional level policy frameworks, binding Metropolitan Development Plans, to localized efforts towards spatially integrating isolated developments using transportation investments (Paulley and Peddler 2000).

3.4 Lessons Learned: Policy Pointers

A global review of measures that encourage mixed use development or density intensification near public transport stations, cooperation of land use and transport planning authorities, along with policies to improve public transport and restrict use of personal vehicles was conducted, which revealed the following:

Clarity in roles and responsibilities has proved imperative for achieving effective integration in the above cases (McEldowney et al. 2003). The use of sustainable accessibility as an indicator for spelling out interdependencies between land use and transport is rampant in global practices. Participation, both external (between community and decision making authority) and internal (among various decision-making authorities) has been stressed in global cases for achieving full policy integration (Salas-Olmedo 2008). Many global practices showcase the use of high-end models to analyse the impact of land use on transport and vice versa, and planning and decision-making tools like a master planning checklist for sustainable transport in due developments, development guidelines, and so on, to give the much-needed technical and scientific robustness to the process of integration. Global practices also point towards the importance of achieving integration of land use and transport with respect to a broader decision-making approach, supporting tools, organizational structures of town planning and transportation engineering (Hayashi 2004). The need for a monitoring mechanism to measure the impact of integrated policy on indicators like accessibility and VMT has also been expressed by policy-making and implementing agencies executing ILU&T efforts worldwide.

4 Framework for State-level ILU&T Policy for India

Based on the understanding of land use and transport planning in the Indian context and the global efforts to achieve co-benefits through IL&UT, this section describes the policy framework for a state-level integrated land use and transport policy that is meant to serve as a guideline for states in India. It is suggested that each state prepare its own state-level policy in consultation with local plan making authorities. For ease of exposition in the rest of the chapter, the state-level integrated land use and transport policy is referred to as 'the policy' and its framework, strategy and outcome is presented in Fig. 1.

4.1 Objective

In order to arrive at the objectives of the policy, it is important to take cognizance of the objectives of national-level sectoral policies on land use and transportation. Whereas sustainable accessibility is the aim of NUTP, there is no national level

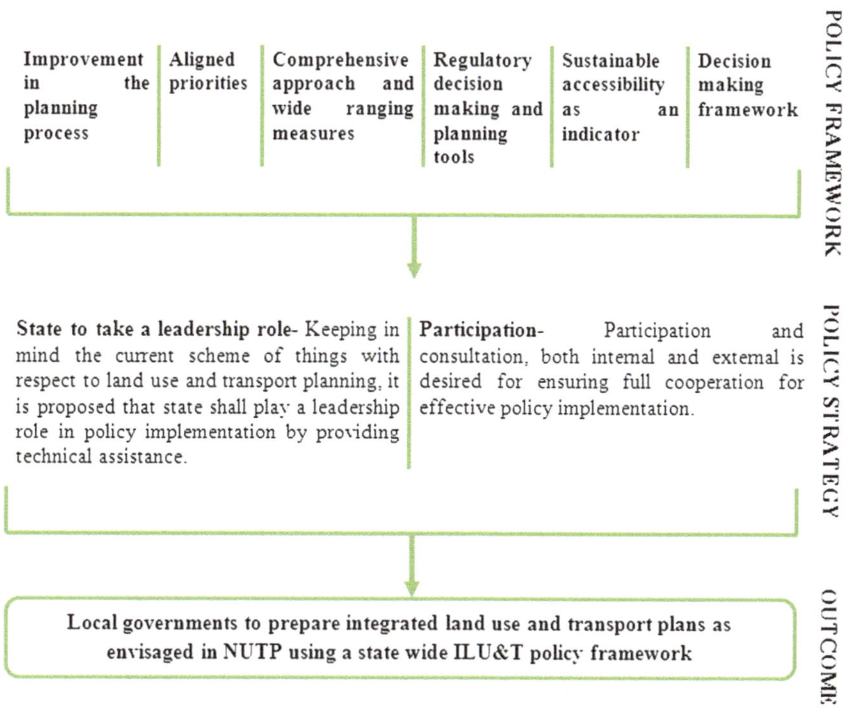

Fig. 1 State-level ILU&T policy

urban land use policy in India. However, the TCPO recognizes land use concentration and mix of uses as key land use policy objectives from the point of view of accessibility (Planning Commission of India 2011). Thus, the objectives of both the policies converge at the aspect of accessibility. Also, if adequately defined, accessibility can be directly related to both the qualities of the transport system (e.g. travel speed) and the qualities of the land use system (e.g. functional densities and mixes) (Bertolini et al. 2005). The NUTP and NMSH recognize the importance of integrated land use and transportation for achieving improved accessibility. Reduction in trip length is also attributed to integrated land use and transportation in NMSH, which eventually leads to reduction in GHG emissions. Hence, the key objectives of policy should be achieving sustainable accessibility and reduced trip lengths which ultimately result in reduced GHG emissions and low-carbon development.

4.2 Multi-pronged Approach

The above-mentioned objectives can be realized by utilizing a multi-pronged approach as described below.

4.2.1 Improvements in the Planning Process

Comprehensive integration in land use and transport can be achieved through legislative improvements. However, introduction of legislative improvements is a time-consuming process and hence the immediate need is to achieve functional cohesion between land use and transportation through improvements in the overall urban planning process. The policy by way of preparing improved guidelines for preparation of Master Plans and CMPs can initiate integration in the plan-making process. Guiding principles for preparation of land use and transportation plans should be based on principles of ILU&T.

4.2.2 Aligned Priorities

The policy should play the crucial role of channelling national priorities to local plans. Hence, the policy should take cognizance and further reiterate the objectives of the NUTP, NMSH, related sectoral policies with respect to urbanization, natural resource management, and so forth. The State Commission on Urbanisation, State Urban Development Agency and Unified Metropolitan Authority (UMTA) at the state level, should facilitate this process. Aligning priorities would result in achieving vertical integration and coordination.

4.2.3 Comprehensive Approach and Wide-Ranging Measures

The complex and dynamic inter-relationships between land use and transport demand a comprehensive approach that addresses the unwanted impacts of each, and an array of measures that aid in successful implementation of the approach. The mix of measures varies with geography, context, local priorities and enabling regulations. For achieving sustainable accessibility and reduction in trip lengths, the approach for the policy should consist of shaping the land use to reduce the demand or need to travel and improving accessibility through a wide range of transport measures. Some of the land use measures to be considered for the above-mentioned approach, also referred to as 'hard measures', are encouraging development near transport infrastructure and 'cities of tomorrow policies' that encourage infill development, urban regeneration, and so on. A combination of 'push' and 'pull' measures, that is, 'pushing' residents from excessive car use through restrictions such as parking management and at the same time 'pulling' users towards environmentally friendly modes by providing an efficient public transport system and favourable conditions for walking and cycling are recommended for inclusion in the policy.

4.2.4 Regulatory Decision-making and Planning Tools

Understanding the complexities in ILU&T planning and decision-making, the policy should mandate the use of tools. Tools can aid in the correct interpretation and operationalization of terms used in policy objectives like 'sustainable accessibility' and hence results in effective implementation. Tools may be prescribed for the following:

(a) Development planning and decision-making—checklist for location of new developments, accessibility profiles/statements for new development and transport infrastructure, decision making tools like high-end land use transport models that analyse the impact of land use on travel patterns and impact of transportation investment on growth, long-term forecasting tools, and so on.
(b) Benchmarking, monitoring and impact assessment—once the policy is in place, monitoring the result with respect to identified parameters can help gauge if the desired impacts are observed or further improvements are desired. The service level benchmarks (SLBs) for urban transport prepared by the Ministry of Urban Development (MoUD) helps in analysing the success of integrated land use and transport measures with respect to various parameters like density, land use mix, availability of infrastructure, and so on. It also measures the level of service of public transport, pedestrian and non-motorized transport (NMT) facilities. The same may be utilized with further improvements.

4.2.5 Defining Sustainable Accessibility Using Land Use and Transport Parameters

Using traditional principles of sustainability, sustainable accessibility may be defined as 'accessibility that is affordable, equitable and environmentally benign'. However, operationalizing this definition through land use and transport parameters would help in easy comprehension, implementation, monitoring, and impact assessment of the policy. The land use and transport parameters corresponding directly to sustainable accessibility are as follows:

(a) *Land use*—mixed land use, development (affordable housing, residential, business, institutional, etc.) around transport infrastructure, high density inner-city areas;
(b) *Transport*—increased use of public transport, improved facilities for non-motorized transport (NMT), like sidewalk, at-grade crossings, cycle tracks, and so on.

4.2.6 Decision-making Framework

It is anticipated that the policy also function as a framework for decision-making with respect to large developments and transport infrastructure. It is to be administered by the state whenever state funding is sought for a major development or transport infrastructure project.

4.3 Strategy

4.3.1 State in a Leadership Role

Keeping in mind the current scheme of things with respect to land use and transport planning, it is proposed that the state shall play a leadership role in policy implementation by providing technical assistance. Considering the lack of transport professionals in local governments, the state should recruit transport professionals and make them available to smaller cities/towns and enable dedicated funding to further strengthen policy implementation by building relevant capacities. Considering that land use planning is an expertise of local development authorities, the UMTA has to facilitate the local development authorities in terms of planning development/land use that adheres to the principles of the policy. The UMTA shall also be critical to policy implementation through its inherent transport functions. Local government along with the development authority, public works department, traffic police, state transport undertaking, and so on are the true implementers of the policy.

4.3.2 Participation

Participation and consultation, both internal and external is desired for ensuring full cooperation for effective policy implementation. Frameworks for integrating both internal and external participation at various stages of policy implementation, that is, decision making, implementation, monitoring, impact assessment, and so forth, should be an integral part of the policy.

4.4 Institutional Mechanism

4.4.1 Policy Formulation

It is proposed that the policy be formulated by a task force at the state level, housed in the UD department, consisting of representatives from other state-level departments having direct or indirect role in land use and transport planning like TCPO, transport department, local government department, pollution control board, state transport undertakings, city-level UMTA, and so on. Further, an ILU&T cell can be established in the state UD department to coordinate regional and local efforts in this regard.

4.4.2 Policy Implementation

Policy implementation rests on the basic assumption that as required by the NUTP, each city shall constitute an UMTA. The UMTA shall be the coordinating agency

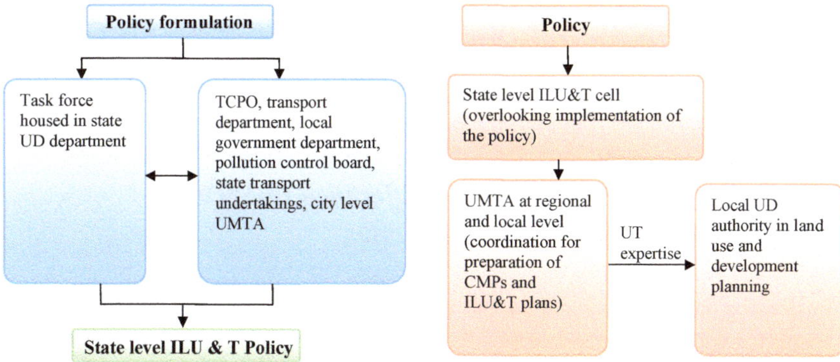

Fig. 2 ILU&T Policy formulation and implementation framework

for integrated land use and transport planning at local level. Having expertise in urban transport (UT), the UMTA shall advise development authorities responsible for development and land use planning on matters relating to transport. This can be further institutionalized by making amendments in relevant acts. The UMTA shall also be responsible for preparation of city-level integrated land use and transport plans. Along with this, the UMTA shall assist in policy implementation by assisting city governments in preparing development and transport infrastructure proposals that adhere to the decision-making framework in the policy. In case of smaller cities wherein constituting an UMTA is not feasible, the same functions shall be performed by the TCPO by recruiting transportation planners (Fig. 2).

5 Challenges in Policy Implementation

Efforts towards integrating land use and transport have traditionally faced a lot of challenges, all extensively argued, that can be summarized as follows.

5.1 Institutional and Financial Barriers

Whereas there is just one agency responsible for land use planning at local level, that is, TCPO or UD authority, urban transportation functions are distributed among centre, state and local governments. Central government is the apex body in terms of policy making; it is also involved in provision of suburban rail service in cities. State governments independently control local land-use policies, motor vehicle and sales tax rates, bus transport systems, and policies for private sector participation (Singh 2005). There is a lack of one coordinating agency and hence transport figures as the secondary responsibility of all the above-mentioned agencies.

The creation of an UMTA at local level is expected to address this but it lacks financial and decision-making powers which may hinder achievement of coordination as anticipated.

5.2 Horizontal Conflict due to Difference in Sectoral Priorities and Planning Processes

Considering no improvement in the existing planning process for land use and transport, major horizontal conflicts could arise due to differences in sectoral priorities and planning processes. The direct conflicts between land use planning and transport planning agencies create a barrier for integration. These conflicts arise because there are separate departments following separate interests, objectives or directives as discussed in Sect. 12.2.

5.3 Lack of Dedicated Budget for Formulation and Implementation of an ILU&T Policy

The NUTP provides for 50% assistance from central government to the local bodies for preparation of ILU&T plans. However, there exists no such allocation for preparation of guiding documents or policies. Considering the level of sophisticated tools and methods required for integrating land use and transport, a dedicated fund for such efforts is very important.

5.4 Lack of Trained Urban Transport Professionals at Local Level

It has been established that any effort towards ILU&T has to be implemented at the local level. However, considering that urban transportation has always been a state subject, there is a lack of trained urban transportation expertise at local level (Ministry of Urban Development of India 2011). Also, urban transport planning professionals, as a norm, are not employed by the cities (Planning Commission of India 2011).

5.5 Attitudinal and Cultural Barriers

ILU&T measures rely heavily on a modal shift to public transport/NMT modes and restriction on the use of private automobiles. In India, using public transport and

NMT represents lack of financial resources and hence has a negative socio-economical connotation attached to it. This, combined with the current lack of appropriate public transport/NMT infrastructure in the country may prevent a modal shift and disruption in achievement of objectives of ILU&T policy. As Indian cities experience socio-economic development, the rate of ownership of private automobiles is expected to rise due to the elevated status attached to it. Thus, curbing the usage of private automobile poses a serious threat to ILU&T policy.

5.6 Lack of Affordable Housing Around Key Public Transport Infrastructure

As mentioned above, the presence of a healthy mix of housing options is essential for successful implementation and sustainability of transit-oriented development projects. In India, high real estate values around the public transport infrastructure have made these areas almost inaccessible to low and middle income sections. Unless regulatory measures are employed to achieve a healthy mix of housing options, effectiveness of such measures cannot be guaranteed.

6 Opportunities Facilitating Implementation of ILU&T Policy in India

6.1 Dedicated Policies and Linked Funding

There is strong support for integrated land use and transport planning from central government through dedicated policies like NUTP that promote cities to prepare integrated land use and transport plans by providing for 50% of costs incurred in preparation of the plans.

6.2 Institutional Reforms

The institutional reforms through the 74th Constitutional Amendment Act (CAA) and JNNURM have made for better coordination and implementation of an ILU&T policy. The establishment of UMTA guarantees mainstreaming transportation planning expertise which can be easily combined with land use/development planning and decision-making at local level.

6.3 Sustainable Development a Cross-sectoral Priority

Sustainable development is now a cross-sectoral priority. The idea has trickled from policy to schemes to programmes to implementation tools. This has led to a shift in focus from mobility of vehicles to accessibility for people, and infill development instead of suburbanization and sprawl. Further, the increased concern over climate change due to GHG emissions has generated spotlight on non-motorized transport (NMT) and resource efficient spatial development like transit adjacent/oriented development, and so on.

6.4 Large NMT Share and Urban Form

The share of NMT in Indian cities is as high as 40% (Wilbur Smith Associates 2008). Restrictive land use polices have evolved into progressive ones that encourage increased density and mixed land use due to increased market pressure. This type of urban form has a potential to achieve 20% reduction in VMT (McKinsey Global Institute 2010). However, achieving increased density and mixed land use is not enough; it has to be supported by improvements in NMT and public transport for it to yield sustainable results.

7 Conclusion

Most effective integration is realized when it is applied to policies, planning methodologies and organization of processes or structures. However, it is important to note that an integration at policy level demands rigorous efforts and is a time-consuming process. Meanwhile, a host of activities in individual sectors may be undertaken in the existing policy environment that can contribute to increased co-benefits in land use and transportation. Some of these are as follows.

7.1 Creating and Strengthening UMTA

The NUTP recommends creation of UMTA in each million plus city for coordinated planning and implementation of urban transport programmes and projects and an integrated management of urban transport systems. It is recommended that UMTA be established in medium-sized cities as well, as a proactive step towards curtailing the negative impacts of uncoordinated transportation planning in these cities. This can also fill the huge gap in terms of transportation expertise at the local level and further initiate land use and transport integration. Efforts should also be made to strengthen already established UMTAs by capacity building in terms of training, making available modelling and decision-making tools to analyse impacts of transportation decisions/projects on land use, environment, and so on. UMTA may also be strengthened by providing statutory power for decision-making and fund allocation.

7.2 Amendments in the State Town and Country Planning/Municipal Acts

The land use and transport planning aspects for a city are governed by various town and country planning/urban development authority/municipal acts. However, none of these acts recognize the importance of integrating land use and transport at development planning level. Much change can be achieved by internalizing impacts of transport on land use and vice versa in the planning process. Structural changes in the various committees that determine land use to include transport professionals may fill the gap of transportation expertise at local level and assist in achieving integration.

7.3 Empirical Research to Demonstrate Co-benefits

Although global literature has recognized that integrating land use and transport planning increases accessibility and reduces VMT, the same needs to be further established and quantified with respect to specific measures like increasing density around transit, restricting development that has limited or no access to public transport, and so on, in the Indian context. Also, considering the diversity in the size and nature of Indian cities, a study analysing the impacts of the above-mentioned measures on accessibility and VMT in various sizes (million plus, small medium, large) and types (heritage, industrial, etc.) of cities should be carried out. The findings can be utilized to further advocate specific integration measures in particular type of cities and securing relevant resources for the same.

7.4 Improving Benchmarks Including Accessibility and VMT Parameters

Although quite comprehensive in its current form, the SLBs when customized to specifically report parameters like accessibility and VMT at city or project level over a period of time can serve as an excellent benchmarking, monitoring and impact assessment tool.

7.5 *Parking Regulations and Congestion Charges to Manage Traffic Demand*

In order to ease congestion on roads, local authorities are adopting parking standards in terms of minimum parking required for a development with no upper limit. However, for parking regulations as an integrated land use and transport measure, it is essential that the development authority stipulate maximum parking standards for each type and mix of land use and development. Parking regulations including parking fees have tremendous potential to curb the use of personalized vehicles. Similarly, congestion charges dependent upon certain land use zones of the city at certain timings could reduce or optimize the traffic demand.

7.6 *Encourage Affordable Housing Around Public Transport Nodes*

Many cities in India are now encouraging transit oriented development (TOD). However, TOD creates negative externality of increased land costs around public transport nodes due to premium location. Thus, these areas become unaffordable for the low and middle income groups that heavily rely on public transport and hence are responsible for its sustainability as well. Making some percentage of affordable housing mandatory for new developments around public transport nodes guarantees increased ridership and hence proactively contributes in sustaining an ILU&T policy.

Cities in India are poised to accrue co-benefits from low carbon and sustainable urban development. With the current policies pushing in the same direction, an ILU&T policy with a dedicated objective to improve accessibility and reduced trip lengths shall not only render sustainability and low-carbon urban development, but also inform and guide future decision-making in similar matters.

References

Adeel, M., Yeh, A. G., & Zhang, F. (2016). Transportation disadvantage and activity participation in the cities of Rawalpindi and Islamabad. *Pakistan. Transport Policy, 47*(4), 1–12. doi:10.1016/j.tranpol.2015.12.001.

Bertolini, L., Clercq, F. L., & Kapoen, L. (2005). Sustainable accessibility: A conceptual framework to integrate transport and land use plan-making. Two test applications in the Netherlands and a reflection on the way forward. *Transport Policy, 12*(3), 207–220.

Bongardt, D., Breithaupt, M. & Creutzig, F. (2010). *Beyond the fossil city: Towards low carbon transport and green growth.* http://www.mcc-berlin.net/~creutzig/Bongardt_Breithaupt_Creutzig_2010.pdf. Accessed January 5, 2015.

Chotani, M.L. (2010). *A critique on comprehensive mobility plan for the City.* In urban mobility india conference and exhibition. http://www.urbanmobilityindia.in/Upload/Conference/440c630d-877d-4552-a62e-3a3ac96ec00f.pdf. Accessed January 11, 2015.

European Commission (2003). Achieving sustainable transport and landuse with integrated policies. *TRANSPLUS.*

Gakenheimer, R. (2006). Transportation and land use in the developing world: Planning in a context of controversy. *International congress on human development Madrid 2006.* http://www.reduniversitaria.es/ficheros/RalphGakemheimer(i).pdf. Accessed January 19, 2015.

Hayashi, Y. (2004). *Integrated land use transport planning and policies for sustainable cities and regions.* Stuttgart: FOVUS-Network for Mobility, University of Stuttgart.

Henar, S. O. (2008). *Spatial and transport planning integrated policies: Guidelines for Northwest Spain.* Oxford: Transport Studies Unit Oxford University Centre for the Environment.

Hrólfsdóttir, H. (2008). *Integrated land use and transportation policies: What can Reykjavík, Iceland and other cities learn from Portland, Oregon?* Montreal: McGill University.

Hugo, P., Nijkamp, P., & Banister, D. (2001). Mobility and spatial dynamics: An uneasy relationship. *Journal of Transport Geography, 9*(3), 167–171.

ICF Consulting (2005). *Handbook on integrating land use considerations into transportation projects to address induced growth.* http://onlinepubs.trb.org/onlinepubs/archive/NotesDocs/25-25(3)_FR.pdf. Accessed January 9, 2015.

ICF International (2009). *Corridor approaches to integrating transportation and land use.* http://onlinepubs.trb.org/onlinepubs/archive/notesdocs/NCHRP08-36(86)_FR.pdf. Accessed February 2, 2015.

Jones, D. (2009). *Integrating land use and transportation policy in California: The legislature's response.* http://elq.typepad.com/currents/2008/09/currents35-12-jones-2008-0915.html. Accessed February 7, 2015.

Khurana, M.L. (2011). *Need for national urban land policy.* International cooperative alliance, Asia and Pacific. http://ica-ap.coop/sites/ica-ap.coop/files/articles_25.pdf. Accessed February 19, 2015.

Klaus, S., & Wegener, M. (2004). Evaluating urban sustainability using land-use transport interaction models. *European Journal of Transport and Infrastructure Research, 4*(3), 251–272.

Litman, T. (1995). Evaluating transportation land use impacts. *World Transport Policy & Practice, 1*(4), 9–16.

Macario, R., Carvalho, D., & Fermission, J. (2005). Achieving sustainable transport and land use with integrated policies. In C. A. Brebbia & L. C. Wadhwa (Eds.), *Urban transport XI—Urban transport and the environment in the 21st century* (pp. 351–360). Southampton: WIT Press.

McEldowney, M., Scott, M., & Smyth, A. (2003). Integrating land use planning and transportation policy formulation in the Belfast metropolitan area. *Irish Geography, 36*(2), 112–126.

Mckinsey. (2010). *India's urban awakening: Building inclusive cities sustaining economic growth.* New Delhi: Mckinsey Global Institute.

Meyer, M., & Miller, E. (2000). *Urban transportation planning* (2nd ed.). New York: McGraw-Hill Education.

Ministry of Urban Development of India. (2011). *Parameters for the National mission on sustainable habitat—Report of the sub-committee on urban transport.* New Delhi: Ministry of Urban Development, Government of India.

Misra, T. (2015). *Chennai's floods were a perfect storm of climate change and failed urban planning.* CityLab. http://www.citylab.com/weather/2015/12/chennai-flood-climate-change-urban-planning-cop21-paris/419349/. Accessed February 14, 2015.

MoEF. (2015). *India's intended nationally determined contribution: Working towards climate justice.* New Delhi: Ministry of Environment & Forests, Government of India.

Oregon Department of Transportation. (2011). *Oregon highway plan policy 1F amendments.* Oregon: Oregon Legislature.

Osborne, J. C. (2012). Indicators that matter : Measuring transportation performance in Ahmadabad. Unpublished thesis report for Masters in City Planning at MIT, pp. 1–132.

Paulley, N., & Peddler, A. (2000). *TRANSLAND: Integration of transport and land use planning*. Berkshire: TRANSLAND.
Planning Commission of India. (2011). *Recommendations of working group on urban transport for 12th five year plan*. New Delhi: Planning Commission of India.
Planning Commission of India (2013). 12th five year plan (2012–2017): Faster, more inclusive and sustainable growth. India.
Planning Institute Australia (2008). Integrated landuse and transportation policy position statement. Kingston, Australia. http://www.planning.org.au/documents/item/281. Accessed February 9, 2015.
Robinovitch, J. (1996). Innovative land use and public transport policy: The case of Curitiba. *Brazil. Land Use Policy, 13*(1), 51–68.
Singh, S. K. (2005). Review of urban transportation in India. *Journal of Public Transportation, 8* (1), 5.
Stohlgren, T. J., Chase, T. N., Roger, A. P., Timothy, G. F. K., & Baron, J. (1998). Evidence that local land use practices influence regional climate, vegetation, and stream flow patterns in adjacent natural areas. *Global Change Biology, 4*(5), 495–504.
Swamy, H. M. S., & Bhakuni, N. (2014). Towards integrated land use transport plan. *Journeys, 5,* 25–36.
Taylor, I. & Lynn, S. (2011). Thriving cities: Integrated land use and transport planning. http://www.urbantransportgroup.org/system/files/20112706ptegThrivingCitiesReportforWebFINAL.pdf. Accessed February 22, 2015.
Ward, M., Dixon, J., Sadler, B. & Wilson, J. (2007). *Integrating land use and transport planning. Research report 333*. New Zealand: Wellington: Land Transport.
Wilbur Smith Associates. (2008). *Traffic and transportation policies and strategies in urban areas in India*. New Delhi: Ministry of Urban Development, Government of India.
Woodbridge, M. (2015). *Cities and the sustainable development goals*. http://localizingthesdgs.org/library/232/ICLEI-Briefing-Sheets-02-Cities-and-the-Sustainable-Development-Goals.pdf. Accessed February 13, 2015.

Chapter 13
Climate Resilience in Urban Planning

Divya Sharma and Raina Singh

Abstract Urbanization as a process, and at its current pace, is posing new opportunities and challenges for growth and development of urban areas and regions in the country. These include acute gaps and constraints to address growing demands on infrastructure, services, housing and other facilities. Additional threats like that of climate change are aggravating these issues. It is unequivocal that climate action would impact the lives of hundreds of millions of people in urban centres. There is a need to address current risks and to begin building climate resilience into urban fabrics and systems to likely future risks. This chapter discusses one such initiative for mainstreaming urban climate resilience carried out in Gorakhpur city, India under the ACCCRN programme.

Keywords Climate resilience · ACCRN · Solid waste management
Gorakhpur · Urban planning · Institutional analysis · Policy

1 Introduction

The twenty-first century is the urban century. This statement holds good particularly for the developing countries that are witnessing a sea change in terms of urbanization. More and more people are now living in urban areas and urban centres are increasingly catering to the economic development of regions and countries. Over

---

This paper is an outcome of a study conducted by the authors at The Energy and Resources Institute, India with support from the Rockefeller Foundation.

D. Sharma (✉)
Urban Policy and Planning, Oxford Policy Management Limited, Oxford, UK
e-mail: divya.sharma@opml.co.uk

R. Singh
Center for Urban Planning and Governance, The Energy and Resources Institute, New Delhi, India
e-mail: raina.singh@teri.res.in

M. Sethi and J.A. Puppim de Oliveira (eds.), *Mainstreaming Climate Co-Benefits in Indian Cities*, Exploring Urban Change in South Asia,
https://doi.org/10.1007/978-981-10-5816-5_13

90% of urban growth is occurring in developing countries, which adds an estimated 70 million new urban residents each year. It is predicted that the developing world will be housing about 80% of the urban population of the world by the year 2030.[1] By this time more than 55% of the population of Asia will be urban.[2] At present, 42% of the world's population lives in Asia generating 80% of the GDP. It is also pertinent to add here that half of the world's slums are in Asia. The urban growth scenario in India displays a similar pattern. As per the latest census (Census of India, 2011), India experiences increased urbanization on account of a large change —a shift of about 80–140 million people of middle tier villages into small and medium urban centres classified as class IV to VI towns. More than 377 million people, constituting 31.16% of the total Indian population (Census of India, 2011) live in these urban centres. Urbanization as a process, and at this pace, is posing new opportunities for growth and development of urban areas and regions, leading to the economic growth of the country. However, this is not without challenges. Developing countries, particularly like India, are facing acute financial and capacity constraint, to address growing demands on infrastructure, services, housing, and other facilities. Additional threats like that of climate change are increasingly gaining attention. Recent extreme events and the loss of lives, property, livelihoods and infrastructure caused by them have caught wide attention. IPCC's Fourth Assessment Report (AR4) predicts a high likelihood of increase in extreme events and changed weather situations in the future.

The development discourse in India has started to address climate change issues in various sectors, including urban development, into the policy-making process—the State Action Plans for Climate Change, the National Mission on Sustainable Habitat (MoUD 2008), and the National Solar Mission, to name a few. However, if one considers the varying degree of potential threats and impacts of climate change across countries, regions and cities, the task seems stupendous and the actions taken still miniscule. Cities like Seattle, Singapore, Cape Town, London, Tokyo and New York are amongst those who have already initiated action towards mitigating and adapting to potential climate change impacts. These cities are not only bringing in policies to support climate action in their cities but are also equipping relevant regulations, planning norms and standards to ensure action and measurable outcomes over time.

In the recent past, several cities in India have also started to think about climate adaptation or resilience initiatives with support from multilateral and bilateral agencies and international funding agencies. Some of these programmes include the Asian Cities Climate Change Resilience Network (ACCCRN) and 100 Resilient Cities (100 RC) supported by the Rockefeller Foundation, Climate Change Resilient Development (CCRD) Program of USAID, and CDKN's Climate Compatible Development programme, among others.

---

[1]United Nations, Department of Economic and Social Affairs, Population Division: World Urbanization Prospects, the 2011 Revision. New York 2012.

[2]ADB Competitive Cities in the twenty first century.

It is unequivocal that climate action would impact the lives of hundreds of millions of people in urban centres. There is a need to address current risks and to begin building climate resilience into urban fabrics and systems for future risks. Making provisions for likely future climate-related risks in new and the present urban development would not only have long-term resilience benefits but will also steer sustainability in the development agenda. India's Intended Nationally Determined Contribution (INDC) identifies the importance of 'immediate action for addressing climate vulnerability and adaptation actions at sub-national levels'. Goal number 11 of the Sustainable Development Goals (SDGs) that aims to 'Make cities and human settlements inclusive, safe, resilient and sustainable' also identifies this need. Therefore, what most urban centres in developing countries need is a development programme, into which measures for climate-change adaptation and mitigation are integrated.

2 The Asian Cities Climate Change Resilience Network and Intervention in Gorakhpur City

Asian Cities Climate Change Resilience Network (ACCCRN) is one such initiative with a mandate to help medium-sized cities in Asia to assess their risks and vulnerabilities to potential climate change impacts and develop their resilience strategies. ACCCRN was launched in 2008 and is funded by the Rockefeller Foundation as part of a US$59 million 7-year initiative aimed at building climate change resilience. The initiative started off with ten cities in Asia, amongst which three were from India—Surat, Indore and Gorakhpur. This initiative aims to catalyse attention, funding, and action in three primary areas:

1. Testing local approaches to build resilience and targeting the poor and the vulnerable community in the process
2. Dissemination, promotion of knowledge so generated in the process
3. Targeting business community, policy-makers and funding agencies to encourage investment in climate resilience.

The initiative follows an iterative approach involving stakeholder engagement and collaboration to understand what is climate change and what it means for cities, what it means to specific sectors and people, who is vulnerable and why. The key feature of the process has been adopting a participatory approach to vulnerability analysis with the aid of local stakeholders. It is this 'shared learning dialogue' that culminates in a city resilience strategy.

While ACCCRN defines climate resilience in a holistic manner as 'the capacity of an individual, community, or institution to dynamically and effectively respond to shifting climate impact circumstances while continuing to function and prosper', the initiative was principally adaptation focused. The process also involved identifying adaptation projects in the project cities to be implemented with support from the Rockefeller Foundation. A total of 32 city projects were identified in 10 cities across South Asia; of which 12 projects are based in India.

In its later phases, the process also saw replication and scaling up of the initiatives to other cities that included Guwahati, Shimla, Mysore, Bhubaneswar, Leh, Panaji, Shillong, Saharsa, Jorhat, Bashirhat, Pune, Kochi and Ludhiana.[3] Wherein cities like Guwahati, Shimla, Mysore and Bhubaneswar replicated the complete process of vulnerability assessment and preparation of resilience strategies, engagement in Pune, Kochi and Ludhiana focused on city competitiveness and making the business case for adaptation. Lately, it was realized that the long-term benefits of this exercise could be reaped only when the resilience strategies are implemented and they become an integral part of the urban planning process. This was also seen as an important step towards the commitment of the ACCCRN programme to inform policy.

A pilot attempt was carried out by TERI in the city of Gorakhpur after conducting a detailed institutional and regulatory review of the urban planning process (as well as the State of Utter Pradesh), and looked at the means to mainstream climate resilience. It is pertinent to note here that Gorakhpur had been one of the three core cities within the ACCCRN programme in India and its resilience strategy had been developed by the Gorakhpur Environment Action Group (GEAG)—an NGO based at Gorakhpur. TERI conducted its exercise while keeping this resilience strategy as the base and made detailed recommendations on mainstreaming climate resilience in Gorakhpur.

3 Need for Such a Review

It is well established now that given the cross-cutting nature of climate change impacts across economic, geographic, administrative boundaries and time scales, the adaptation strategies should essentially be formulated as part of a broader policy framework for development (UNDP-UNEP PEI 2011).

It is also suggested through various studies that this has to be at all relevant levels of the policy/regulatory framework namely national, sub-national (state/region), local and sectoral levels and should consider policy, financing, implementation and monitoring as the areas of intervention in the long run. Climate change would impact various sectors such as agriculture, water, and infrastructure, and so mainstreaming climate change action would not be a function of any one ministry or department (MoUD). It will have to be included in all sectoral planning and decision-making besides involving the sub-national bodies like states, regional planning boards and so on. Further to this, the impacts would occur at a local level affecting the livelihood, energy use, vulnerability patterns, health of individuals and

[3]Web portal of ACCCRN (last accessed on 9 February 2016): http://acccrn.net/country/india.

community. Hence the responses that are mainstreamed at national and sub-national levels ought to be rooted in local conditions as well as built into local laws and regulations.

Mainstreaming starts with finding the entry points for climate-related responses within the existing framework of governance and policy. This basically entails understanding the linkages between climate change and development, the government, political, institutional and regulatory mechanism. This, however, has to be substantiated with climate knowledge, risks and vulnerability analysis results and socio-economic analysis. Hence regulations and policy documents need to be required in the light of climate change and additional measures and policies as required need to be introduced. This review should cover official documents of policies, legislations, sectoral studies and information on current institutional structures. It should also be based on consultations with relevant government institutions at all levels of the government, citizens and experts in the field (UNDP-UNEP PEI 2011).

Such an assessment of the government, institutional and regulatory contexts would go a long way in understanding and highlighting the need to improve the existing setting to integrate climate change resilience measures throughout the government mechanisms (UNDP-UNEP PEI 2011). A review puts the resilience efforts into an advantageous position by helping to identify suitable entry points, while simultaneously identifying a potential break through champions of the mainstreaming effort as well as the appropriate implementation agencies.

4 Methodology

The process involved a detailed assessment of the vulnerability report, the geo-hydrological study report of Gorakhpur among others. In addition, GEAG shared the climate scenario report prepared by the Institute for Social and Environmental Transition (ISET), USA. A review of the same was carried out by the climate modelling team to understand the future implications of climate change on the city. Various questions on the same were taken up with GEAG and ISET and further clarifications were sought. The team made a presentation to the Divisional Commissioner of Gorakhpur and sought his support for the study. An advisory note from the Divisional Commissioner was then released to all the important stakeholders at the city level to support the study and a nodal person was appointed to assist various engagements in the city during the course of the study.

The process took around 8 months to complete, which included several rounds of discussion with key stakeholders in the city to make it an iterative process. It was ensured that the concerns, experience and learning of the city stakeholders were well documented. This was substantiated with an in-depth analysis of the regulatory-institutional set-up at the level of the Gorakhpur city as well as at the level of the State of Uttar Pradesh. Final recommendations were presented to Divisional Commissioner of Gorakhpur, Municipal Commissioner and all city officials for detailed discussions. Later, an 'Action Points' document was submitted to the city which is being pursued for adoption by the city.

5 Study Area and Approach

Gorakhpur is one of the fastest growing urban centres in Uttar Pradesh, located in the *Terai Belt* of Eastern Uttar Pradesh. The city is 147 km^2 in area and divided into 70 administrative wards. The topography of the city is largely plain with a marginal slope from North to South. There are many water bodies in the city, the Ramgarh Tal is the biggest, situated in south-eastern Gorakhpur.

The city is grappling with insufficient infrastructure facilities and basic services due to tremendous pressure from the influx of the rural population. This has led to proliferation of slums and there are about 110 slums in the city accommodating 33% of the total population (GEAG 2010). The city is already facing problems of flooding, water logging, temperature extremes, power shortage, poor quality of water and increased incidence of water and vector borne diseases. All these pressures are likely to be aggravated by potential climate change impacts which were analysed while preparing the resilience strategy for Gorakhpur.

'Water logging' was identified as the most prominent risk to city systems due to future and predicted climate change (refer to Fig. 1). It then prepared an implementation strategy and a supporting regulatory framework (taking into account adaptation measures suggested by GEAG) that reinforced these sectors to withstand present development pressures and to become resilient towards any future climate impacts. The recommendations draw from various stakeholder consultations made during the process of the study with the city officials.

The key sectors requiring actions to build resilience were identified as: solid waste management; drainage and sewerage system; drinking water; ecosystem conservation and flood management; public health; urban planning; and energy.

6 Institutional Analysis

The analysis contributed to the overall institutional set-up that drives urban development in the cities in India. The sectors were divided into urban planning, environment and disaster management, housing and infrastructure for ease of understanding. The section also presented the corresponding institutions for Uttar Pradesh state and for Gorakhpur city. The assessment included a national, state and city level framework that is present in India right now. The institutions were studied for their functions, and the legislations/regulations under which they were established. An assessment of the reach and importance of the institution in question was also established in terms of its role in bringing in climate resilience.

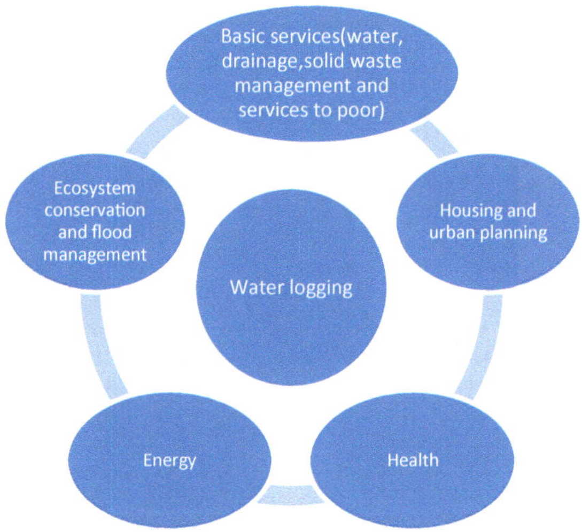

Fig. 1 Options to address risk of water logging

6.1 Sectoral Analysis and Detailed Action Points

The resilience strategy had outlined a few sectors that formed the basic core of the resilience building process in Gorakhpur. A detailed assessment of each sector in terms of the current situation of the city in that sector was made and actions points which the city could take up in the short term, medium term and long term to build and mainstream planned resilience into the city were recommended. The sectoral action points were provided for energy, housing, urban planning, public health, ecosystem conservation and flood management, urban services (including drinking water, sewerage and sanitation and solid waste management). Figure 2 shows one example of how this was carried out in the case of solid waste management.

An Example of the Action Points Proposed for Solid Waste Management
The present waste dumping sites are located in low lying areas prone to water logging, which is hazardous and may contaminate the soil as well the ground water. These sites should be assessed and closed down in a phased manner. The location of the proposed central SWM plant should be demarcated and buffered within the land use plan of the city. The municipal body should also consider potential climate impacts in making decisions for siting.

Since 72% of the developed area in Gorakhpur is residential in nature this offers an opportunity for adopting decentralized community waste management models. These could be managed at the level of 8 zones for which the zonal plans are proposed to be prepared under the provisions proposed in the development plan of the city.

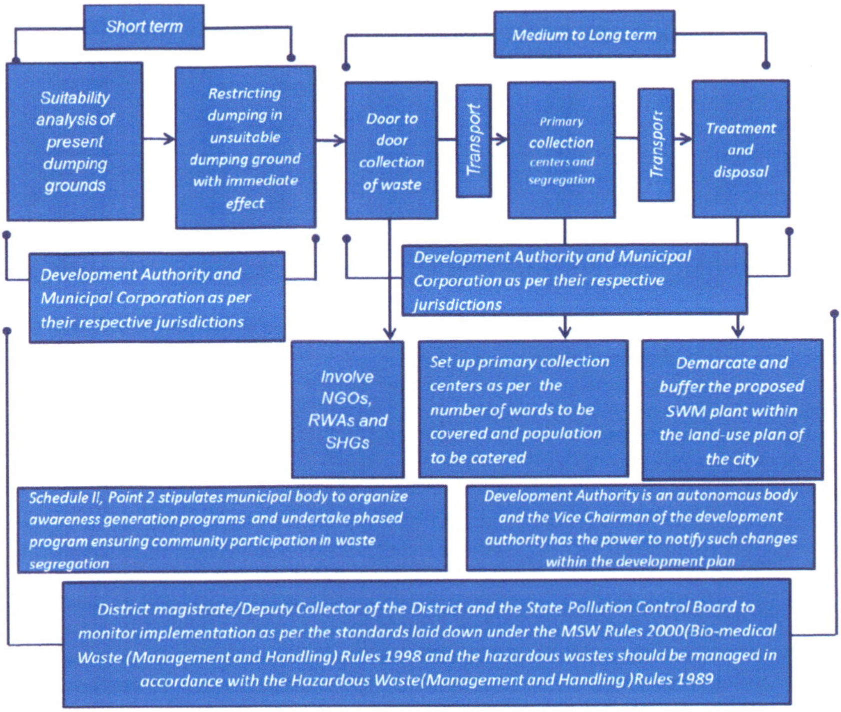

Fig. 2 TERI's proposal on resilient solid waste management

Refer to MSW Rules 2000 for detailed stepwise methodology to manage solid waste under the following categories:

Door to door collection and segregation—Schedule II, Point 2 of the MSW Rules 2000, stipulates municipal body to organize awareness generation programmes and undertake phased programme ensuring community participation in waste segregation, hence alternatively, such an arrangement could be made in the city.

1. **Primary collection centres**—Primary collection could be set up at zonal levels depending upon the wards covered under each zone. Each zone could have more than one primary collection centre depending on the population to be catered. The upkeep and operation of the primary collection centre including waste segregation would essentially be the responsibility of the municipal corporation.
2. **Treatment and disposal**—The non-biodegradable, inert waste and waste not suitable for recycling or biological processing should be sent to the land fill site.

The **CPHEEO Manual on 'Solid Waste Management'** should be referred while designing a detailed solid waste management plan for Gorakhpur city.

This was followed by detailed enabling mechanisms which enlisted not only the current laws and regulations that can help the city to take up the action plan proposed but also gave details of the guidelines that are applicable and available to make this happen. A detailed action plan was thus prepared and shared with the city of Gorakhpur for implementation. The particular section on solid waste management is given in Table 1. Similar action points were developed for other sectors as well.

6.2 Overall Recommendations and Key Findings

While the detailed sectoral interventions covered implementation mechanisms for the resilience strategy, the institutional regulatory review and concurrent stakeholder consultations revealed the challenges and opportunities to mainstream adaptation and resilience into the urban development framework in the city of Gorakhpur. One of the key recommendations has been enforcement of the 74th CAA to give a legal backing to ward level plans. The study also recommended including a chapter on climate change resilience in the Master Plan of the city. Capacity, accountability, coordination and efficiency of institutions were identified as areas requiring foremost attention to ensure sustainable and resilient urban systems. Capacity building and awareness generation of the state agencies/ULB/ and local community can play a major role for building resilience. For instance, capacity building of the government health set-up in the city by assessment of required infrastructure and manpower can help build an effective Public Health Surveillance and Management System.

Another key recommendation was to revisit the drainage (storm water drainage) project sanctioned under UIDSSMT to allow for disintegration of storm water drains appropriately from the sewer drains when they are sanctioned for. Exploring the option of decentralized solid waste management systems was also proposed. The application of ICT for building resilience and strengthening municipal database management systems was highlighted. On these lines, in the case of drinking water, establishing multiple channels for data collection and reporting on water quality was suggested. For urban planning and land use change issue, mapping, demarcation of green areas and water bodies in the city to regulate encroachment and reclamation was suggested. It was noted that preparation and enforcement of a Flood Management Plan which includes measures for preparedness, response, relief and recovery, can go a long way in addressing the enhanced risk of water logging due to climate change. Implementation of Sustainable Urban Drainage Systems (SUDS) in residential areas could further help in this respect. Table 2 summarizes

Table 1 Action plan for solid waste management in the city

City	Actions	Institutions	Supporting regulation/policy
Medium term	Phasing out current dumping grounds, particularly those in the low lying areas of the city	Municipal Corporation of Gorakhpur	
	Facilitating door to door collection of waste in the city	Municipal Corporation of Gorakhpur with help from NGOs, CBOs, SHGs, RWAs and community	MSW rules (2000)
	Setting up primary collection systems with segregation facility	Municipal Corporation of Gorakhpur and Development Authority	MSW rules (2000)
	Consider decentralized solid waste management system for peri-urban areas.	Municipal Corporation of Gorakhpur and Development Authority	MSW rules (2000) CPHEEO guidelines on solid waste management
	Campaign to create public awareness to keep the city clean	Municipal Corporation of Gorakhpur and Development Authority	
	For cleaning open drains: Special drives for cleaning the drains pre-monsoon and during the monsoon	Regular cleaning drives to be ascertained and notified by Municipal Corporation of Gorakhpur	Rules and guidelines of UP pollution control board
	Banning use of polythene Prepare a project for plastic recycling and implement[a]	Municipal Corporation of Gorakhpur	
	Strict enforcement	Overseen by Divisional Commissioner, Gorakhpur	

(continued)

Table 1 (continued)

City	Actions	Institutions	Supporting regulation/policy
Long term	Prepare detailed SWM scheme	Municipal Corporation of Gorakhpur, Development Authority	MSW rules (2000) CPHEEO guidelines on solid waste management
	Make an application for grant of authorization for setting up waste processing and disposal facility including landfills from the State Pollution Control Board	Municipal Corporation	Section 4 MSW rules (2000)
	Development, operation and maintenance of SWM	Municipal Corporation	
	Manage bio-medical waste (waste from hospitals)	Development Authority, Municipal Corporation and Health Department	Bio-medical waste (Management and Handling) rules 1998
	Manage hazardous waste	Development Authority, Municipal Corporation and GIDA	Hazardous waste (Management and Handling) rules 1989

Note The District Magistrate or the Deputy Commissioner of the district shall have the responsibility of enforcement of the provisions of the MSW rules. The State Pollution Control Board shall monitor compliance of the standards regarding ground water, ambient air, leachate quality and the compost quality as laid out under the schedule II, III and IV of the MSW Rules (2000)

[a]As suggested by the Divisional Commissioner during the second consultation in Gorakhpur on 2 June 2012

Table 2 Sector specific recommendations

Sector	Institutions	Regulations	Community
Housing	• Need for stringent compliance mechanism if the suggested changes in the building by-laws happen	• Revision/customization of Building Construction and Development Regulations (2008)	• For implementation of SUDS, improvised septic tanks and rainwater recharging/storing schemes, community buy-in few demonstration projects will be essential • Awareness programme on case studies for community
Urban planning	• Ensure Implementation of Master Plans • Develop a mechanism that evaluates and monitors the implementation (this is partially carried out now at the time of revision) • Resilience measures to be included in the Master Planning process • Strengthening technical capacity of the institutions to do so	• Bringing in sustainability, environmental and climate change related issues within the purview of a single Act (an Urban Development Act that talks of urban development in totality) • Master Plan to be guided by the new proposed Act • Implementing 74th CAA to ensure planning at grass roots level	• Ensure participation within the Master Planning process by developing a transparent and easy mechanism to include people in the decision-making system/planning
Basic services	• Ensuring inter-institutional coordination and integration • Developing monitoring and evaluation mechanisms • Database management and data sharing • Developing capacity of institutions to deliver quality services	• Enforcement of 74th CAA (service provision and user charges) • Bringing in law that sets standards of service delivery and prescribes measurement of service level • Regulations setting capacity requirements particularly targeting urban local bodies	• Public awareness programmes through involvement of NGOs, schools, volunteers, etc. on behavioural issues
Ecological conservation and flood management	• Strengthening of the ULB by adoption of 74th CAA (Schedule XII devolves powers to ULB for environment, climate change and disaster risk reduction issues)	• Need to enforce the provisions of various regulations and guidelines—UP Draft Environmental Policy, Parks, Playgrounds and Open Spaces (Preservation and	• Public awareness programmes through involvement of NGOs, schools, volunteers, etc. on climate change,

(continued)

Table 2 (continued)

Sector	Institutions	Regulations	Community
	• Technical and financial capacity building	Regulation) Act (1975), Model UP Zoning Regulations, etc. • Enforcement of 74th CAA to give a legal backing to ward level plans (currently under preparation) • Ward resilience plans should take climate change related future vulnerability into account	conservation of natural resources, pollution abatement, etc. • Participation of local community in formulation and implementation of various plans for better enforcement
Public health	• Setting up of a surveillance system for public health management • Capacity building of the government health set-ups in the city by assessment of required infrastructure and manpower	• Draw up a city health and sanitation policy guided by the provisions of the national sanitation policy, the Air Pollution Act, the Environment Act and the Water Pollution Act	Public awareness programmes on preventive sanitation and health measures—role of NGOs, schools, youth, etc.1

the actions at three levels, namely institutions, regulations and community under the broad sectors taken up in this study. These were then substantiated further with detailed recommendations on each one of these broad recommendations under short-term, medium-term and long-term interventions.

7 Discussion/Measuring Climate Resilience in Urban Planning

While detailed sector-wise recommendations were made on institutional and regulatory mechanisms, the study also addressed underlying issues that need equal attention. One of the foremost was to understand the institutional and regulatory landscape and political economy of the city in order to develop urban governance mechanisms that enable climate change resilience in the development process.

For instance, in Gorakhpur, unauthorized land use conversion can be legalized by private builders through the payment of an 'impact fee'. Cost-benefit analysis that favours long-term ecosystem conservation and public benefit over immediate financial gains is a farfetched dream. Moreover, such decisions are driven by an underlying political agenda at both the centre and state level which shape the governance systems and usually change with change in the political seat of power. The UN Habitat's 2011 Global Report on Human Settlements, titled *Cities and Climate Change: Policy Directions*, notes that urban areas with weak governance systems—as a result of political instability, exclusion of climate change from the political agenda, or lack of governmental resources—are especially vulnerable to climate change impacts. This holds true for Indian cities as well. In addition, it is also characterized by jurisdictional overlap; multiplicity of institutions which makes it difficult to implement adaptation actions in a holistic manner.

Though urban climate governance is at a very nascent stage in Indian cities, where it exists, there are several political factors that shape the opportunities and constraints for urban climate governance. There are issues of leadership (individual and organizational), questions of opportunity (windows of opportunity), the framing of the costs and benefits of acting on climate change, and underlying structures and processes of political economy (UN Habitat 2011). Since resilience is intrinsically linked with local governance, in order to plan for urban climate resilience the governance systems must operate on the principles of decentralization, accountability and transparency, responsiveness and flexibility, participation and inclusion. Moreover, the relationship between the arenas of authority is critical in shaping the capacity to govern climate change.

I. *Decentralization*: The 12th Schedule of the 74th Constitutional Amendment Act (CAA) delegates the subject of urban planning, including town planning to the Urban Local Bodies (ULBs)—Municipal Corporations and Municipalities. Unfortunately, the CAA has not been implemented fully in all the states and the provisions have not been adopted in their true spirit.

· There is hardly any ULB, which follows all the 18 functions defined in Schedule XII of the Act. It has also been observed that ULBs do not have the capacity (neither financial nor technical) to implement these provisions. For instance, many cities have not yet adopted mandatory reforms under Jawaharlal Nehru National Urban Renewal Mission (JNNURM).[4] In cities where ward committees are established, it is reported that ward councillors don't have sufficient powers (especially financial) to develop urban infrastructure and strategies for adaptation. At the city-region scale, another key issue is the fragmentation of urban governance across multiple authorities. This can be seen as a challenge of horizontal coordination. Unfortunately, a fragmented approach prevails over an integrated approach, which is required especially for land use/urban planning, basic service provision and so on. Multiplicity is further aggravated by PPPs. These modes are preferred by central and state governments and sometimes included as prerequisites for projects (JNNURM). There is no doubt about their success, but currently these do not integrate resilience arrangements. In fact, no climate change related arrangements exist at city level.

II. *Policy Implications*: The National Action Plan on Climate Change (NAPCC) steered by the Prime Minister's Council on Climate Change addresses the climate change related critical concerns of the country. NAPCC identifies measures that not only promote the development objectives of the country but also includes measures and strategies to address climate change. Through its 8 Missions dealing with long-term and multipronged strategies that would eventually lead to achieving the adaptation and mitigation goals of the country, the NAPCC fosters understanding of climate change adaptation and mitigation, energy efficiency and resource management. The National Mission on Sustainable Habitat (MoUD 2008), one of the 8 missions under the NAPCC, seeks to bring in sustainability of habitats through energy efficiency in buildings, urban planning, improved management of solid and liquid waste, public transport, and climate change and disaster mitigation and adaptation. The MoUD, Government of India is the nodal agency for formulation and implementation of the sub-missions. Besides this, the Ministry of Environment and Forests (MoEF) has asked all Indian states to develop action plans to define how they intend to undertake activities and programmes aimed at climate change adaptation and mitigation. These State Action Plans on Climate Change (SAPCC) should be in line with the objectives of the NAPCC and ensure its implementation at state level. However, as of now there are no policy obligations for cities to follow the same. On similar lines, the recently launched urban development schemes of the Government of India, including the Smart Cities Mission and AMRUT

[4]As per the reforms progress section in the official website (http://jnnurm.nic.in/reforms.html).

(Atal Mission for Urban Renewal and Transformation), also identify disaster resilience as one of the key action points,[5] though it is very early to make a conclusive statement on their implications and outcome since these schemes are still in their initial stages of implementation.

III. *Responsiveness and Flexibility*: At present there is no law that sets standards for service provision, let alone accounting for climate change impacts on services. Hence there is a need to introduce one that stipulates standards of service delivery and holds the municipal corporation responsible for quality and efficiency of services.

IV. *Accountability and Transparency*: Achieving this becomes a challenge due to the multiplicity of authorities. The JNNURM reforms include enacting the Community Disclosure Law (CDL) to bring in transparency in the government's transactions. However, this is applicable for JNNURM cities only. Other cities should also be encouraged to enact CDL. The disclosure law not only brings in accountability and transparency within the governance mechanisms, but in a way also promotes inclusion and participation from the citizens as they tend to understand the systems of the city management. In terms of climate action, the CDL would be useful to provide real-time information to citizens on the actions taken by the government and the preparedness of the government to face any unforeseen climate related situation.

V. *Participation and Inclusion*: Community participation law (CPL), a mandatory reform under the JNNURM[6] refers to making appropriate provisions in the state-level municipal statute(s) for the establishment of three tiers of decision-making in a municipality, namely, the municipality, the ward committee, and the *Area Sabhas.*[7] It aims to involve citizens in municipal functions like setting priorities, budgeting provisions, exerting pressure for compliance of existing regulations. Though applicable to JNNURM cities, this model law can be adopted by states for non-JNNURM cities as well as to promote community participation. Climate action in a city would particularly depend upon the local knowledge and would benefit by involving the community in the planning for climate resilience. The plight and needs of the vulnerable community—for example, the people effected by floods, droughts, Tsunamis or other climate related impacts—will have to be essentially incorporated within the climate resilience strategy to enable designing of practical coping strategies. The urban poor form the target group of MoHUPA policies. It has formulated the Rajiv Awas Yojana (RAY)—a scheme that targets provision of low-cost housing to the urban

[5]MoUD, 2015. Smart Cities Mission Guidelines (Available at: http://smartcities.gov.in/writereaddata/SmartCityGuidelines.pdf) and AMRUT Guidelines (Available at: http://amrut.gov.in/writereaddata/AMRUT%20Guidelines%20.pdf).

[6]The JNNURM makes it mandatory for states to either enact a separate CPL or make appropriate amendments to their existing municipal laws.

[7]*Area Sabhas* would consist of all registered voters of a polling booth in urban areas.

poor. MoHUPA is also responsible for implementation of the sub-mission on Basic Services to Urban Poor (BSUP) under JNNURM. But, it has done little to ensure inclusion of the poor and marginalized groups in decision-making, monitoring and evaluation. It needs to go a step further to include the urban poor, which is a key to improving the living conditions of these vulnerable groups.

VI. *Relationship between Different Levels of Authority*: The UN also notes that *municipalities are more or less coherent and have varying degrees of autonomy from international policies and from regional and national governments, the relationship between these arenas of authority is critical in shaping the capacity to govern climate change* (UN Habitat 2011). While Centre transfers power to State, more responsibilities need to be transferred to the local level as well. If the Centre directly offers assistance (city development funds) to cities by directing funds available through donor agencies, this may speed up the process. But the next question that arises is will this then dilute the powers of the state?

Capacity, accountability, coordination and efficiency of institutions need foremost attention to ensure sustainable and resilient urban systems. It is also essential that mechanisms and institutions are aligned not only to ensure quality and reliable services but also to account for future vulnerabilities like that of climate change. Multiplicity of organizations in a single sector as well as overlapping jurisdictions of the development authorities and the municipal corporation creates a lot of confusion and affects the quality of services. Due to this, there is no single agency which can be held accountable in case of non-delivery or poor quality of service provision. The government must come up with a mechanism to introduce and ensure inter-departmental coordination. Apart from these, the study also suggested key recommendations for integrating climate resilience into the city level urban planning framework. Some of these are discussed below.

1. The study revealed that there is no dearth of guidelines that cities could use to plan the systems effectively; however, there is definitely lack of technical capacity and manpower that needs to be looked at.
2. It was seen that there are several laws that address issues of urban planning, disaster preparedness, housing, environment, and so on. However, to make these regulations implementable, a city-level charter of activity could be prepared that draws from all of these regulations and comes up with a clear plan of action with defined responsibilities.
3. The urban development Acts of the state should be revised to incorporate all sustainability measures required for urban centres at one place, ensuring implementation of its provision in totality. The Acts must bear implementable guidelines that specify important questions like who will do what (institutional mechanisms), how this will be done (technical capacity and know-how), where will the funds come from (economic and financial considerations) and necessary time lines and schedules to be able to monitor results. In a nutshell, a complete

ecosystem should be derived in the guidelines that can help implementation of the law.

4. The National Mission on Sustainable Habitat has come up with legal habitat standards which the city would have to implement. A chapter on climate change impacts in the Master Plan would prove to be a beneficial instrument in drawing up an action plan for implementing the above standards. Cities and States could tie up with experienced agencies to conduct an impact assessment exercise for their respective cities. For JNNURM cities, City Development Plans (CDPs) could incorporate the resilience chapter and provide for funding options under the investment plans.
5. There are four types of ICT tools, namely, geographic information systems (GIS), e-governance, early warning systems (including telemetry), and wireless communications commonly adopted by local governments worldwide for helping their cities adapt to the effects of climate change. Gorakhpur can adhere to some of the below listed ICT-based recommendations:

   - Strengthen its e-governance systems (municipality database management). One such proposal has been made in Sect. 3.3 on drinking water, which recommends the setting up of a database management system at the JalKal Department or the State Ground Water Department office at Gorakhpur.
   - Build capacity of universities and institutes already applying ICT (like Gorakhpur University) as well as introduce it at other levels.
   - ICT tools, such as online mapping and mobile phone-based applications can be applied to strengthen the coordination mechanisms of relief agencies during disasters.
   - Topographical features (low-lying areas) could be mapped with rainfall patterns and weather predictions for establishing early warning systems. The GIS is a recognized tool for this.
   - ICT shows potential to enable participatory governance and transparency through facilitating sharing and updating of information between government and citizenry as well as allowing citizen monitoring and reporting of environmental status.

8 Conclusion

This study was one of its kind within the Asian Cities Climate Change Resilience Network (ACCCRN), where ways and means to mainstream climate resilience in a project city were looked at. While the study provided detailed recommendations for implementation of each of the sector specific strategies identified within the resilience strategy, it also supplemented the same with short-term, medium-term and long-term action points that the city could adopt and systematically approach the climate resilience objective.

The study conducted several stakeholder discussions to understand the challenges on the ground and the will and capacity of the city government in pursuing a resilience strategy. While the inputs received from the stakeholders were very enriching and encouraging, the discussions revealed many challenges to the actual adoption of the entire resilience strategy in the absence of a mandate from the state government and a specific policy at the national level. However, there was a unanimous consensus on the efficacy and importance of the National Mission on Sustainable Habitat to achieve climate resilience goals in urban areas.

While the study revealed that the cities are keen in taking up a climate resilience agenda into their development planning system, it could be reasonably concluded that each city would have to contextualize the resilience activities depending on their socio-economic and geographical characteristics. The governance, regulatory mechanisms and institutions would have a major role to play in steering interest within the citizens and initiating actions on this.

Notes

For detailed information on climate resilience initiatives in Gorakhpur, please refer to:

- Gorakhpur city resilience strategy prepared by GEAG and ISET: *Towards a resilient Gorakhpur.*

(http://www.acccrn.org/sites/default/files/documents/GorakhpurCityResilienceStrategy_ACCCRN_Jan2011_small_0.pdf)

- The synthesis report of TERI's work on mainstreaming the city resilience strategy in Gorakhpur, prepared by TERI: *Mainstreaming Climate Resilience in Urban Areas—A case of Gorakhpur City.* (http://www.acccrn.org/sites/default/files/documents/Gorakhpur%20report_Synthesis.pdf)
- A film titled '*Tales of Gorakhpur*' produced by TERI's in-house Film & TV Unit. It captures the work and insights of all ACCCRN partners who have been working in Gorakhpur for the last several years. It showcases TERI's yearlong work in Gorakhpur. (*View this film at:* http://www.youtube.com/watch?v=og41J9tP1Hk)

Acknowledgements The authors are grateful to The Rockefeller Foundation for granting this study to TERI. Of particular mention is the support from the Gorakhpur Environmental Action Group, for the help extended to TERI to carry out the project in Gorakhpur city. We are thankful to Mr K. Ravindra Naik, Divisional Commissioner, Gorakhpur for supporting the study and steering interest and participation from the city officials during the course of the study. We extend our gratitude to all the members of the Consultation Committee formed within the project for their time, inputs and expert advice to make this study practical and implementable for the city and the State Government. Finally, our sincere thanks to the senior colleagues and the other team members associated with this project within TERI.

References

Ministry of Urban Development. (2008). *National mission on sustainable habitat.* New Delhi: Ministry of Urban Development, Government of India.

Municipal Solid Waste (Management and Handling) Rules. (2000). New Delhi: Ministry of Environment & Forests, Government of India.

Habitat, U. N. (2011). *Global report on human settlements 2011: Cities and climate change.* London/Washington DC: United Nations Human Settlements Programme & Earthscan.

Uttar Pradesh Building Construction and Development Regulations. (2008). *Uttar Pradesh Building Construction and Development Regulations.* Lucknow: State Government of Uttar Pradesh, India.

Bibliography

Air (Prevention and Control Act). (1981). New Delhi: Ministry of Environment & Forests, Government of India.

Brown, A., Dayal, A., & Del Rio, C. (2012). From practice to theory: Emerging lessons from Asia for building urban climate change resilience. *Environment and Urbanization, 24*(2), 531–556.

CIRIA. (2007). *Improving the flood performance of new buildings—Flood resilient construction.* London: Department of Communities and Local Government.

CPHEEO-Central Public Health and Environmental Engineering Organization. (2000). *Manual on solid waste management.* New Delhi: Ministry of Environment and Forests, Government of India Disaster Management Act 2005, Government of India.

District Disaster Management Authority Gorakhpur. (2011). *Gorakhpur district disaster management and risk reduction plan 2011, supported by State Government of Uttar Pradesh and Government of India.* Gorakhpur: UNDP Disaster Risk Reduction and Urban Risk Reduction Project.

Environmental (Protection) Act. (1986). New Delhi: Ministry of Environment & Forests, Government of India.

France, R. L. (2003). *Wetland design: Principles and practices for landscape architects and land-use planners.* New York: Norton.

Gorakhpur Environment Action Group. (2010). *Towards a resilient world.* Colorado, USA: The Rockefeller Foundation and Institute for Social and Environmental Transition.

Gorakhpur Environment Action Group and Department of Geography, D.D.U Gorakhpur University. (2009). *Geo-hydrological study of Gorakhpur city.* Supported by The Rockefeller Foundation and Institute for Social and Environmental Transition, USA. Gorakhpur: Gorakhpur Environment Action Group.

Gorakhpur Municipal Corporation and Gorakhpur Environment Action Group. (2009). *Vulnerability analysis—Gorakhpur city.* Supported by The Rockefeller Foundation and Institute for Social and Environmental Transition, USA. Gorakhpur: Gorakhpur Municipal Corporation.

Hazardous Waste (management and handling) Amendment Rules. (2003). New Delhi: Ministry of Environment & Forests, Government of India.

ITPI-Institute of Town Planners, India. (1996). *Urban development and plan formulation and implementation (UDPFI) guidelines.* New Delhi: Ministry of Urban Affairs & Employment (erstwhile; now known as Ministry of Urban Development), Government of India.

IWMI-Tata Water Policy Program. (2002). Bringing pumps to people. *Water Policy Briefing, 2,* 1–6.

Model Town and Country Planning Act. (1960). *Model Town and Country Planning Act.* New Delhi: Ministry of Works & Housing, Government of India.

River Board Act. (1966). New Delhi: Ministry of Water Resources, Government of India.

Srinivasan, R. K., & Suresh-Babu, S. V. (2008). *Do-it-yourself: Recycle and reuse wastewater*. New Delhi: Centre for Science and Environment.

Town and Country Planning Department and Gorakhpur Development Authority. (2005). *Gorakhpur master plan-2021*. Lucknow: State Government of Uttar Pradesh, India.

UNDP-UNEP PEI. (2011). *Mainstreaming climate change adaptation into development planning: A guide for practitioners*. Nairobi, Kenya: UNDP-UNEP Poverty—Environment Initiative.

Uttar Pradesh Avas Evam Vikas Parishad Adhiniyam. (1965). *Uttar Pradesh Avas Evam Vikas Parishad Adhiniyam*. Lucknow: State Government of Uttar Pradesh, India.

Uttar Pradesh Municipal Corporation Act. (1959). *Uttar Pradesh Municipal Corporation Act*. Lucknow: State Government of Uttar Pradesh, India.

Uttar Pradesh Parks, Playgrounds and Open Spaces (Preservation and Regulation) Act. (1975). Uttar Pradesh parks, playgrounds and open spaces (preservation and regulation) act. Lucknow: State Government of Uttar Pradesh, India.

Uttar Pradesh Special Area Development Authorities Act. (1986). *Uttar Pradesh Special Area Development Authorities Act*. Lucknow: State Government of Uttar Pradesh, India.

Uttar Pradesh Urban Planning and Development Act. (1973). *Uttar Pradesh Urban Planning and Development Act*. Lucknow: State Government of Uttar Pradesh, India.

Uttar Pradesh Water Supply & Sewerage Act. (1975). *Uttar Pradesh Water Supply & Sewerage Act*. Lucknow: State Government of Uttar Pradesh, India.

Vandana Menon Architects. (2010). Nizamuddin East: Colony as community-water harvesting initiative, presentation in training program on Rain water harvesting by the Centre for Science and Environment, New Delhi.

Water (Prevention and Control of Pollution Act). (1974). New Delhi: Government of India.

Wetlands (Conservation and Management) Rules. (2010). New Delhi: Ministry of Environment & Forests, Government of India.

Chapter 14
Rights-Based Approach to Realize Co-benefits in Delhi

Magali Dreyfus

Abstract An important actor in the urban reform agenda in India is the judiciary. In fact, through public interest litigation, the Supreme Court (SC) and the High Courts have received claims from citizens, directly related to city development. Although the basis of the claims in front of the SC was the violation of a fundamental right, the judiciary was able to step into sectors such as land use, transport as well as cross-cutting issues such as air pollution. The case law of the SC has been particularly important for the city capital, Delhi. Based on this example, this chapter shows how the SC, starting from the violation of the fundamental right to life, ended up making detailed policy and technological orders to local public authorities in order to tackle the problem of air pollution. It shows the outcomes of these orders and co-benefits. Finally, it discusses the limits and opportunities of this rights-based approach for city development.

Keywords Rights-based approach · Public interest litigation
Vehicular pollution · Air pollution · Health and climate co-benefits

1 Introduction

Engaging local governments in climate policies is a challenge. In fact, the benefits associated to greenhouse gas (GHG) emissions mitigation are global and are neither immediate, nor certain, at the local level. Adaptation is yet more relevant as it contributes to reducing the vulnerability of a territory and its population. Identifying the potential co-benefits of public policies may therefore be an important driver for local authorities to develop climate strategies. Surprisingly in India, an example of this operation can be found in the case law of the Supreme Court (SC) through the use of a rights-based approach (RBA). The highest jurisdiction of the country has established a link between the right to life enshrined in the Fundamental Rights

M. Dreyfus (✉)
CNRS, CERAPS, 1 place Deliot, BP 629, Lille Cedex 59024, France
e-mail: magali.dreyfus@univ-lille2.fr

M. Sethi and J.A. Puppim de Oliveira (eds.), *Mainstreaming Climate Co-Benefits in Indian Cities*, Exploring Urban Change in South Asia,
https://doi.org/10.1007/978-981-10-5816-5_14

section of the Indian Constitution and city development issues, such as the local transport policy. The SC addresses in particular the issue of air pollution, a suitable medium to generate climate co-benefits. In this process, the SC acted both as a law enforcing authority, asking the central, state and local public authorities to implement legislations, and as almost an executive agency, issuing very detailed and technical orders to improve public policies.

This chapter examines different issues raised by the activity of the SC. First, how did the SC establish a link between the right to life and urban development issues? Second, to what extent do these decisions generate urban climate co-benefits? Finally, what governance lessons can be drawn from the SC's case law that are relevant for policy innovation and to develop co-benefits strategies at an early stage of the decision-making? The different sections of this chapter address these questions. Section 2 describes the process that connects the violation of a fundamental right, namely the right to life, with the local development agenda. Then Sect. 3 focuses on the case of air pollution in Delhi, which appears to be one of the most comprehensive and contentious areas of activity of the SC, and Sect. 4 examines the outcomes of these decisions. Finally, Sect. 5 discusses the value of this RBA and Sect. 6 provides a short conclusion.

2 From the Right to Life to City Development Issues

The connection between the right to life and city development issues made by the SC has its origin in procedural law and aligns with the RBA. In fact, by easing access to the courts through public interest litigation (PIL) and a wide interpretation of Article 21 of the Indian Constitution, the SC allowed city dwellers to have their voice heard by the top judicial institution. As a result, city development issues tend to become an important topic in public debates.

2.1 A Rights-Based Approach to the City

In development studies, the RBA is a way to put human rights at the centre of development and public policies. Rights can be understood as claims (of one person or group, on another person, group or institution) that have been legitimized by social structures and norms (Moser 2001). These human rights go beyond traditional civil rights and can include economic, social, political (e.g. participation) and cultural dimensions. The right to a healthy environment and its corollary, the right to life, fall within this category. The implementation and protection of these rights create obligations for public authorities: they should respect and not interfere with the enjoyment of these rights; they should also protect them, that is, prevent third parties from interfering in their enjoyment. Finally, they should be proactive in ensuring the implementation of these rights. Hence the RBA is associated with claim-making on the state to secure services (Mitlin and Patel 2005). It also means that strong legal frameworks are necessary to allow citizens to ask for the realization of their rights.

Now the RBA is relevant to city development as access to essential services such as housing, education, clean water availability, waste management, transport and so forth are a major issue. In that case, local governments are responsible for the provision of these services and citizens should refer to them in the first place. A city with rights should be then both a process and an outcome for its sustainable development (Jonsson 2004). This corresponds to the definition included in the 2004 World Charter for the Right to the City, adopted by the World Urban Forum, a body set up by the United Nations. Article 1 paragraph 2 of the Charter provides that, 'the Right to the City is defined as the equitable usufruct of cities within the principles of sustainability, democracy, equity, and social justice [...] this assumes the inclusion of the rights [...] to social security, public health, clean drinking water, energy, public transportation, and other social services; to food, clothing, and adequate shelter; to quality public education and to culture'. The activity of the SC of India in Delhi and other cities is largely in compliance with this approach. Most of the times, the judiciary reminds public authorities of their duties and urges them to take action on the basis of existing norms. At the same time, its case law puts fundamental rights, such as the right to life, at the centre of the urban process.

2.2 *Public Interest Litigation*

In India, public interest litigation (PIL) (see Table 1) developed in the 1970s and 80s under the influence of activist judges. PIL opens access to the higher courts through a relaxed right to stand ('locus standi'). The power of PIL lies in its freedom from traditional judicial proceedings. Originally this process was aimed at allowing the representation of the poorest share of the population in courts. It is based on the reasoning that the persons who 'by reasons of poverty, helplessness or disability or socially or economically disadvantaged position [are] unable to approach the court for relief, can ask or be represented by any member of the public to maintain an application for them' (Baviskar 2012). Thanks to this, an individual through a mere written letter can start a legal suit.[1] Moreover, citizens are welcome to represent the interests of the poor and oppressed (representative standing).[2] They can also take action to ask public authorities to fulfil their duties (citizen standing). PIL can be pursued either in the High Courts or the SC. It allows bypassing ordinary legal proceedings. In front of the High Courts, citizens base their claim on a legal wrong (Article 226 of the Constitution of India). To reach the SC, claimants must allege the

[1]*Bandua Mukti Morcha* v. *Union of India*, AIR 1984, S.C. 802.

[2]*S.P. Gupta* v. *Union of India* (1981) Supp SCC 87–233.

Table 1 Public interest litigation

Supreme court		High courts
Article 32 Constitution of India '1. *The right to move the SC by appropriate proceedings for the enforcement of the rights conferred by this Part is guaranteed.*'		**Article 226 Constitution of India** '1. *Notwithstanding anything in article 32, every High Court shall have power, throughout the territories in relation to which it exercises jurisdiction, to issue to any person or authority, including in appropriate cases, any Government, within those territories directions, orders or writs, including writs in the nature of habeas corpus, mandamus, prohibition, quo warranto and certiorari, or any of them, for the enforcement of any of the rights conferred by Part III and for any other purpose.*'
Article 32 empowers the Supreme Court to issue writs. Article 32 is itself a fundamental right, since it is included in Part III on Fundamental Rights of the Constitution. It can be suspended during an emergency period. Article 32 empowers the Supreme Court to issue the writs only when the Fundamental Rights are violated or threatened. Through Article 32 a claimant can go straight to the SC.		Article 226 empowers the High Courts to issue writs. Article 226 is not itself a fundamental right. It has a wider scope than Article 32 as it is does not apply only to fundamental rights.
Who can file a case?		
Anybody on behalf of group of people affected 'Representative standing'	People whose rights are affected	Any public spirited person 'Citizen standing'

Source Author

violation of a fundamental right (Article 32 of the Constitution of India[3]) (Rajamani 2007). On this basis, major progress in the fields of education, prisoner's rights and the protection of the environment were achieved (Rosencranz and Jackson 2003).

Soon, environmental groups started to introduce PILs to stop activities causing environmental hazards (e.g. the oleum gas leak case in 1994) and subsequent health problems. Since the process requires that the claimant bases his/her pleas on the violation of a fundamental right, they referred to Article 21 of the Constitution, which provides the right to life. They stated that this involved a right to a healthy environment. The SC has endorsed this approach and developed an important case law on the basis of Article 21. Thus slowly the SC started monitoring civic issues in the urban domain.

2.3 *The Wide Interpretation of the Right to Life*

Article 21 of the Indian Constitution enshrines the right to life. It provides that 'no person shall be deprived of his life or personal liberty except according to procedure established by law'. The SC interpreted this provision broadly, triggering a growing number of PIL submissions. Among them an important number related to the life conditions and welfare of city dwellers. In the course of ten years, the SC identified a series of social rights extending the scope of Article 21. First the SC stated that, in combination with Article 32 of the Indian Constitution, Article 21 includes the right to food, clothing and shelter.[4] The SC also recognized that the right to life encompasses the right to livelihood.[5] Parallel to that, in the specific area of healthcare, the judges admitted that the right to life means the right to live with human dignity, which means that people are entitled to the protection of health.[6] About ten years later, it reaffirmed this right stating that the right to health and medical care is a fundamental right,[7] creating obligations for public authorities. For instance, the Court reckons it is the primary duty of a welfare state to ensure that medical facilities are adequate and accessible to provide treatment.[8]

At the same time, the SC acknowledged that Article 21 includes the right to a healthy environment. The SC observed that 'the right to live [...] includes the right of enjoyment of pollution-free water and air for full enjoyment of life'.[9] It

[3]Article 32 of the Constitution of India: '1. The right to move the SC by appropriate proceedings for the enforcement of the rights conferred by this Part is guaranteed'.

[4]*Francis Coralie Mullin* v. *Union Territory of Delhi* 1981 (1) SCC 608.

[5]*Olga Tellis* v. *Bombay Municipal Corporation* AIR 1986 SC 180.

[6]*Bandhua Mukti Morcha* etc. v. *Union of India and Others* AIR 1984 SC 802.

[7]*Consumer Education and Research Centre* v. *Union of India* (1995)3 SCC 42.

[8]*Paschim Banga Khet Mazdoor Samity and Others* v. *State of West Bengal* 1996(4) SCC 37.

[9]*Sybhash Kumar* v *State of Bihar* AIR 1991 SC.

reaffirmed it in several decisions and applied it to several sectors.[10] For instance, the High Court of Allahabad ordered the local development agency to enforce the city master plan and create a public park to facilitate the enjoyment of the right to a healthy environment.[11] This context was favourable for the SC to take action against air pollution in Delhi.

3 The Air Pollution Case in Delhi

Environmental activists have tapped the opportunity of reaching the courts to submit their claims and question city development related issues. Judicial activity was particularly intense in the capital city, Delhi, where air pollution has been a steady problem. Yet air pollution is a cross-sectoral issue affecting public health and climate change. So the SC took action in two major sectors: land use and transports.[12]

3.1 Air Pollution, Health and Climate Co-benefits

Air pollution is a cross-cutting issue, which requires action in several policy sectors: health, transport, energy, infrastructure, land use and so forth. It has a significant impact on health. In fact, air pollution is the cause of numerous deaths. In 2012 about 7 million people died worldwide of air pollution related disorders, making it the biggest environmental health risk (WHO 2014). While many local air pollutants like CO_2 and NO_x contribute to global climate change, it is also acknowledged that global warming exacerbates air pollution. In fact, higher temperatures increase the concentration of some particles matters and levels of ozone in the air and consequently increase health problems (deaths, hospital care, asthma, bronchitis and lung problems). Reducing GHG emissions can therefore produce co-benefits by downscaling air pollution and the health impacts associated to it (Cifuentes et al. 2001). It requires action in sectors where GHG emissions are the highest, and where public authorities can take significant action. These are, according to the UNFCCC guidelines on national inventories: energy (including transport), industrial processes, solvent and other products, agriculture, land-use and forestry, waste and

[10] *Virendra Gaur and Others* v *State of Haryana* 1995 (2) SCC 577. Several High Courts took decisions: *D.D. Vyas* v. *Ghaziabad Development Authority* AIR 1993 All 57; Allahabad High Court, Kerala High Court, *Antony* v. *Commissioner, Corporation of Cochin* (1) KLT 169.

[11] *D.D. Vyas* v. *Ghaziabad Development Authority* AIR 1993 All 57.

[12] There is also an important case law regarding the waste management sector (see Rajamani 2007).

others.[13] The SC's decisions aiming at reducing air pollution were based on several of these sectors (industry, land use, transport). Its case law therefore goes beyond the reduction of air pollutants. It also has positive impacts in GHG emission reduction although this does not appear as a prime objective in the decisions.

3.2 *Land Use Planning: Industrial Pollution*

In the 1980s, environmental activists and lawyers filed several cases before the SC, arguing that local authorities had failed to protect the citizens from the harmful effects of pollution (Sharan 2013). They focused on industrial pollution caused by polluting industries located within the city's jurisdiction. The claimants were seeking the displacement of hazardous industries out of the city. At that time, the 1962 first Master Plan of Delhi split industries in different categories and established zoning regulations on this basis. In particular, industries were divided between 'conforming/non-conforming industries' for those not complying with the proposed zoning regulations in a particular zone, and 'noxious/non-noxious industries', depending on their polluting levels. But the plan remained unenforced. The revised version of the Master Plan of 1990 was not implemented either. However in the 1990s, the SC started issuing orders to enforce the master plan. It urged the government to notify several noxious and non-conforming industries to shut down and move out of the city. Thousands of industrial units were affected.[14] The SC thus became a key actor in the reshaping of Delhi's urban space (Bhan 2009). Likewise the City Master Plan became a major tool in tackling air pollution. It highlights the importance of development planning and land use regulations for that purpose.

3.3 *The Transport Sector: Vehicular Pollution*

But the degradation of air quality caused by vehicular pollution has been the major area of concern and action of the SC. Indeed, this is a greater source of pollution, as it accounts for two-thirds of city air pollution (Sharan 2013). The problem was

[13]Updated UNFCCC reporting guidelines on annual inventories following incorporation of the provisions of decision 14/CP. 11, FCCC/SBSTA/2006/9. Available at: http://unfccc.int/documentation/documents/advanced_search/items/6911.php?priref=600003988#beg.

[14]*M. C. Mehta* v. *Union of India*, Petition N. 13381, 1984. This happened in different phases and ordered the closure and/or displacement of hazardous and heavy industries, water-polluting industries as well as non-conforming industries. Thousands of industrial units were concerned (Baviskar 2012).

pointed out by different agencies and civil society organizations. The publication in 1996 of a report written by an advocacy group, the Centre for Science and Environment (CSE), titled 'Slow Murder: the Deadly Story of Vehicular Pollution in India' was highly mediatized for the scaring data released (CSE 1996). This report dwelled on the results of a research undertaken by the World Bank. It showed that the annual health costs of air pollution in 1995 were superior to more than US $2,102 in millions. Moreover, the Energy and Resources Institute estimated that in 1998, the incidence of mortality and morbidity due to PM_{10} exposure cost 2.5 million premature deaths and about 88,500 crore Indian rupees to 4,25,000 crore annually (Pandee et al. 2006). Against this background, M.C. Metha, an environmental lawyer, asked the Court to direct various public authorities to implement the Air (Prevention and Pollution) Act of 1981, which already contained guidelines to tackle air pollution. The SC therefore asked Delhi public authorities and the central government to take action. In particular, an amendment to the Air (Prevention and Pollution) Act in 1986, led the central government to establish the Environment Pollution (Prevention and Control) Authority (EPCA), a new advisory body (Narain and Ruth 2005; Sharan 2013).[15] At the same time, both the central government ('White Paper on Pollution in Delhi with an Action Plan') and the Delhi state government (National Capital Territory of Delhi (NCDT)) developed air pollution action plans. The NCDT Plan included measures such as the use of better technology for new vehicles, the suspension of the registration of three-wheelers, the removal of all public bus fleets, and so on (Narain and Ruth 2005).

In addition, the SC decided on 28 July 1998, precisely on the basis of the EPCA recommendations, that all pre-1990 auto-rickshaws should be replaced. As for post-1990 auto-rickshaws and taxis, the decision stated that they had to be retrofitted in order to run on clean fuels. Then, the Court ordered the withdrawal of buses that were more than eight years old, and the conversion of the rest of the city fleet to compressed natural gas (CNG) (Pandee et al. 2006). The Court also mandated fuel quality specifications such as the reduction of sulphur content in diesel supplied in Delhi from 1 to 0.25% (Narain and Ruth 2005). So in the case of vehicular pollution, the SC went several steps farther than for industrial pollution. It did not merely ask for the enforcement of existing laws, but asked for the adoption of new norms, stating what they should provide. The outcomes of this process for the city's development need to be discussed.

[15]Government of India, Ministry of Environment and Forests, 1998 Constitution of the Environment Pollution (Prevention and Control) Authority of the NCR, Delhi. Available at: http://envfor.nic.in/legis/ncr/ncrauthority.pdf.

4 Environmental Outcomes of the Court's Decisions

The immediate measures adopted by the Delhi state government and other implementing local authorities, on the basis of the SC's orders were beneficial but their efficiency was challenged by economic development and the SC triggered a new set of reforms.

4.1 Air Pollution Reduction

It is challenging to assess the benefits of the measures adopted on the basis of the SC's decisions. First, some of the noxious industries are still operating within the city. Second, for industrial as well as vehicular pollution, it must be borne in mind that the pollution has not totally disappeared. Most of it has been displaced, out of the city borders, but still affects the health of city dwellers and workers (Sharan 2013).

The bus fleet in Delhi was one of the main sources of air pollution in the capital. In fact, buses were old and poorly maintained. In addition, they were functioning on petrol or diesel and this was combined with bad traffic conditions (low average speeds, frequent stops, long idling times, etc.; Pandee et al. 2006). Therefore the decision of the NCDT following the SC's orders, to impose CNG on the bus fleet and rickshaws, had a significant impact. Compared with 1997, the concentration of carbon monoxide in 2002 fell by 32%, sulphur dioxide by 39% and particles matters also decreased (Pandee et al. 2006; Siemiatycki 2006). Moreover, the introduction of CNG brought some co-benefits. It reduced CO_2 emissions, even though this was not the primary goal of the policies (Sharma and Tomar 2010).

Yet these good results did not last long, mostly because the number of private vehicles grew significantly. Also some national subsidies for diesel favoured the sale of diesel vehicles. As a result, the nitrogen dioxide levels have increased by 30% since 2002. During the same decade (2002–2012), vehicle numbers have increased by as much as 97%, contributing to pollution load and direct exposure to toxic fumes (EPCA 2014). In the end, in 2014, a World Health Organization (WHO) report shows Delhi has the world's highest annual average concentration of $PM_{2.5}$. In other type of pollutants, Delhi rates better but still stands at the very bottom of the cities on the world scale (WHO 2014).

4.2 New Set of Reforms

With air pollution levels rising again, so did the number of health problems. Senior advocate Harish Salve stated in 2014 that 3000 children die annually from air pollution.[16] Against this background, the SC issued new notices to the Government of India as well as to different Indian states (Haryana, Uttar Pradesh, Rajasthan and Delhi), to compel them to adopt new measures for better air quality.

As in the former set of measures adopted in the 1990s, the SC goes very much into detail in its decision. It refers to the conclusions of a new report from the ECPA acknowledging the gravity of the air pollution problem. Once again the focus of the priority measures identified by EPCA is transport. Some measures aim at improving public transports such as running 10,000–11,000 buses within a year, implementing an appropriate network of dedicated pedestrian and cycling lanes with a high degree of safety, and increasing public transport services. Other measures relate to fiscal policies: promoting a CNG programme through a long-term favourable taxation policy, imposing an additional 30% environment compensation charge on private diesel cars, removing entry taxes on public transport buses across borders in the capital region, refraining from the practice of imposing higher taxes on public transport buses as well as increasing parking charges. Finally, they also set emissions standards consisting in implementing Bharat Stage IV emissions standards nationwide by 2015, moving to Euro V and Euro VI by 2021, and implementing a daily air quality index and health advisory for public information immediately (CSE 2014). Although notices are not binding per se, they are recommendations which public authorities are very much encouraged to follow to avoid future potential sanctions from the Court.[17]

5 Discussion: A Rights-Based Approach to City Development

The SC of India has played an important role in Delhi's governance. This is quite unique in the world. It is therefore interesting to see what lessons in terms of governance can be learnt. Some aspects are particularly contentious, like participation, balance of interests, and expertise and decision-making.

[16]http://timesofindia.indiatimes.com/home/environment/pollution/Impose-30-cess-on-diesel-cars-panel-tells-Supreme-Court/articleshow/30180391.cms.

[17]The SC applied some sanctions in the past for the non-enforcement of its orders. For instance, after the deadline for the conversion of all buses to CNG, the Court ordered private diesel buses to pay a fine of 500 Rupees per day (Rajamani 2007). It also chided the Delhi government on various occasions for non-enforcement of its orders on the closure and displacement of industrial activity (Baviskar 2012).

5.1 Participation

Access to the courts is one on the essential elements of principle 10 of the 1992 Rio Declaration on Environment and Development, also called the 'Environmental Participation Principle',[18] (endorsed by the UN General Assembly). This principle has three dimensions: right to information, participation in decision-making and access to judicial and administrative courts. Indeed, through the opening of the right of standing ('locus standi') and the wide interpretation of some fundamental rights, the SC of India has allowed conscious citizens to represent collective interests, which could not be heard beforehand in courts. To that regard, the Indian judiciary have provided an illustration of how the principle of effective access to judicial proceedings, enshrined in international conventions such as the Rio Declaration or regional conventions as the 1998 Aarhus Convention on Access to Information, Public Participation in Decision-Making and Access to Justice in Environmental Matters, can be implemented. Yet the extent of the opening of the SC to citizens has been discussed.

5.2 Balance of Interests of All City Dwellers

The decisions of the courts and the level of participation that they actually entail have been discussed. While PIL allowed some citizens to be represented in courts for the first time, others largely remained out of the process. Indeed, although access to court is eased from a procedural point of view, other charges and costs still impede the participation of the poorest. For instance, even if lawyers do not charge for their service, petitioners would need to be present continuously to increase their chances of success, which involves taking work leaves and transport costs. They would also need to find support from big solicitor firms to support their claims. Lack of legal literacy is not really compensated by PIL as these are often filed on their (affected party's) behalf but not by them (Rajamani 2007). Participation is therefore mediated by middle-class activists, which might not always be fully aware of the poor's needs. Baviskar (2012) highlights the distress of workers left without job after the SC's orders in the industrial pollution cases, were implemented. The Court valued more the right to a healthy environment over the right to livelihood. This led to important protests (Baviskar 2012). Moreover, the legitimacy of the

[18]Principle 10 Rio Declaration on Environment and Development: 'Environmental issues are best handled with participation of all concerned citizens, at the relevant level. At the national level, each individual shall have appropriate access to information concerning the environment that is held by public authorities, including information on hazardous materials and activities in their communities, and the opportunity to participate in decision-making processes. States shall facilitate and encourage public awareness and participation by making information widely available. Effective access to judicial and administrative proceedings, including redress and remedy, shall be provided.'.

judiciary to adopt measures for the entire city despite their lack of representativeness (they are not elected) is also an issue (Rajamani 2007), and some authors suggest that judges may be sensitive to middle-class concerns but less to the poorest dweller's (Baviskar 2012; Bhan 2009). These pitfalls in the effectiveness of participation cannot be ignored for the peaceful and equitable development of the city.

5.3 Expertise and Decision Making

In the vehicular pollution case, the order of the SC to impose the conversion of the entire city fleet to CNG was contentious.[19] Some experts were promoting other fuels and technologies, as well as more flexible norms (Rajamani 2007). The discussion between the Court, the ECPA and other Committees went back and forth for about five years. The discussion was also very vivid outside of the Court in the media. The SC, on the basis of the EPCA report recognized that they were no real clean fuels but also declared that fuels such as CNG, liquefied petroleum gas (LPG) and propane were less polluting. It also rejected the arguments that the supply and costs of infrastructures to support the CNG fuelling would be complicated and costly. In fact, the Government of India appointed an expert committee to make recommendations over the air pollution problem in Delhi. R.A. Mashelkar, director general of the Council of Scientific and Industrial research (CSIR), headed the committee. The Mashelkar Committee shared with the Court and the ECPA the objective of protecting Delhiite's health, and the idea that vehicular pollution was one of the causes of air quality degradation. But its recommendations were different. While ECPA and the SC marshalled the introduction of CNG for the bus fleet, the Mashelkar Committee recommended a multi-fuel and multi-technological approach with competent public authorities defining the appropriate norms (Sharan 2013). By rejecting these recommendations, the SC challenged directly the expert committee appointed by the Government of India, and the question of its legitimacy to define the transport policy was raised. Some authors observed that the SC legitimately made its decisions on the basis of the precautionary principle. This principle enshrined in the Rio Declaration of 1992 specifically mentions the 'lack of full scientific certainty shall not be used as a reason for postponing cost-effective measures to prevent environmental degradation' (Principle 15). Therefore the impossibility to establish clearly the benefits of CNG over diesel could not restrict the SC to search for an innovative solution. Emissions norms had already been tested and failed to improve the air quality. So it has been argued that the SC rightly ruled over this issue (Rajamani 2007; Sharan 2013).

---

[19]*M. C. Mehta* v. *Union of India and Others*, SC, 28 July 1998 (written petition n. 13029, 1985).

6 Conclusion

This chapter demonstrates how a RBA is a way to put fundamental rights at the centre of urban development and thus achieve co-benefits. The SC of India on the basis of international environmental principles and the Indian Constitution has, through its activism, provided one of the most noteworthy examples of the concrete outcomes a RBA may offer. Although this chapter focused on Delhi, the case law shows a process which can be replicated in other Indian cities and states, and also a reasoning which can be inspiring for lower courts.

Yet the process is not enough in itself. Indeed although the Indian judiciary has been active for about 30 years now, the initial positive results have diminished with time. Today, air quality in Delhi remains one of the worst in the world and the Indian judges still struggle with public authorities over the question of air pollution. Indeed, recently, the National Green Tribunal[20] put out a new plan of action for the Delhi administration and the Delhi government, which includes the ban of vehicles over 15 years old, the installation of air filters in the city's public spaces and the ban on parking on roads and lanes.[21] Although these measures are binding, their efficiency and enforceability are questioned (like limited scoped, resources needed for enforcement, counter-effect, etc.).[22] Moreover, the legitimacy of the judiciary to influence decision-making and its capacity to consider the interests of all urban stakeholders, even those not represented in courts, is still debated.

To overcome these challenges, the RBA needs to be considered and included in an early stage of the decision-making process, and action in front of the courts, which comes later, should only make sure policies are implemented, instead of being a catalyst for public action. To this regard, early participation of communities and private actors is vital to highlight interests and needs which might be otherwise overseen. Policy-makers will thus anticipate potential co-benefits and the rights and duties of all different stakeholders aligning with the latest recommendations from the Paris Climate Agreement of 2015, which provides that:

---

[20]The National Green Tribunal (NGT) was established in 2010, on the basis of Article 21 of the Constitution. The National Green Tribunal (NGT) Act provides that it was established 'for the effective and expeditious disposal of cases relating to environmental protection and conservation of forests and other natural resources including enforcement of any legal right relating to environment and giving relief and compensation for damages to persons and property and for matters connected therewith or incidental thereto'. Its activity is expected to release the SC from the high number of PILs filed in front of it..

[21]NGT, application n. 21 of 2014, *Vardhaman Kaushik* v. *Union of India and others*, 4 December 2014, available at: http://www.greentribunal.gov.in/Writereaddata/Downloads/21-2014%28PB-I%29OA-4-12-2014_1v.pdf.

[22]http://www.downtoearth.org.in/content/ngt-slams-delhi-government-worsening-air-pollution-gives-directions-curb-vehicular-emissions.

acknowledging that climate change is a common concern of humankind, Parties should, when taking action to address climate change, respect, promote and consider their respective obligations on human rights, the right to health, the rights of indigenous peoples, local communities, migrants, children, persons with disabilities and people in vulnerable situations and the right to development, as well as gender equality, empowerment of women and intergenerational equity.[23]

References

Baviskar, A. (2012). Public interest and private compromises: The politics of environmental negotiation in Delhi, India. In J. Eckert, B. Donahoe, C. Strumpell and Z. Ozlem Biner (Eds.), *Law against the state, ethnographic forays into law's transformations* (pp. 171–201). Cambridge: Cambridge University Press.

Bhan, G. (2009). This is no longer the city I once knew: Evictions, the urban poor and the right to the city in millennial Delhi. *Environment and Urbanization, 21*(1), 127–142.

Cifuentes, L., Borja-Aburto, V. H., Gouveia, N., Thurston, G., & Davis, D. L. (2001). Hidden health benefits of greenhouse gas mitigation. *Science, 293,* 1257–1259.

CSE. (1996). *Slow murder: The deadly story of vehicular pollution in India. State of the environment series*. New Delhi: Centre for Science and Environment.

CSE India. (2014, 11 February). Supreme Court admits EPCA report on priority measures to control air pollution. http://www.cseindia.org/content/supreme-court-admits-epca-report-priority-measures-control-air-pollution. Accessed 23 Dec 2016.

EPCA. (2014). Report on priority measures to reduce air pollution and protect public health, in the matter of W.P. (C) No.13029 of 1985, http://cseindia.org/userfiles/EPCA_report.pdf. Accessed 23 Dec 2016.

Jonsson, U. (2004). *A human rights-based approach to programming*. New York: UNICEF.

Mitlin, D., & Patel, S. (2005). *Re-interpreting the rights-based approach—A grassroots perspective on rights and development*. Paper presented at the Winners and Losers from Rights-Based Approaches to Development, Manchester University, Manchester, UK, February 21–22, 2005. http://www.sed.man.ac.uk/research/events/conferences/documents/Winners%20and%20Losers%20Papers/Mitlin.pdf.

Moser, C., Norton, A., Conway, T., Ferguson, C., & Vizard, P. (2001). *To claim our rights: Livelihood security, human rights and sustainable development*. London: Overseas Development Institute.

Narain, U., & Ruth, G. B. (2005). *Who changed Delhi's air? The roles of the Court and the executive in environmental policy making, Discussion Paper*. Washington, D.C: Resources for the Future.

Pandee, S., Singhal, S., Jaswal, P., & Guliani, M. (2006). Urban environment. In A. Rastogi (Ed.), *India infrastructure report 2006—Urban Infrastructure*. New Delhi: Oxford University Press.

Rajamani, L. (2007). Public interest environmental litigation in India: Exploring issues of access, participation, equity, effectiveness and sustainability. *Journal of Environmental Law, 19*(3), 293–321.

Rosencranz, A., & Jackson, M. (2003). The Delhi pollution case: The of India and the limits of judicial power. *Columbia Environmental Law*, 223–254.

[23]Draft Paris Agreement, 12 December 2015, FCCC/CP/2015/L.9/Rev.1 http://unfccc.int/files/meetings/paris_nov_2015/application/pdf/cop_auv_template_4b_new__1.pdf.

Sharan, A. (2013). One air, two interventions: Delhi in the age of environment. In A. Rademacher and K. Sivaramakrishnan (Eds.), *Ecologies of urbanism in India. Metropolitan civility and sustainability* (pp. 71–90). Hong Kong: Hong Kong University Press.

Sharma, D., & Tomar, S. (2010). Mainstreaming climate change adaptation in Indian cities. *Environment and Urbanization, 22*(2), 451–465.

Siemiatycki, M. (2006). Message in a Metro: Building urban rail infrastructure and image in Delhi, India. *International Journal of Urban and Regional Research, 30*(2), 277–292.

WHO. (2014). Ambient (outdoor) air pollution in cities databases 2014. http://www.who.int/phe/health_topics/outdoorair/databases/cities/en/. Accessed 23 Dec 2016.

Chapter 15
Mainstreaming Co-benefits in Urban Policy, Governance and Finance

A. Narender and Mahendra Sethi

Abstract This chapter concludes the book by examining the potential for mainstreaming a climate co-benefits approach in urban policy, governance and finance to address climate impacts and reducing greenhouse gas emissions in India. The cities have a key role to accrue co-benefits as they have jurisdictional and functional responsibility for planning and provision of services. The five-year plans, sectoral policies and projects comprise the urban policy framework in India and past efforts have not paid much attention to climate co-benefits. It is only in the last decade that the programmes and policies have started paying attention to the climate co-benefits in their development agenda. There is a greater scope for mainstreaming a climate co-benefits approach into these policies and ensure their effective implementation at the state and Urban Local Body (ULB) levels. The governance and financing capacities of implementing agencies can play a key role in effective implementation of urban sector policies. The 12th Five Year Plan and the High Power Expert Committee Report have made suggestions for strengthening the governance and financing framework for promoting urban development, which needs to be implemented on a priority basis. The chapter presents the need and approach to internalize a climate co-benefit approach, discusses the key issues and suggests the way forward to mainstream co-benefits into the urban policy, governance and financing framework in India.

Keywords Global climate framework · Urban policy · Governance
Finance · Climate Co-benefits · Research and policy imperatives
Recommendations

This book started with a dialogue significant to several developing and developed economies, but a challenge that particularly hounds their urban areas. It is the most crucial dichotomy of today's cities that forms the premise for this research, that is,

A. Narender
Consultant to Various Organizations, Hyderabad, India

M. Sethi (✉)
National Institute of Urban Affairs, New Delhi, India
e-mail: mahendrasethi@hotmail.com

M. Sethi and J.A. Puppim de Oliveira (eds.), *Mainstreaming Climate Co-Benefits in Indian Cities*, Exploring Urban Change in South Asia,
https://doi.org/10.1007/978-981-10-5816-5_15

while cities are meant to tirelessly function as engines of growth, guzzlers of energy, absorbing additional manpower migrating from rural areas, at the same time they are expected to have minimal emission footprints by pursuing sustainable strategies. In view of the New Urban Agenda adopted during Habitat III Conference in Quito, the intention of this book was to discuss how we can develop policies and instruments to boost the capacity of societies, in generating climate, environmental and development oriented co-benefits in cities to synergistically achieve the local and global goals of sustainable development. The edited volume aimed to build on the deliberations in this multi-disciplinary area, and recognize pathways to identify and generate large urban co-benefits. Accordingly, the analysis was structured into four main thematic areas: (1) concepts and theories behind cities and climate co-benefits; (2) contextualizing co-benefit issues across spatial scales and sectors; (3) sectoral analyses of co-benefits in energy, transport, buildings, waste and biodiversity; and (4) innovations and reforms needed to promote co-benefits in the urban context. Making deliberations in the last thematic area, this chapter ties up the research, and concludes with suggestions and recommendations that could mainstream climate co-benefits in urban policy, governance and finance.

1 Aligning with the Global Climate Framework

The conventional strategies for economic growth and development in urban areas in the past have disregarded their impact on environment and climate change. The recent efforts to promote measures for adaptation and mitigation of climate change have been pursued independently of the growth and development strategies resulting in limited impacts. The climate co-benefits approach refers to pursuing economic growth and development objectives along with co-benefits to address climate change impacts effectively. Hence, it is increasingly being realized that mainstreaming the climate co-benefits approach with the development approach could have a greater positive impact on environment and climate. Although there are various challenges in the estimation of potential co-benefits in the inter-disciplines of cities and climate change (Sethi 2015), urban policy, governance and financing framework in India is a key enabler of both development objectives and adapting and mitigating climate change and offers a huge potential to internalize the climate co-benefits approach. The urban policy framework consists of various sectoral policies such as transport policy, sanitation policy, housing policy, and so on, and efforts have been initiated to integrate the climate co-benefits approach into them. A number of service delivery improvement projects in the areas of water, sanitation, solid waste management, and so forth, bear potential to promote the climate co-benefits approach. The governance at the national, state and city level need to be effective to implement these policies, projects and reforms so as to improve technical capacity and performance of local governments and service delivery agencies. The existing financing framework for urban development needs to be sensitive to incentivize and promote climate co-benefits in policies and governance measures.

The cities and their urban local bodies (ULBs) can play a key role in mainstreaming climate co-benefits in urban development. City governments in many countries are responsible for or have significant influence over areas and activities that produce GHG emissions. Consequently the municipal governments have the potential to bring about co-benefits, particularly in the areas of transportation, waste management, buildings and construction (Puppim de Oliveira 2013). It has been further suggested to strengthen the autonomy and capacities of local governments, better regulation and enforcement, improved institutional coordination, integrating various sectoral policies, reforming public budget allocation process, engaging non-governmental actors, awareness and advocacy and promoting innovations as the key means of mainstreaming climate co-benefits into urban development.

According to the UN Habitat Global Report 2011 on Cities and Climate Change, municipal authorities are potentially important actors in addressing climate change as they have jurisdictional responsibility for key processes impacting climate change, they can act as laboratories for testing solutions and provide a key interface for engaging the private sector and civil society (UN Habitat 2011). Reviewing the global experience of cities that have implemented climate change adaptation and mitigation initiatives, the report observed that though climate change remained firmly on the urban policy agenda, its implementation has been marginal and ineffective due to policy fragmentation and lack of governance capacity (UN Habitat 2011). Some of its observations in this matter could be summarized as:

1. International and national policies have provided overall framework for municipal policies and national policies have served as direct drivers for municipal actions
2. Municipalities with specific competencies for direct provision of waste, energy and transport can have significant capacity to address climate change
3. Municipalities with limited powers could achieve GHG emissions by integrating policy goals across different levels of government, full utilization of their powers, effective institutional coordination with other departments and partnering with the private sector and civil society
4. Effective implementation and enforcement of environmental regulations and standards can promote climate change adaption and mitigation
5. Building expertise, scientific knowledge and skills among public sector, private sector and civil society are a prerequisite
6. Ability and autonomy to adopt innovative strategies for mobilization of internal revenues and greater access to external sources of funding is needed
7. Presence of effective leadership, individual political champions and policy entrepreneurs, committed individuals and a degree of commitment to action can play a key role

8. Bundling climate change mitigation with other potential social or environmental benefits can have a positive impact
9. Effective management of conflicting agendas and ensuring equity while addressing climate change is needed.

According to the 'Guide to Climate Change Adaptation in Cities', city governments are responsible for delivery of a wide range of services such as land use and zoning, water, sanitation, housing, transportation, and so on, and hence are better positioned for adaptation of climate change impacts. Many of these services are vulnerable to disruption due to climate change impacts and cities can develop increased adaptive capacity and strategies to reduce vulnerability (World Bank 2011). The suggestions made to develop climate change adaptation and mitigation strategies in cities include, inter alia: instituting and enforcing strong management principles to address problems related to governance such as corruption, lack of transparency and weak administration; establishing a dedicated climate change unit in the mayor's office or in the environment agency, an inter-agency task force of the city government or a wider stakeholder group beyond the city government; starting a climate leadership team within the government and partnerships with civil society. The World Bank also recommends understanding and assessing climate change impacts; developing city adaptation plans, policies and actions; setting performance indicators, evaluating and prioritizing potential adaptation actions; and seeking financial resources from a wide variety of sources such as local, national, international, public and private sources.

The World Economic Forum projects that by 2020, about US$5.7 trillion will be needed annually for green infrastructure investments in the developing countries. There are several bilateral and multilateral mechanisms for funding from developed countries. The Green Climate Fund, the operating entity of the United Nations Framework Convention on Climate Change (UNFCCC) along with the Climate Investment fund makes up the major share of multilateral aid available for climate change for the developing world. In addition, there is the Special Climate Change Fund, Adaptation Fund and official development assistance from regional and bilateral organizations like the European Union, Organization for Economic Cooperation and Development, Germany and the UK. The negotiations in Bonn, followed by those in Marrakech, Morocco yielded no distinct road map to systematically target US$100 billion of annual climate change finance by 2020. While it is generally understood that climate finance could play a crucial role in assisting developing countries in making the transition to more environmentally sustainable systems and addressing developmental priorities, there is no specific international framework to directly finance climate action in cities or sub-national authorities.

The IPCC 2014 WG III Report has made several observations on urban policy, governance and financing for adaptation and mitigation of climate change (IPCC

2014b) that include: reducing basic service deficits, improving housing, and building resilient infrastructure systems that could significantly reduce vulnerability and exposure in urban areas; urban adaptation benefits from effective multi-level urban risk governance; alignment of policies and incentives; strengthened local government and community adaptation capacity; synergies with the private sector, and appropriate financing and institutional development; and increased capacity, voice, and influence of low-income groups and vulnerable communities and their partnerships with local governments to benefit adaptation.

The IPCC 2014 WG Report advises for the following measures for mitigation of urbanization impacts on climate change (IPCC 2014b):

1. Mitigation options vary by urban trajectories and are expected to be most effective when policy instruments are bundled. Infrastructure and urban form are strongly interlinked and lock in patterns of land use, transport choice, housing and behaviour. Effective mitigation strategies involve packages of mutually reinforcing policies, including co-locating high residential with high employment densities, achieving high diversity and integration of land uses, increasing accessibility and investing in public transport and other demand management measures.
2. The largest mitigation opportunities with respect to human settlements are in rapidly urbanizing areas where urban form and infrastructure are not locked in but where there is often limited governance, technical, financial and institutional capacities. The bulk of urban growth is expected in small to medium-size cities in developing countries. The feasibility of spatial planning instruments for climate change mitigation is highly dependent on a city's financial and governance capability.
3. Thousands of cities are undertaking climate action plans, but their aggregate impact on urban emissions is uncertain. There has been little systematic assessment on their implementation, the extent to which emission reduction targets are being achieved, or emissions reduced. Current climate action plans focus largely on energy efficiency. Fewer climate action plans consider land-use planning strategies and cross-sectoral measures to reduce sprawl and promote transit-oriented development.
4. Successful implementation of urban-scale climate change mitigation strategies can provide co-benefits. Urban areas throughout the world continue to struggle with challenges, including ensuring access to energy, limiting air and water pollution, and maintaining employment opportunities and competitiveness. Action on urban scale mitigation often depends on the ability to relate climate change mitigation efforts to local co-benefits.

Urban policy, governance and financing frameworks also have relevance to achieving the United Nations Sustainable Development Goals particularly to Goal

11, 16 and 17. While Goal 11 aims at making cities inclusive, safe, resilient and sustainable (see Box 1 for details), Goal 16 aims to promote peaceful and inclusive societies for sustainable development, provide access to justice for all and build effective, accountable and inclusive institutions at all levels. This goal also aims to ensure responsive, inclusive, participatory and representative decision-making at all levels. Another key goal of SDGs is Goal 17, which aims to strengthen the means of implementation and revitalize the global partnership for sustainable development through measures such as strengthening domestic resource mobilization to improve domestic capacity for tax and revenue collection, targeted capacity building, multi-stakeholder partnerships, effective public-private and public-civil society partnerships and availability of high quality data, monitoring and accountability (United Nations 2015). Strengthening urban policy, governance and finances on these lines can have a significant impact on achieving SDGs 16 and 17. Meanwhile, Goal 13 (Take urgent action to combat climate change and its impacts) acknowledges that the UNFCCC is the primary international, intergovernmental forum for negotiating the global response to climate change.

Box 1: SDG Goal 11: Make Cities Inclusive, Safe, Resilient and Sustainable

- By 2030, ensure access for all to adequate, safe and affordable housing and basic services and upgrade slums
- By 2030, provide access to safe, affordable, accessible and sustainable transport systems for all, improving road safety, notably by expanding public transport, with special attention to the needs of those in vulnerable situations, women, children, persons with disabilities and older persons
- By 2030, enhance inclusive and sustainable urbanization and capacity for participatory, integrated and sustainable human settlement planning and management in all countries
- Strengthen efforts to protect and safeguard the world's cultural and natural heritage
- By 2030, significantly reduce the number of deaths and the number of people affected and substantially decrease the direct economic losses relative to global gross domestic product caused by disasters, including water-related disasters, with a focus on protecting the poor and people in vulnerable situations
- By 2030, reduce the adverse per capita environmental impact of cities, including by paying special attention to air quality and municipal and other waste management

- By 2030, provide universal access to safe, inclusive and accessible, green and public spaces, in particular for women and children, older persons and persons with disabilities
- Support positive economic, social and environmental links between urban, peri-urban and rural areas by strengthening national and regional development planning
- By 2020, substantially increase the number of cities and human settlements adopting and implementing integrated policies and plans towards inclusion, resource efficiency, mitigation and adaptation to climate change, resilience to disasters, and develop and implement, in line with the Sendai Framework for Disaster Risk Reduction 2015–30, holistic disaster risk management at all levels
- Support least developed countries, including through financial and technical assistance, in building sustainable and resilient buildings utilizing local materials

The UNFCCC Conference of Parties (COP) convened in December 2015 to conclude the Paris Agreement, advocating various actions for strengthening capacity building, transparent action and support and finance for implementing climate change policies and strategies by member countries (UNFCCC 2015), which have relevance for the urban context as well. The important ones are identifying capacity gaps and needs and recommending ways to address them; promoting the development and dissemination of tools and methodologies for implementation of capacity building; fostering global, regional, national and sub-national cooperation; identifying and collecting good practices, challenges, experiences, and lessons learned; education, training and public awareness, public participation and public access to information; facilitating improved reporting and transparency over time; financial resources provided to developing countries to enhance the implementation of their policies, strategies, regulations, action plans and climate change mitigation and adaptation actions; global funding arrangements such as the Green Climate Fund, Least Developed Countries Fund, Special Climate Change Fund and Global Environment Facility to support the agreement; enhance coordination and delivery of resources to support country-driven strategies through simplified and efficient application and approval procedures. Although the above suggestions are broad and provide a framework for developed countries and international institutions to support country specific capacity building and financing requirements for implementing climate change adaptation and mitigation actions, by the same token these are applicable for cities in India too.

In pursuit of the Paris Agreement, the Government of India communicated the Intended Nationally Determined Contribution (INDC) to the UNFCCC (MoEF 2015). India's INDC has assigned a key role for both urban sector and building sector. The INDC noted that rapid urbanization in the country will be a notable trend in the coming years and the urbanization levels are going to increase from 30% at present to 40% by 2030. The INDC also noted that half of India is yet to be built by 2030 highlighting the opportunity to make the process sustainable and efficient. It further observed that the broad policy framework for environment and climate change in India comprises the National Environment Policy (NEP) (2006), National Action Plan on Climate Change (NAPCC) (Government of India 2007), State Action Plan on Climate Change (SAPCC), Energy Conservation Act 2001, National Policy for Farmers, National Electricity Policy (NEP), Integrated Energy Policy (IEP), National Mission for Enhanced Energy Efficiency (NMEEE), Energy Conservation Building Code (ECBC), Green Rating for Integrated Habitat Assessment (GRIHA) and so on. A majority of these policies are directly or indirectly related to cities. The Government of India is implementing several schemes for strengthening cities through the Ministry of Urban Development (MoUD) which also aims at making them climate resilient. These schemes include the Smart Cities Mission, Atal Mission for Rejuvenation and Urban Transformation (AMRUT), National Heritage City Development and Augmentation Yojana (HRIDAY), Swachh Bharat Mission and National Urban Transportation Policy. Actions in the areas of water, health and disaster management also have relevance to cities.

The key instruments identified by the INDC for capacity building are the setting up of INCCA (Indian Network for Climate Change Assessment) for knowledge sharing and collaboration comprising of 127 institutions, creation of climate change centres by the Department of Science and Technology especially for Himalayan region, formulation of National Training Policy which prescribed earmarking 2.5% of salary budget by each and every department for training including for climate change and sustainable development and launching of the Skill India Programme for imparting skills to 400 million persons in various sectors including climate change and sustainable development by 2022. The INDC articulated for a framework for diffusion of cutting edge technology and research and development in future technologies. The other capacity building initiatives include formation of thematic knowledge networks, more intensive state centric knowledge and awareness, training programmes in sectors and so on.

With regard to financial resources for climate change and sustainable development, INDC observed that most of them come from budgetary sources of departments for various schemes and programmes and this is applicable for urban programmes as well. However, measures are introduced in recent times to mobilize financial resources through a mix of market mechanisms, fiscal and regulatory instruments. The sources of national funds include, cess on coal, national adaptation fund, increase in taxes and reduction in subsidies on petrol and diesel, tax free infrastructure bonds and so on (MoEF 2015). The INDC articulated for mobilizing domestic and external funds from developed countries. In addition, it has estimated a requirement of US$206 billion at 2014–15 prices to implement climate change adaptation and mitigation actions during 2015–2030 in various sectors. The estimates by NITI Aayog are much higher at US$834 billion till 2030 at 2011 prices. The INDC further estimated that at least US$2.5 trillion are required at 2014–15 prices to meet India's climate change actions between now and 2030. More specific policies and governance and financing framework related to the urban sector and sustainable habitat in India are presented in sections below.

2 National Urban Frameworks

2.1 Urban Policy Framework

Urban policies in India have evolved mainly through the focus of successive five-year plans (FYPs). India never had an independent urban development policy or strategy. The role of national government through the Ministries of Urban Development and Housing and Poverty Alleviation was to provide a catalytic, facilitating and guiding role to states and cities through policy, strategy, programmes and schemes under the FYPs. Until recently, urban policy within the provisions of FYPs was piecemeal, scheme and project based and not strategic or programme based. The focus of various schemes was to provide housing or basic infrastructure without addressing the issues of governance, municipal finance and capacity of ULBs. The issues related to climate change and environment were not paid attention in any of the FYPs until the 12th FYP for the period 2012–17.

A series of national housing policies were formulated from time to time. The first National Housing Policy was formulated in 1988 followed by revisions in 1994 and 1998. This was followed by National Housing and Habitat Policy (NHHP) was formulated in 2007 (Ministry of Housing and Urban Poverty Alleviation 2007). These policies have articulated the goal of providing *shelter for all* and emphasized increasing the supply of serviced land, provision of basic services and promoting healthy environment. The ultimate goal of the NHHP 2007 was to ensure sustainable development of all urban human settlements, duly serviced by basic civic amenities for ensuring better quality of life to all urban citizens. There was no conscious effort in these policies to integrate the climate co-benefits approach into the housing and habitat agenda. A number of schemes such as the Basic Services for the Urban Poor, Integrated Housing and Slum Development Programme and *Rajiv Awas Yojana* were formulated for provision and improvement of housing and services to the slum dwellers and the urban poor. These policies carried the potential to achieve climate co-benefits as they convert informal housing stock into formal housing stock with adequate water, sanitation and other services. These policies could have also promoted energy efficiency and reduce climate change impacts. Unfortunately these schemes were executed only as slum development schemes without considering their potential for climate co-benefits.

The Government of India through MoUD is implementing several programmes and policies for the development of cities. The Jawaharlal Nehru National Urban Renewal Mission (JNNURM) and Urban Infrastructure Development Scheme for Small and Medium Towns (UIDSSMT) were two such programmes launched in 2005 for promoting infrastructure, services and governance reforms with a total outlay of Rupees one lakh crore (1000 billion) over a seven-year period. The MoUD formulated the National Urban Sanitation Policy (NUSP) in 2008, providing a framework for states and cities to formulate strategies and action plans for achieving universal access to sanitation (MoUD 2008a, b, c). The MoUD also brought out service level benchmarks for water supply, sanitation, solid waste management and storm water drainage sectors in 2008, which focused on sustainable and efficient service delivery. The Municipal Solid Waste Management Rules 2000, now the updated Solid Waste Management Rules 2016, provide an effective framework for sustainable management of waste including recycling and scientific disposal. These programmes have made reference to measures that have the potential to yield environmental benefits. However, there was no conscious effort to integrate climate co-benefits into these schemes.

MoUD has brought out the National Urban Transport Policy (NUTP) in 2006, 2014. The key objectives of the NUTP include encouraging integrated land use and transport planning and investing in transport systems that encourage greater use of public transport and non-motorized modes. The other objectives of the policy include more equitable allocation of road space, establishing regulatory mechanisms, introducing intelligent transport systems, addressing concerns of road safety, reducing pollution levels, building capacity, promoting the use of cleaner technologies, raising finances through innovative mechanisms, associating private sector and taking up pilot projects. The policy has suggested reserving lanes and corridors exclusively for public transport and non-motorized transport. The NUTP has suggested a number of supply-side and demand-side initiatives for reducing climate change impacts and promoting sustainable habitat in Indian cities. But these measures are meant to improve the efficiency of transport system and addressing the local environmental problems rather than mainstreaming climate co-benefits.

The Government of India has formulated several policies and initiatives in the energy and building sector for reducing GHG emissions. These include the Energy Conservation Act 2001, Energy Conservation Building Code 2007, National Mission for Enhanced Energy Efficiency, National Building Code and a rating system for green buildings. These policies are closer to integrating climate co-benefits into the urban policy. A key issue with these policies is their effective implementation at the state and ULB level to realize the potential benefits.

The 12th FYP formulated a strategy for urban development for the period 2012–17 based on the recommendations of the High Powered Expert Committee set up by the MoUD which submitted its report in 2011. It also identified the desired outcomes of planned, inclusive and sustainable urban development: affordable housing, sustainable livelihood and enterprises, universal access to water and sanitation, quality and affordable public transport, and a clean and healthy environment. The 12th FYP has further identified the necessary enablers to achieve the desired outcomes: strengthen local governance systems, integrate planning organizations and processes, build capacity across all levels, financially empower ULBs, and promote innovation in urban management. The 12th FYP has emphasized sectoral policies and projects for improved governance and service delivery. These policies and projects have the potential for climate co-benefits but the same was not emphasized. Thus, most of the urban sector policies have not realized the potential of integrating a climate co-benefit approach. Currently the only framework available for this purpose at the national level is the National Mission for Sustainable Habitat (NMSH) launched by the MoUD in 2008 under the NAPCC. The strategies articulated by the NMSH for sustainable urban habitat include extension of the energy conservation building code; better urban planning; urban transport management and a modal shift to public transport; recycling of material and urban waste management including solid and liquid waste and storm water management; better water management including water resource management and drinking water supply; promotion of energy efficiency in residential and commercial sectors; and energy efficient and green buildings.

A key practical issue with regard to the above strategies is that these ought to be implemented by the states and cities and there is no plan of action or institutional mechanism to achieve the same. The year 2015 saw the official launch of some of the most widespread and promising initiatives in modern India unveiled for cities by the MoUD, Government of India, namely the Smart Cities Mission (SCM), AMRUT, HRIDAY and last but not the least, Housing for All. The schemes collectively pose to be a reckoning and unprecedented force of such a scale, in transforming the living conditions of so many city dwellers in a planned manner probably in the entire history of human civilization.

The much anticipated SCM aims to make 100 major cities of the country smart through a competitive process of selection. The Government has committed to collaborate with states and local bodies in implementation of the scheme and dedicated Rs. 48,000 crore (480 billion) for five years. As the chapter on 'Livable and Smart Cities' in this book reveals, in spite of having no internationally accepted definition of a Smart City or a national urbanization policy in India, the Smart Cities objectives nonetheless hold immense potential to achieve multiple benefits of sustainability, systems efficiency, economic growth, participatory governance and better quality of life. AMRUT, an upscale version of JNNURM (2005–12), would sponsor 500 cities having 1 lakh (0.1 million) population and more people with the aim to (i) ensure that every household has access to a tap with assured supply of water and a sewerage connection; (ii) increase the amenity value of cities by developing greenery and well-maintained open spaces (e.g. parks); and (iii) reduce pollution by switching to public transport or constructing facilities for non-motorized transport (e.g. walking and cycling). All these outcomes are valued by citizens, and their indicators and standards have been prescribed by the MoUD in the form of service level benchmarks. The SCM includes components of public private partnership, capacity building and urban reforms. In addition, SCM is also developing its own set of indicators to benchmark cities and guidelines for inspiring innovative smart solutions at the grassroots.

HRIDAY aims at holistic development of heritage cities. With a duration of 27 months and a total outlay of INR 500 crores (5000 million), the scheme is being implemented in a mission mode in 12 identified cities with cultural significance, namely: Ajmer, Amaravati, Amritsar, Badami, Dwarka, Gaya, Kanchipuram, Mathura, Puri, Varanasi, Velankanni and Warangal. It supports development of core heritage infrastructure projects, including revitalization of areas around heritage assets identified/approved by the Ministry of Culture, Government of India and state governments. Meanwhile, *Housing for All* is an ambitious scheme that aims to construct 2 crore (20 million) houses in urban areas in the next seven years. Each house in the country is stipulated to have basic facilities of 24-h power supply, clean drinking water, a toilet and connection by road. In addition to the above scheme, the Government of India formulated the Swachh Bharat Mission (SBM) that became operational on 2 October 2014, the birthday of Mahatma Gandhi. The scheme that will be implemented over five years, aims at elimination of open defecation, conversion of insanitary latrines to pour flush latrines, eradication of manual scavenging, 100% collection, treatment and scientific disposal of solid waste, bringing behavioural change among people to promote healthy sanitation practices, generate citizen awareness on sanitation and public health linkage, strengthen urban local bodies to formulate, implement and maintain schemes and create private sector participation in the provision and management of sanitation. The objective of SBM is to provide sanitation and household toilet facilities for all 4041 statutory towns in the country which accommodate about 31% urban population (as per 2011 Census). The components of the scheme include construction of individual, community and public toilets, solid waste management, information, education and communication, and capacity building.

2.2 Urban Governance Framework

Urban governance constitutes the leadership, managerial and technical capacity of institutions and staff engaged in managing urban development. Good urban governance is considered as a precondition for promoting equitable and sustainable growth of cities. In 1999, UN Habitat launched Global Campaign for Good Urban Governance which focused on articulation of principles and initiation of strategies in various countries through country-specific campaigns. The aim of this campaign was to eradicate poverty by increasing the capacity of state and local governments and raise awareness on good urban governance in countries across the world. It advocated seven norms of good urban governance, namely sustainability, subsidiarity, equity, efficiency, transparency and accountability, civil engagement and citizenship, and security. A National Campaign on Good Urban Governance was launched in India as part of the Global Campaign by the MoUD in September 2001 with support from UN Habitat. It led to formulation of a National Action Plan on Good Urban Governance (NAPGUG) that identified key working themes of decentralization, municipal finance, urban environment, integration of the poor and marginalized, transparency and civic engagement, municipal management and capacity building. The suggested strategies for promoting a sustainable environment included integrated waste management systems, waste water recycling, non-conventional energy sources, implementing solid waste management rules 2000, preparing an environmental status report, community participation in planning and delivery of services, and so on. JNNURM in 2005 also emphasized promoting good governance by introducing reforms in the functioning of urban departments at the state and local level. The states and cities were required to implement 23 reforms for accessing funds under JNNURM falling into three categories, namely state level reforms, ULB level reforms and optional reforms, as shown in Table 1.

It needs to be underscored that three initiatives have played a key role in promoting good urban governance in India. These are the 74th Constitutional Amendment Act (CAA) 1992 for ULBs, NAPGUG 2001 and the JNNURM Programme in 2005. NAPGUG has created awareness on principles and norms of good governance and highlighted various good practices. It has also led to implementation of a few initiatives on a pilot basis. The JNNURM programme had a significant impact on governance and accountability in states and cities (Narender 2013). It also imparted training to various municipal functionaries to enhance their technical and managerial capacity. However, it is observed that there is not much focus of these initiatives in mainstreaming environmental or climate co-benefits in promoting good urban governance. The High Powered Expert Committee (HPEC) Report constituted by the MoUD has identified the several basic prerequisites of a well-functioning local governance framework (MoUD 2011), the significant ones being *functional autonomy*—functions of local governments vis-à-vis state governments, and their entities must be unambiguous; *financial autonomy*—own revenue and inter-governmental transfers, must match local requirements and should be accompanied by the necessary autonomy to expand these resources; *functional competency*—functionaries must be competent to discharge the local functions effectively seeking an ongoing process of training and knowledge dissemination must be built into the system of governance; *functional outcomes*—authority for

Table 1 Implementation of 23 reforms within JNNURM

State level reforms	ULB level reforms	Optional reforms
1. 74th CAA (transfer of 12th schedule of functions), constitution of district planning committee (DPC) and metropolitan planning committee (MPC) 2. Integration of city planning and delivery functions and water supply function 3. Reform in rent control 4. Stamp duty rationalization to less than 5% 5. Repeal of Urban Land Ceiling and Regulation Act 6. Enactment of community participation law 7. Enactment of public disclosure law	8. E-governance set up 9. Shift to accrual based double entry accounting 10. Property tax 11. Levy of reasonable user charges 12. Internal earmarking of funds for services to urban poor 13. Provision of basic services to urban poor	14. Introduction of property title certification system 15. Revision of building by-laws to streamline the approval process 16. Revision of building by-laws for mandatory rain water harvesting 17. Earmarking 25% developed land in all housing projects for EWS/LIG 18. Simplification of legal and procedural framework for conversion of agricultural land 19. Introduction of computerized process of registration of land and property 20. By-laws on reuse of recycled water 21. Administrative reforms 22. Structural reforms 23. Encouraging public private partnership

approving and disbursing money for approved projects, must match the finances allotted within a framework of transparency, accountability, community participation; and *social accountability* must be ensured.

Meanwhile, the 12th FYP has suggested the following strategies for strengthening urban governance in India (Planning Commission 2012):

1. Convergence of all central and state schemes at the municipal level
2. Setting up of an independent utility regulator at the state level to monitor service levels and adjudicate disputes related to delivery and pricing of services
3. Empowering Mayor's office and extending the Mayor's term to five years
4. Publishing citizen charter by every municipality
5. Adoption of information technology and e-governance by all municipalities
6. Clarification of roles and responsibilities of ULBs and parastatals and making service delivery by all agencies accountable to ULB
7. Institutionalizing participatory development processes by establishing Area Sabhas and Ward Committees for decentralized governance
8. Adoption and management of prudential financial management in ULBs through appropriate legislation
9. Putting in place a robust monitoring system through outcome-based approach and results framework document

10. Establishing Lokayuktas/Ombudsman at state and local level for conflict and dispute resolution
11. Reforming the urban planning process, town planning laws and master plan preparation by adopting a whole city approach, long-term planning and participatory process
12. Preparation of city development plans and financial plans by all cities
13. Constitution of metropolitan planning committees and district planning committees and restructuring of metropolitan development authorities to focus on metropolitan plan preparation

2.3 Urban Finance Framework

Municipal finances are a key determinant of urban infrastructure. Expenditures towards provision of urban infrastructure require capital investments for asset creation. These are largely undertaken by the ULBs comprising of revenue expenditure, operation and maintenance (O&M) charges, establishment charges and capital expenditures. The poor state of municipal finances requires that most of the revenues are committed towards revenue expenditure with little available for capital expenditure resulting in severe service backlogs over the past few decades.

The revenue base of ULBs consist of own revenues, shared taxes/assigned revenues, plan and non-plan grants and loans. The own revenues are from taxes and non-taxes. The key tax sources include property tax, vacant land tax and advertisement tax. The non-tax revenue comprises of trade licences and user charges. Historically the revenues from these sources are low due to poor data base, low coverage, low rates, non-revision of rates and poor administration. The assigned revenues or shared taxes are those which are collected by ULB on behalf of the state government and given share due to their effort. These include professional tax and stamp duty charges. The plan and non-plan grants are transfers from state government and centre. The transfers from state government have been ad hoc and not regular. The 74th CAA mandated constitution of State Finance Commissions (SFCs) to devise regular and formula based allocations to ULBs but this has not been systematically followed. The central government allocations include transfers from Central Finance Commissions and grants through programmes and schemes under FYPs. Though ULBs were permitted to take loans, this was not often resorted to, as they had to obtain permissions from respective state governments. All these factors have contributed to evolution of a poor revenue base of ULBs in India over the years. The problem has worsened due to growth of population of ULBs causing increased demand for services and devolution of additional functions to ULBs by the 74th CAA without corresponding devolution of finances, a feature labelled as vertical imbalance. Fighting global warming and pursuing climate co-benefits is one such aspect that is new to the traditional functional portfolio of ULBs.

In addition, the poor state of municipal finances in India is considered as responsible for the severe backlogs in provision of urban infrastructure. It is felt that the requirements of finances for urban infrastructure have gone up significantly over the years such that major reforms in the municipal finance system are required to mobilize large funds to ensure service delivery as per the prescribed norms. Several studies and reports by expert committees in recent years have estimated the financial requirements for urban infrastructure and articulated strategies for mobilizing the same. A summary of different estimates of urban infrastructure investment requirements is presented in Table 2.

The HPEC Report (MoUD 2011) has made the following recommendations for strengthening municipal finances.

- More broad-based revenue sharing by states with ULBs with appropriate amendments to Constitution and other measures namely:
  - Introducing Local Body Finance List along with Union and State Lists
  - Power to ULBs to levy property tax, professional tax, entertainment tax, advertisement tax and retain whole proceeds
  - Sharing of a pre-specified percentage of revenues from all taxes on goods and services
  - Formula based and regular allocations
  - Strengthening SFCs
- Comprehensive and time bound reform of property tax system including measures such as setting up of Property Tax Board, self-assessment and area based taxation, revision every five years, IT based collection system, and so on
- Reforms in user charges to meet O&M costs, debt servicing, depreciation and surplus to meet a part of capital expenditure
- Establishing a Municipal Service Regulator with the responsibility of revising user charges regularly
- A good inter-governmental transfer system to ensure regular and formula based allocations
- A new and improved JNNURM programme with an allocation of 0.25% of GDP per annum over a 20-year period
- Facilitating municipal borrowing by creating a Regulatory Guidelines Handbook for Municipal Borrowings
- Promoting PPPs in urban infrastructure and service delivery
- Land-based financing instruments such as betterment charges, impact fees, development charges, use of floor space index (FSI), and so on

The 12th FYP (Planning Commission 2012) suggested important strategies for strengthening municipal finances that mandate to create robust tax and non-tax based revenue streams for ULBs; attract private capital; incorporate a Local Bodies Finance List through a Constitutional Amendment facilitated by MoUD providing for taxes such as property, entertainment, professional, motor vehicle, advertisement, stamp duty, user charges, trade licences, land based instruments, and so forth.

Table 2 Financial estimates of urban infrastructure by various committees/reports

Source	Sectors	Period	Estimates (Rs crores)
Committee of Ministers constituted by the Central Council of Local Government (1963)—also known as Zakaria Committee	Water supply, Sewerage, Storm water drains, Urban roads, Street lighting	–	211.3 (2113 million) at 1960–61 prices (annual)
India Infrastructure Report (2006)	Water supply, Sewerage, Solid waste management, Urban roads	1996–2006	56,000 (560 billion) at 1995–96 prices
Mohanty et al. (2007)	Water supply, Sewerage, Solid waste management, Storm water drains, Urban roads, Urban transport	2004–14	630,000 (6300 billion) at 2004–05 prices
McKinsey Global Institute (2010)	Water supply, Sewerage, Solid waste management, Storm water drains, Urban roads, Mass transit, Affordable housing	2010–30	$1.2 trillion or Rs 56 lakh crores (56 trillion)
MoUD (2011)	Water supply, Sewerage, Solid waste management, Storm water drains, Urban roads, Urban transport, Traffic support infrastructure, Street lighting, Renewal and redevelopment (including slums), Other sectors	2012–31	3,918,670 (39.2 trillion) at 2009–10 prices

It should provide for regular revision based on scientific principles; allocate 25% of goods and services tax to urban and rural local bodies as suggested by HPEC Report; enable ULBs to generate resources by using FSI and other land value based instruments such as betterment fee, land use conversion charges, impact fee, development charges which could be parked into a city development fund that could be used for urban infrastructure.

The plan recommends for putting in place a comprehensive and transparent system for land monetization; levy user charges for all measurable services covering O&M costs, debt servicing costs, depreciation and minimum profit that should be regulated by a municipal service regulator; establish a comprehensive approach to People–Pubic–Private-Partnerships (PPPPs) to be able to raise 13–23% of investment for urban infrastructure under the 12th FYP; set up a ring-fenced state/city level development fund with revenues from innovative sources to ULBs exclusively for urban infrastructure. The fund could have two parts where part one could be used for transport projects while the other part can be used for shelter and other projects; empower ULBs to leverage municipal bonds including pooled financing and create a handbook specifying lending instruments, roles and responsibilities of ULBs, states, centre, and so on; strengthen SFCs and municipal information system (MIS) at local level with assistance under the capacity building component of urban schemes.

As evident, the existing municipal finance framework focuses on enhancing conventional municipal services such as water supply, sanitation, solid waste management and has no focus on addressing climate change mitigation and impacts. A new approach to municipal finance allocations may be required to mainstream a climate co-benefits approach. The above reforms and strategies are needed to enhance the revenue base of ULBs, at the same time the municipal finance framework should incentivize adoption of a climate co-benefit approach while implementing policies and projects.

3 Issues in Mainstreaming Climate Co-benefits Approach in Urban Policy, Governance and Finance

The policy, governance and financing framework for urban areas in India in the past have not paid adequate attention to climate co-benefits and sustainable habitat issues. The focus was more on sectoral policies and projects towards service delivery which in turn adopted conventional approaches. These either adversely impacted climate change and environment or at best ignored climate co-benefits. The governance arrangements defined by leadership, organizational, managerial and technical capacity of states and cities was not adequate to undertake the conventional functions, let alone address the emerging issues related to climate change. The municipal finances have been totally inadequate to provide even the basic minimum level of services. Consistent efforts to reform the sectoral policies and projects, governance and municipal finances have been initiated only during the past one decade. These were aimed at addressing the existing weaknesses and promoting efficiencies and improvements in governance and service delivery and not aimed at mainstreaming a climate co-benefits approach.

The urban policy in India is sectoral and project based and is articulated mainly through FYPs. A number of urban projects related to water supply, sanitation, solid waste management, housing, slum development and transportation were implemented through this instrument without focusing on the environment and climate change issues. The housing policies also did not pay much attention to sustainable habitat. Formulation of the Municipal Solid Waste Management Rules in 2000, followed by their revision through the Solid Waste Management Rules 2016, is probably the first sectoral initiative integrating the climate co-benefits approach. The JNNURM Programme formulated in 2005 and implemented over 2005–12 has also focused to a limited extent on promoting environment friendly initiatives in cities. On the other hand, the Integrated Energy Policy 2006 and National Urban Transportation Policy 2006 brought some thought on climate co-benefits. Significant attention has been paid to climate change and habitat issues only after the formulation of the NAPCC and preparation of eight mission documents, including the NMSH by MoUD in 2008. NUHHP 2007, NUSP 2008 and service level benchmarking indicators in 2008 also aimed at promoting better management of urban habitat that encouraged states and cities to formulate environmentally sustainable policies. It can be reasonably argued that climate co-benefits could be efficiently integrated within the urban development paradigm if intermittent research and policy aspects are sufficiently addressed.

3.1 Research Imperatives

In order to realize climate co-benefits in urban development and local action, it is imperative to scientifically evaluate causations and impacts of climate change against development initiatives, as well as to build capabilities in cities to operate with this new paradigm of urban development (co-benefits approach). In this process there are still several grey zones, or research gaps of the conceptual, methodical, empirical and governance kind, that pose to conceal some of the co-benefits. In order to deal with the unresolved issues in effective application of co-benefit concepts and tools, there is a need to further focus on: (a) assimilation of mitigation sectors and adaptation risks, vulnerabilities and hazards; (b) greater application of spatial-temporal and real-time analysis; (c) utilization of available national/regional datasets and city indicators; (d) outputs that could directly orient governance strategies, through greater science-policy convergence; and (e) mandatory application of assessment tools by city authorities.

In fact, there should be assessment tools to guide policy at every scale of intervention—institutional, societal, administrative, and so on—but more so in the urban domain. Co-benefit models need to back-cast capped emission levels and accordingly establish a co-relation with urbanization, development thresholds, mid-term and long-term goals of urban planning, within the country or region. Then at the city-region scale, there are technologies that measure spatial causation and impacts real time, with accurate ground positioning. In this regard, a spatial and temporal dataset for cities and city-regions could be very helpful to assess their climate co-benefits. For instance, greater empirical research in the area or land use and transport could better demonstrate potential co-benefits. It has been concluded that service level benchmarks when customized to specifically report parameters like accessibility and vehicle miles travelled at city or project level over a period of time can serve as an excellent benchmarking, monitoring and impact assessment tool.

3.2 Development Policy

3.2.1 Energy

A superimposed scenario of economy, energy-generation and consumption, and environment (notably emissions), collectively termed as E3 Nexus, increasingly exhibits rapid growth with mutual inter-dependency. This emphasizes the immediacy to take measures in low-carbon technology and economic or policy instruments to produce climate co-benefits in different energy generating and consuming sectors. Green alternatives are particularly inclined towards promoting renewables, enhancing energy efficiency, minimizing losses, avoidance or reduction in consumption, fuel efficiency, and so on. Certain measures like introducing a light emitting diode, energy efficient appliances and demand side management are low-hanging fruits. Meanwhile, policy alternatives should focus on a mix of alternatives like energy pricing, carbon-tax, cap-and-trade, subsidies and regulation. Whereas national policies do spell out technical and policy measures to deal with the situation, comprehensive planning, modelling of energy and action by the state governments in India in improving performance standards would go a long way in realizing climate co-benefits locked in the interphase of energy, economy and environment at the local level.

3.2.2 Transport

Transport co-benefits such as emissions savings (climate pollutants and traditional air pollutants) may not necessarily form a big portion of the monetized savings, but can provide that needed increment to make projects more feasible. Moreover, a more holistic analysis of the costs of the 'baseline' scenario (e.g. accounting for the infrastructure requirements to accommodate the growth in private motorized travel in cities) will provide additional arguments for choosing alternatives that promote public transport and non-motorized transport. Broader benefits—which are more difficult to quantify and, moreover, harder to monetize—should also be taken into consideration, such as improved quality of life, equity and social cohesion. The call for the adoption of a co-benefits approach in the transport sector is a call for the adoption of a more holistic approach to assess and evaluate investments to move towards sustainability. There is also a pressing need for mainstreaming integrated land use and transport planning at the state level. Structural changes in various acts, and committees that determine land use to include transport professionals may fill the gap. Initiatives like developing maximum parking standards for each type, mix, of land use and development, encouraging congestion charges, affordable housing around transit oriented development could generate multiple environmental, economic and development co-benefits. A similar approach of integrating land use and transport themes in training of urban, regional, housing and transport planners in academic institutions is highly required. A special training course could be organized in this regard focusing on the basic principles, best practices, methods and tools to prepare and evaluate proposals, estimate multiple co-benefits of development and environment, project work, etc.

3.2.3 Buildings and Construction

In spite of India being one of the first few countries to develop a comprehensive framework for promoting energy efficiency in green buildings that can ensure climate co-benefits, only a few states have adopted the energy conservation building code (ECBC). Many urban local bodies are yet to formulate or amend the building by-laws. The sustainable habitat principles articulated by the NBC 2005 and URDPFI Guidelines 2014 should be strictly adopted in plan formulation and implementation by the state and city governments. National urban policies and programmes that finance various services have a huge potential in facilitating implementation of energy efficiency and green building policies in Indian cities. The Government of India's Smart City covering 100 cities are required to prepare a Smart City Plan which should incorporate the climate co-benefits framework and strategies for implementation. Similarly, guidelines for AMRUT and *Housing for All* should also make it conditional for adoption of ECBC and green buildings by cities. There is a need to create awareness and build capacities among the officials of ULBs on these issues. Methods for assessing the climate co-benefits in the building sector and informal housing sector should be put in place. In addition, a compliance evaluation system for ECBC and green buildings should be executed at the city and state level to assess the progress, measure the impact and capture the co-benefits.

3.2.4 Municipal Waste

With regard to realizing waste related co-benefits in urban areas, Indian cities face major technical and managerial challenges in moving to 3R (reduce, reuse, recycle) and 'waste to energy' paradigm. The chapter on waste to energy in this book demonstrated some cases for initiating action on converting waste to energy but most of these efforts are small-scale, fragmented and project-level initiatives that have not been integrated into national policy frameworks. Existing management for the current system with continuous research and development in collaboration with external agencies and enhanced number of plants with tertiary treatment process (promoting good initiatives that happen in one part of the city to other parts) can be a learning example for many other municipal authorities in India. As evident, only few states/UTs have set up processing and disposal facilities and even those plants are either closed or underperforming, which is a major cause of concern. The challenge is to address the higher cost of technologies (bio-chemical waste to energy and thermo-chemical waste to energy), lack of awareness, lack of a focused national policy and not having skilled manpower, particularly at the ULB level, to make this kind of action more common in India. This requires the serious attention of policy-makers to disseminate learning and integrate them into national policy frameworks by which these practices could well expand into other regions and bring further positive co-benefits.

3.2.5 Biodiversity

Co-benefits arising from urban biodiversity, that is, forestry or agriculture, may seem insignificant, particularly because of limited theoretical and empirical studies in this area. But systematic evaluation of concerned impacts and co-benefits through a strategic environmental assessment (SEA) could bridge this gap. The internationally accepted benefits of SEA are that it allows for a wider consideration of impacts and alternatives, is a proactive tool that can be used to support strategic action formulation for sustainable development, can increase the efficiency of tiered decision-making (including strengthening of project EIA), allows for a systematic and effective consideration of the environment at higher tiers of decision-making and allows a greater level of public participation and consultation. It needs to be underscored that India's National Environment Policy provides for the incorporation of SEA in the plan-making process in the country. Even the 11th and 12th FYP and the Regional Plan for NCR 2021 provides for SEA type assessments.

4 The Way Forward: Recommendations to Mainstream Climate Co-benefits in the Urban Sphere

The states and cities have exclusive jurisdiction to enforce climate adaptation and mitigation strategies in the urban sector, which play a major role in reducing the overall impact of global warming. Mainstreaming co-benefits in urban policy, governance and financing framework is key to achieve this goal. The following suggestions would be pertinent in this regard.

4.1 Research, Monitoring and Reporting

There should be innovations in co-benefit assessment research, monitoring and reporting to guide decision-making at every possible scale of intervention—national, regional and the city level itself. An integrated urban tool is highly needful that takes into account risks, vulnerabilities and hazards on cities and also impacts of city-based GHG emissions from various activities on climate. The tool should be technically competent to downscale national scenarios on urbanization, GHG emissions, climate variability, and so forth, as regional-level inputs, and at the same time upscale local urban and user behavioural data collected through surveys to macro-scale. The tool should also be able to quantify co-benefits based on simulation of different urban factors particularly user choices, income levels, spatial planning, density, type of built-form and their energy needs, city structure, land use mix and transportation/mobility. Sethi (2017) is one such research that attempts to correlate GHG emissions emanating from urban areas with diverse spatial and non-spatial factors using regression modelling for over 41 Indian cities from different bio-physical locations. To begin with, the results would be pertaining to several other global cities too.

The states in coordination with the member departments should develop indicators, benchmarks, guidelines, manuals and good practices related to climate co-benefits in the urban sector and disseminate them among the city governments. This would serve as a checklist for monitoring and reporting. Some examples of measuring co-benefits from urban areas include energy efficiency in water supply and street lighting, designing energy efficient buildings, water audit, rainwater harvesting, wastewater recycling and reuse, composting and energy from waste, tree planting, preparedness for disasters such as flooding, fire, diseases, and so on. In this regard there are many lessons and learning opportunities from international experiences of other developing countries that could be applied in India. For example, the experience on urban waste composting in Indonesia could be important to India to reduce the large-scale urban waste that goes untreated (to the tune of 81% of total municipal solid waste generated), and at the same time produce compost for urban gardens and municipal parks, or for agriculture, as well as generate local jobs in waste collection. Similarly, as this research affirms with fresh evidence from India, there are huge co-benefits in the inter-disciplines of land use–transport, health and GHG emissions. India could improve cities and reduce GHGs with simple technologies already in use and showing effective results elsewhere. The low costs of some of those initiatives could help their use in Indian cities almost immediately. Boosting the exchange of experiences for Indian cities among themselves and with cities worldwide could also help the dissemination of good practices in India for cities facing similar development challenges.

4.2 Creating Horizontal and Vertical Linkages for Potential Co-benefits

Aligning global environment and locally pertaining urban issues forms the bedrock of achieving co-benefits. The GHG emission from urban areas is large, generated mainly from motor vehicles, industries, energy use in buildings and from municipal waste. Therefore, the policies and programmes of the government should focus on sectors that will substantially bring down the GHG emission levels. This in turn depends on governance and financing capacities of ULBs. Efforts are being made by the government programmes to strengthen urban governance and several government reports themselves have made recommendations towards the same. These measures need to be integrated with the capacities and systems to formulate and implement a climate co-benefits approach at the local level. ULBs should consciously identify and create climate co-benefits in various sectors and projects while preparing local level plans such as city development plan and city sanitation plan. There is a need to strengthen the urban policy framework in India articulated through various sectoral policies for integrating and mainstreaming climate co-benefits approach consciously as presented in Table 3.

Thus urban policies and schemes having potential for mainstreaming a climate co-benefits approach can be grouped into two categories. Category one policies and programmes have minimum focus on climate co-benefits and there is greater scope to include those in future. These include FYPs, housing, slum, water supply and sanitation schemes. Category two policies and schemes have adequate focus on climate co-benefits but require greater application and implementation at the state and ULB level. These include NMSH, SCM, NUTP, AMRUT, building sector policies and schemes and solid waste management schemes. Even category one policies require application and implementation at the state and city level once they are strengthened.

There are several plausible cross-linkages between different transects. As this research shows, immense gains could be made by pricing carbon from thermal energy and transportation emissions and incentivizing biodiversity and forestation measures. Similarly, government-funded urban schemes like smart cities could become an important vehicle to aggregate market demand in the renewable sector, which could substantially bring down their capital and operational costs. The sudden reduction in LED (light-emitting diode) lamp prices in the market due to policy-led aggregation of demand and negotiations with supply side manufacturers is a perfect example of how out of the box decision-making could generate climate co-benefits by cheaply offsetting GHG emissions. While public finance could be used as a seed fund, private companies could make locality level projects sustainable by investing in equipment, facilities and manpower to introduce smart meters, electric mobility, energy storage devices, and so on. With some persuasion

Table 3 Potential to mainstream climate co-benefits approach in urban policy framework

Sector	Policy	Existing focus on climate co-benefits	Potential to strengthening climate co-benefits in future policy
Overall urban policy	12th Five Year Plan, National Mission for Sustainable Habitat 2008, Smart Cities Mission, AMRUT, HRIDAY	Moderate focus in Five Year Plan through suggesting implementation of NMSH and need for clean and better environment in cities Adequate focus in NMSH and Smart City Mission	Mainstreaming climate co-benefits approach in the overall urban agenda, in each and every sectoral policy and project under the Five Year Plans is needed Effective implementation of NMSH and Smart City Mission is required
Housing	National Housing and Habitat Policy 2007, Housing for All	Emphasis on sustainable housing with use of renewable energy	Suggest and incorporate compulsory adoption of ECBC 2007, NBC, green building rating into the policy Incentives of higher FAR, TDR for low emissions
Transport	National Urban Transport Policy 2006	Emphasis on public transport, non-motorized transport, integrating land use with transport and use of clean fuels	Continue the same focus with greater awareness and policy-making and implementation at the state and ULB level
Energy sector	Integrated Energy Policy 2006, Electricity for All, National Smart Grid Mission, Energy Conservation Act 2001, ECBC 2007	Increasing supply from renewable energy sources Emphasis on energy conservation, green building rating	Installation of model WTE and renewable energy projects by ULBs. Transfer of electricity functions to ULBs or make distribution companies accountable Cap and trade market with reduction targets for GHG emissions Strengthening regulatory and market instruments for REC and RPO Polluter pays, penalties on high emissions from captive

(continued)

Table 3 (continued)

Sector	Policy	Existing focus on climate co-benefits	Potential to strengthening climate co-benefits in future policy
			plants/diesel generator sets Phased overhaul of municipal infrastructure in industrial estates
Water supply	Service level benchmarks and water supply schemes and projects	Moderate focus with expansion and sustainability of service with 24 by 7 water supply, reduction of non-revenue water, cost recovery etc.	Greater scope to mainstream climate co-benefits by designing projects and schemes with climate co-benefits across the project life cycle and effective implementation of service level benchmarks
Sanitation	National Urban Sanitation Policy 2008, Service Level Benchmarks, Swachh Bharat Mission	Moderate focus for improved sanitation	Greater scope to address entire sanitation value chain through better septage collection, transportation, treatment and disposal practices along with universal access and open defecation free cities
Solid waste management	MSW Rules 2000, SWM Rules 2016 and Service Level Benchmarks	Adequate focus on collection, transportation, treatment and scientific disposal	Strengthen implementation and financing framework in cities particularly related to 100% treatment (composting, waste to energy etc.) and scientific disposal (landfill)
Slums	Housing for All, State Specific schemes	Focus only on security of tenure, provision of housing and services	Greater scope to integrate co-benefits approach through ECBC 2007, NBC, use of renewable energy and green building concept

(continued)

Table 3 (continued)

Sector	Policy	Existing focus on climate co-benefits	Potential to strengthening climate co-benefits in future policy
Biodiversity	Respective Master Plans, EPA 1986, Biodiversity Act 2002, NBAP 2008, NMSH, GIM, JFM, NAP, Redd-Plus policy	Co-benefits to be identified while protection of areas of significant ecology, EIA of development projects using SEA methodology	Better linkages with poverty eradication, energy access and air-pollution mitigation Climate Action Plans of cities include greening of urban areas in tune with the biodiversity Master Plans of urban areas have to provide areas for open spaces/recreation preferably including existing pockets of green Creating network of parks and roadside greens

and incentivization, large public entities that are guzzlers of fossil fuels could be brought to the forefront of sustainable transformations. For instance, the mandatory procurement of renewable energy by thermal plants can further increase. Similarly, railways, special economic zones, industrial estates, government office districts, and so on could invest in waste to energy plants in collaboration with municipal authorities that ensure continual and quality waste streams. India has an ecosystem where mini-*ratna* public sector undertakings and corporate social responsibility within corporate sector acts as an instrument of change.

Our research also reflects that it is important to integrate climate resilience in urban planning. While it mandates detailed recommendations for implementation of sector specific strategies identified within the resilience framework, it supplements the same with short-term, medium-term and long-term action points that cities could adopt to systematically approach the climate resilience objectives. There is a unanimous consensus on the efficacy and importance of the NMSH to achieve climate resilience goals in urban areas. While cities seem to be keenly interested in taking up a climate resilience agenda into their development planning system (as it means channelizing additional climate funds for essential development works), it is felt that each city would have to contextualize activities depending on their socio-economic and geographical characteristics. The governance, regulatory mechanisms and institutions would have a major role to play in steering interest within the public sphere and initiating actions on this.

Thus, formulating urban sector policies and projects by mainstreaming climate co-benefits is needed along with their effective adoption and implementation at the state and ULB level to make them effective. But apart from exploring the recognized policy sectors, as above, this research establishes with conviction and profound evidence that there are several other initiatives, approaches and measures that could promote climate co-benefits in cities. It could be through an active and larger role played by legislative and judicial bodies, media and civil society, social entrepreneurs, urban planning agencies, government commissions and probably some newly formed institutions.

4.3 Creating Institutions at the State and City Level

As suggested by NMSH, all states should create an apex body on climate change headed by the Chief Secretary with Principal Secretaries from Energy, Environment, Urban Development, Rural Development, and so on, as members. The apex body should be responsible for preparing a state-level strategy and its execution in a convergent and coordinated fashion with the urban sector policies. It would also define the implementation, monitoring and evaluation arrangements across development sectors, institutions and cities for mainstreaming climate co-benefits. All states should prepare a climate co-benefits strategy articulating a long-term, medium-term and short-term action plan covering key urban sectors and specifying roles and responsibilities of institutions and funding arrangements. The state level co-benefit strategy should be prepared within the framework of NMSH. It ought to fix long-term (5–10 years), medium-term (2–5 years) and short-term (0–2 years) targets for potential climate co-benefits within inter-linkages of major sectors such as energy, buildings, water, wastewater, solid waste and transportation.

On the pattern of a state-level apex body, all cities, particularly with five lakh (0.5 million) and above population should set up a city-level committee consisting heads of major ULBs and agencies for key sectors such as energy, power, health, disaster management, and so on. This committee should be responsible for preparing a city-level co-benefits strategy and integrate it with the national and state-level strategies to ensure coordination and address regional issues. The committee would also articulate a long-, medium- and short-term action plan covering all the key sectors. The strategy should undertake assessment of the current status and impact of climate change and formulate and implement action plans with climate co-benefits. Each city should also prepare a ward-level action plan to accrue climate co-benefits by involving various stakeholders and adopting participatory processes. All cities irrespective of their population size would need to integrate city-level climate co-benefit strategies into master plans, city development plans and/or city sanitation plans. Each and every local development activity or urban plan should ensure integration of climate co-benefit strategies. For example, all schemes and projects dealing with water, sanitation, solid waste management, urban transport, environment, energy, health, and so on, should ensure that climate co-benefit strategies are internalized.

4.4 *Innovative Funding Mechanisms*

In addition to innovative regulatory mechanisms, this research suggests encouraging policy instruments strongly in favour of appropriate carbon and fossil based energy pricing. While it is of the utmost necessity to ensure access and affordable energy to all, particularly the backward areas, nevertheless pricing in general should do away with subsidies and adequately represent the costs incurred in production, transmission, distribution and maintenance. In addition, it is recommended to include the marginal social and environmental costs of conventional energy sources such as coal, petroleum and gas, through carbon cess or tax. While most development sectors are grappling with sudden technological and economic shifts in the prevailing global markets, industry and businesses prefer environments that have the least financial uncertainties. In this regard, public authorities ought to create a conducive policy and financial regimes for energy efficient technologies, clean energy, and so forth, so that private enterprises can plan for the long-term horizon.

Strengthening the urban policy and governance framework for mainstreaming climate co-benefits calls for reinforcing the municipal finance framework. While ULBs should be encouraged to implement measures for enhancing own revenues, central and state governments should be encouraged to enhance budgetary allocations to ULBs. New and innovative sources of financing through strategic partnerships, market finance and leveraging land values should be implemented by ULBs. The centre and state should provide incentives, preferably routed from carbon tax or cess, to the public and private sector for innovative projects that mainstream climate co-benefits and also to strengthen the systems and capacities of ULBs.

At present, the national and state governments are providing funding to cities for implementation of various projects. The project and scheme-based funding should be linked to a mandatory set of measures that promote climate co-benefits. Depending upon local needs, national and state governments could prescribe a priority list of ten initiatives with climate co-benefits across varied sectors. The cities could start implementing two to three measures every year while release of funds for various projects and schemes may be linked to performance against these initiatives. In addition, a dedicated fund for mainstreaming a climate co-benefits framework should be set up at the state level by pooling resources from all the urban departments. The budgetary transfers from national and state governments should incentivize implementation of co-benefit measures by cities. They should provide dedicated budget allocations for climate co-benefits over and above the sectoral or departmental allocations, which should be utilized as a catalyst to attract greater resources particularly from the private sector. This fund should be administered by the proposed apex body in accordance with the provisions of a state-level climate co-benefit strategy.

4.5 Mobilizing Multi-stakeholders in Governance

ULBs should develop institutional processes for engaging civil society and private sector. The local leadership needs to be sufficiently oriented and sensitized towards mainstreaming the climate co-benefits approach. States should be encouraged to develop core capacities to advise cities in mainstreaming a climate co-benefits approach and extend handholding and technical support to ULBs. Currently, the urban local governments have sectoral specialists to deal with transportation, water supply, sanitation and solid waste. They are required to undergo training programmes to update their skills from time to time. There is a need to incorporate climate co-benefit concepts and methodologies through dedicated workshops and modules to enhance awareness and intervening skills on climate co-benefit issues among the officials. The senior and middle-level officials dealing with housing, slums, buildings, transportation, water supply, sanitation and solid waste management sectors should be imparted training in methodologies to identify and include climate change co-benefits in city level plans and projects. Regular skills and learning opportunities should be imparted to all executives and workers involved in operationalizing the climate co-benefit strategy.

In addition to achieving energy efficiency in established urban centres, it is equally important to bring human settlements undergoing rapid rural to urban transformations within the folds of reliable energy access. This research demonstrates that it is necessary to have local social entrepreneurs and their innovative thinking to address social needs and market challenges. Availability of technology or skills alone would not make any considerable difference, unless it is made accessible to the poor and downtrodden communities. Thus, the multiple roles played by social entrepreneurs to accrue several co-benefits are noteworthy. The most significant enabler is to bring the governing public authorities and funding agencies or banks on board. Learning reflects that there were successful cases that were supported by government and its entities. However, there are many initiatives which require awareness, education support and encouragement of local enterprises to scale-up co-benefits.

The civil society members should be involved in the preparation and operationalization of city-level climate co-benefit strategies and in information, education and communication campaigns since they are responsible for effective implementation of such strategies like adopting energy efficient appliances, green buildings, rainwater harvesting structures, solar water-heating systems, and so on. Most city jurisdictions are divided into local wards for better administration. A task group comprising civil society groups and associations should be set up at the ward level to provide inputs to a ward-level plan and implement the same. Targets can be fixed for doable actions like generating clean energy, reduction of energy consumption, improving the groundwater, increasing the green cover and creating green buildings at the ward level. Competition could be arranged between different wards, and the wards with greater performance for implementing climate co-benefit initiatives should be rewarded.

This research has also revealed that a rights-based approach is a valuable medium to incorporate fundamental rights at the centre of urban governance. The Supreme Court of India through the use of the 'right to life' as a legal basis, provides one of the most noteworthy examples of that kind for the city of Delhi. In fact, through the lens of protecting the fundamental right to have a clean environment, the Court has stepped into different sectors such as human health, air pollution, local environment, land use and transport. Thus, it not only pressed upon the implementation of existing norms, but issued detailed orders to be pursued by public authorities. This triggered a process of policy-making on key urban environmental issues (like air pollution in the national capital New Delhi) and ensured that they are taken into account early in decision-making. Each decision increased the pressure over public authorities to take preventive action. Therefore the rights-based approach to generate climate co-benefits in cities is positive in terms of the values it promotes and hence serves as the stepping stone to produce larger co-benefits. It is part of a larger urban governance scheme where ULBs as well as the state and central governments are major actors. Hence, urban India's sustainable development requires structural and incremental transformations, which will largely depend on the involvement and coordination of all the concerned agencies.

In view of the New Urban Agenda adopted during the Habitat III Conference in Quito, the intention of this book was to discuss how communities, cities and nations can develop policies and instruments to boost the capacity of societies, in generating climate, environmental and development oriented co-benefits in cities to synergistically achieve the local and global goals of sustainable development. With empirical investigations, the book built on the deliberations in the multi-disciplinary area of climate change, cities and their local development, and forged pathways to identify and generate large urban co-benefits. It presents a strong case where co-benefits can help us to understand how we make economic, policy and technological transitions that promote more sustainable urban initiatives, cutting across sectors and regions and that lead to significant impacts on the way that our societies use environmental resources and distribute the benefits. The adopted methodology not only evaluated the development sectors, but also stepped beyond to assess their inter-linkages for innovations and reforms that could generate larger co-benefits in urban areas. While the initial assessment mandates quantitative rigour in data collection, evaluation tools and forecasting, the latter part interprets it to influence constructive decision making. This research vividly displays how globally circulated theories, concepts and accorded treaties could be actualized to produce social and environmental benefits at the grassroots level. It demonstrates that within the new post-Paris climate regime and Habitat III era, similar studies could be initiated in other countries too. They would be highly relevant to researchers, policy-makers and decision-makers working in diverse development contexts, but with the common objective of making this world a better place to live, and to leave for future generations.

References

Central Council of Local Government. (1963). *Report of the committee of ministries (Zakaria committee) – Augmentation of financial resources of ULBs*. New Delhi: Government of India.

Government of India. (2007). *National action plan on climate change*. New Delhi: Prime Minister's Council on Climate Change, Government of India.

India Infrastructure Report. (2006). *'Urban Infrastructure' in India Infrastructure report*. New Delhi: Indian Development & Finance Corporation.

IPCC. (2014a). Summary for policymakers. In: C. B. Field, V. R. Barros, D. J. Dokken, K. J. Mach, M. D. Mastrandrea, T. E. Bilir, M. Chatterjee, K. L. Ebi, Y. O. Estrada, R. C. Genova, B. Girma, E. S. Kissel, A. N. Levy, S. MacCracken, P. R. Mastrandrea, & L.L. White (Eds.), *Climate change 2014: Impacts, adaptation, and vulnerability. Part A: Global and sectoral aspects. contribution of working group ii to the fifth assessment report of the intergovernmental panel on climate change*. Cambridge, United Kingdom and New York, NY, USA: Cambridge University Press, pp. 1–32.

IPCC. (2014b). Summary for Policymakers. In: O. Edenhofer, R. Pichs-Madruga, Y. Sokona, E. Farahani, S. Kadner, K. Seyboth, A. Adler, I. Baum, S. Brunner, P. Eickemeier, B. Kriemann, J. Savolainen, S. Schlömer, C. von Stechow, T. Zwickel and J.C. Minx (Eds.), *Climate change 2014: Mitigation of climate change. contribution of working group iii to the fifth assessment report of the intergovernmental panel on climate change*. Cambridge, United Kingdom and New York, NY, USA: Cambridge University Press.

McKinsey Global Institute. (2010). *India's urban awakening: Building inclusive cities, sustaining economic growth*. New Delhi: McKinsey & Company.

Ministry of Housing and Urban Poverty Alleviation. (2007). *National housing and habitat policy 2007*. New Delhi: Ministry of Housing and Urban Poverty Alleviation, Government of India.

Ministry of Urban Development. (2006). *National urban transport policy*. New Delhi: Ministry of Urban Development, Government of India.

Ministry of Urban Development. (2008a). *National mission on sustainable habitat*. New Delhi: Ministry of Urban Development, Government of India.

Ministry of Urban Development. (2008b). *National urban sanitation policy*. New Delhi: Ministry of Urban Development, Government of India.

Ministry of Urban Development. (2008c). *Handbook on service level benchmarking*. New Delhi: Ministry of Urban Development, Government of India.

MoUD-Ministry of Urban Development. (2011). *Report on urban infrastructure and services*. New Delhi: High Power Expert Committee, Ministry of Urban Development, New Delhi.

Ministry of Urban Development. (2012). *Report of the sub-committee on financing urban infrastructure in the 12th plan*. New Delhi: Ministry of Urban Development, Government of India.

Ministry of Urban Development. (2014a). *Guidelines for swachh bharat mission*. New Delhi: Ministry of Urban Development, Government of India.

Ministry of Urban Development. (2014b). *Draft concept note on smart city scheme*. New Delhi: Ministry of Urban Development, Government of India.

Mohanty P. K., Misra B. M., Goyal, Rajan & Jeromi P. D. (2007). *Municipal finance in India: An assessment, Development Research Group, Study No. 26*. Mumbai: Department of Economic Analysis and Policy, Reserve Bank of India.

MoEF. (2015). *India's intended nationally determined contribution: Working towards climate justice*. New Delhi: Ministry of Environment & Forests, Government of India.

Narender, A. (2013). Urban policy under JNNURM and the 12th five year plan in India. *Nagarlok, 45*(1), 1–31.

Narender, A. (2014a). Financing urban infrastructure in India: Trends, requirements, challenges and strategies. *Urban India, 34*(2), 97–121.

Narender, A. (2014b). Ethics and social accountability in remaking urban governance in India: Framework and perspective. In U. B. Singh (Ed.), *Ethics & social accountability in urban governance* (pp. 52–81). New Delhi: Global Research Publications.

Planning Commission. (2011). *Report of the working group on financing urban infrastructure*. New Delhi: 12th five year plan steering committee on urban development & management.

Planning Commission. (2012). *Twelfth five year plan (2012-17)- Economic sectors: Volume II*. New Delhi: Planning Commission

Puppim de Oliveira, J. A. (2013). Learning how to align climate, environmental and development objectives: Lessons from the implementation of climate cobenefits initiatives in urban Asia. *Journal of Cleaner Production, 58*(1), 7–14.

Sethi, M. (2015). Cities and climate change: Challenges in estimation of potential co-benefits. *Nagarlok, 47*(1), 1–17.

Sethi, M. (2017). *Climate change and urban settlements: A spatial perspective of carbon footprint and beyond*. London, UK: Routledge (Taylor & Francis).

UN Habitat. (2011). *Global report on human settlements 2011: Cities and climate change*. London/Washington DC: United Nations Human Settlements Programme & Earthscan

United Nations. (2015). *Transforming our world: The 2030 agenda for sustainable development, A/RES/70/1*. https://sustainabledevelopment.un.org/post2015/transformingourworld/publication. Accessed on December 15, 2015.

UNFCCC. (2015). *Conference of parties, December 2015, FCCC/CP/2015/L.9. United Nations framework convention on climate change*. http://unfccc.int/resource/docs/2015/cop21/eng/l09.pdf. Accessed on January 24, 2016.

World Bank. (2011). *Guide to climate change adaptation in cities*. Washington, D.C: World Bank.

Printed by Printforce, the Netherlands